34/35 Advances in Polymer Science

Fortschritte der Hochpolymeren-Forschung

Edited by H.-J. CANTOW, Freiburg i. Br. · G. DALL'ASTA, Colleferro
K. DUŠEK, Prague · J. D. FERRY, Madison · H. FUJITA, Osaka
M. GORDON, Colchester · J. P. KENNEDY · W. KERN, Mainz · S. OKAMURA, Kyoto
C. G. OVERBERGER, Ann Arbor · T. SAEGUSA, Kyoto · G. V. SCHULZ, Mainz
W. P. SLICHTER, Murray Hill · J. K. STILLE, Fort Collins

With 12 Figures

Springer-Verlag
Berlin Heidelberg GmbH 1980

Editors

ISBN 978-3-662-15388-8   ISBN 978-3-540-38241-6 (eBook)
DOI 10.1007/978-3-540-38241-6

Library of Congress Catalog Card Number 61-642

Typesetting: Elsner & Behrens GmbH, Oftersheim.
2152/3140 – 543210

Cationic Polymerisation

Initiation Processes with Alkenyl Monomers

Alessandro Gandini and Hervé Cheradame

Laboratoire de Chimie des Polymères, Institut National Polytechnique de Grenoble, 44 Avenue Félix-Viallet, F-38000 Grenoble

Table of Contents

Preface

Initiation of cationic polymerisation is arguably the most difficult single elementary reaction step studied in polymer science as a whole. But the publication, in this issue of Advances in Polymer Science, of an unusually comprehensive review, does not rest on the plea of *complexity:* on the contrary the justification lies in the degree of *order* which the authors have induced in the wider field of cationic polymerisation as a whole. Just what it is that cationic initiation initiates, emerges in a better light almost everywhere.

This review has clearly matured on the basis of long and arduous experience. Years ago, Cheradame and Sigwalt demonstrated the occurrence of initiation by direct acid-base reaction between a Lewis acid and a vinyl monomer, while Gandini and Plesch first put pseudocationic polymerisation involving propagation by undissociated ester intermediates on the map. Each of the two authors of this review was, accordingly, trained by one of the pioneers of cationic polymerisation. From them they learned the rigorous vacuum techniques which no other methods have yet been able to displace. The recent explosion of interest in cationic research does not, however, stem from further refinements in the clean handling of sensitive materials. Wistfully, in this respect, the authors record a steady decline in standards. As a result, spectacular improvements due to modern spectroscopic stopped-flow methods, and to instrumental advances in unravelling chemical mechanisms generally, have perhaps been slower than elsewhere to emerge in studies of cationic polymerisation mechanism. Only in 1976 do we find — in the work of Kunitake and Takarabe — 'the first thorough examination of the kinetics of initiation' of a dimerisation (!) based on spectroscopic stopped-flow measurements.

While generous with praise, Cheradame and Gandini are often sceptical and sometimes scathing. Where they themselves merely speculate, they say so clearly, and are worth listening to. Their text bears the imprint of wide experience in chemical mechanistics, far transcending the field under review. Many other reactions are drawn in to help us understand cationic polymerisation: the addition of iodine or acids to olefines, or Friedel-Crafts reactions, are among the many examples. As a result, the authors are able in several places to suggest new mechanistic schemes, and to support often quite comprehensive novel formulations by recalculations based on literature data.

Initiation mechanisms are everywhere elucidated by their integration into the complex overall rate behaviour of cationic polymerisation. Propagation rates determine yields and conveniently multiply the measurable effects of initiation. Termination steps contribute their effects to the jigsaw of cationic polymerisation, and even various kinds of cationic grafting reactions are covered. There can be few reviews punctuated by so many specific proposals for timely research, not only by academics, but also sometimes for possible industrial development.

To combine an impeccably systematic account with imaginitive presentation is not the least merit of this book, which may be studied with profit by chemists who pursue elusive intermediates in any field.

M. Gordon

I. Introduction

Cationic polymerisation is undoubtedly the most complex and controversial of all poly-additions. A thorough understanding of its phenomenology is still lacking, despite several decades of extensive research. Considerable progress has been made since the pioneering work of the forties, yet certain problems which arose from those first fundamental studies still linger today. Basically, these problems are related to the subtle behaviour of short-lived intermediates, a classical situation in physical organic chemistry. The determining contribution of some ingenious experiments, coupled with the advent of more sophisticated probing instruments, have set an irreversible trend towards the ultimate unravelling of many of these difficulties. However, the present stage does not allow the formulation of a comprehensive theory of cationic polymerisation, and in its place one finds different schools of thought, each one with some merits and some obvious weaknesses.

The crucial themes which stimulate present research in the field are sketched out below:
— Unravelling the mechanism(s) of initiation, whereby chain carriers are formed in a system. This topic includes such studies as the mode of proton transfer from a Brønsted acid to a monomer, the steps involved in the interaction of a Lewis acid with an olefin (in the absence and/or in the presence of a cocatalyst), the reaction of stable carbenium salts with various vinylic moieties, etc.
— Establishing a precise knowledge of the intimate structure of these chain carriers. Here, one is faced with the possibility of finding several species coexisting in equilibrium, viz., polarised ester molecules, ion pairs with various degrees of solvent participation, and free ions. All these species, moreover, can be stabilised or activated by one of the components of the polymerisation system.
— Determining the concentration of these various species and their relative contribution to the growth of a polymer chain. The aim of these studies is to arrive at values for the individual propagation rate constants, a task which has so far proved insurmountable.
— Minimising or exploiting rationally the important role of transfer and termination reactions. The well-known difficulty of obtaining high molecular-weight products is perhaps the most serious problem in applied cationic polymerisation, and many potentially useful systems have not gained industrial status due to the preponderance of these reactions and the lack of adequate means to reduce their detrimental effect. On the other hand, following the opportunistic principle, "if you can't beat them, join them", interesting research has been developping on the possibility of synthesising oligomers and cooligomers with useful properties and applications. Further advances in both areas, however, necessarily imply a better understanding of the chemistry of chain-breaking reactions and of the influence of various parameters upon their importance, relative to propagation.

— Searching for the conditions which will provide living systems, i.e., fast initiation and negligible termination and transfer. Naturally, the attainment of this goal is contingent upon the solution of some of the problems listed above. The ultimate rewards for gaining such a fine control over this hostile domain are all too obvious to be discussed here. The recent history of achievements in anionic polymerisation is the only reference needed.

— Acquiring the necessary theoretical and practical mastery over the discipline to be able to predict, at least in a semiquantitative fashion, the behaviour of hitherto untested systems. This requires a deep understanding of the way a large number of physicochemical factors are interrelated. While studying the literature, or working in cationic polymerisation, one is often reminded of the danger of extrapolating certain conclusions reached on a given system to another one, however similar the two might seem. An apparently minor change of conditions (e.g., temperature, or cocatalyst, or solvent) can bring about a drastic and often unexpected modification of the overall response of a system. Predictions are thus still hazardous, which goes to show that the vast amount of information which has been gathered has a panoramic character, but is far from exhaustive in terms of depth and quality. Therefore, the intricate relationships which govern the interplay of the components of a polymerisation system (monomer-catalyst(s)-solvent-temperature-impurities) require further and more thorough studies.

— Controlling the stereoregularity of the polymer growth. Surprisingly, this has been an aspect which, despite its remarkable potentials, has not yet received the attention it deserves. Although it is true that side reactions, and thus the danger of upsetting the regular growth of a chain, are unfortunately commonplace in cationic polymerisation, one cannot overlook that the first stereoregular polymers were obtained using cationic initiators on vinyl ethers.

— Synthesising polymers with predetermined architecture. Of course, living systems would provide an adequate solution to this general problem (cf. anionic polymerisation), but their feasibility is still uncertain. Other techniques have given some encouraging results. The recent advances in cationic grafting and some successful block copolymerisations using ingenious expedients (see below) are examples of such alternative approaches.

The literature on cationic polymerisation is vast and varied. Bibliographies covering the progress in the field up to 1974 are available[1−3]. As for an understanding of its problems and for thorough analyses, discussions and new concepts and theories, it is essential to consult the two classical books edited by Plesch[1−2] and more recent reviews and essays[4−19], and conference proceeding[20−23]. The quality of the latter monographs and collections varies considerably and the long-term relevance of some of them in terms of their effective contribution to the advancement of the field is questionable. We have nevertheless attempted to list all of them for the sake of completeness and to emphasise the abundance of criteria and theories presently available.

Why, then, another review?

We feel that several important considerations justify this venture. First, the specific topic we chose within the general subject of cationic polymerisation, i.e., initiation, is perhaps the most critical and we believe that the comments we have to offer might contribute to the clarification of some issues, or at least to the promotion of helpful debates.

It is our hope that our views will constitute a further step towards comprehensive and unifying theories of initiation. Second, we have attempted to establish a better connection between available evidence from various collateral disciplines and its relevance to the more specialised knowledge on initiation, often too isolated from the advances in other fields. In doing so, we found a considerable amount of pertinent information which seems to have been neglected in previous studies. Third, some interesting new material has been published in the last few years which requires a critical assessment and an evaluation of its real contribution to a better understanding of both the mechanisms of initiation and the nature of the chain carriers. Fourth, recent new theories on initiation, including our own views on this subject, made it necessary to reexamine some of the better older work in order to verify the validity of these new ideas in the face of large variety of experimental data and/or to reinterpret those results.

The scope of the present review is to discuss the various chemical pathways leading to the formation of reactive intermediates capable of inducing the growth of a polymer chain through a cationic or a pseudocationic mechanism. Essentially, therefore, we wish to study how systems achieve their basic function of generating chain carriers, as sketched out in the first of the themes enumerated above. Throughout this review, frequent comments and observations will also be made on problems related to propagation and other reactions. We consider that these inevitable extrapolations from the main topic complement the purpose of the essay rather than breaking its unity. It is in fact impossible to discuss the various modes of initiation without touching upon the qualities and properties of the different active species formed thereby.

We have decides to restrict our coverage to alkenyl monomers because the cationic polymerisation of monomers containing oxygen, nitrogen, sulphur and phosphorus as positively chargeable atoms is much better understood, given the higher stability of these intermediates. Progress recently made with these systems is well documented and reviewed[18,24–27,895,896]. In that field it seems that some clear answers have already been obtained concerning the critical problems listed above. Nevertheless, olefins and other vinylic monomers constitute an extremely important class of compounds, which has been the subject of the *majority* of studies in the area of cationic polymerisation, but for which the controversies over the mechanisms of initiation and propagation are still lively. Our choice, then, focusses upon intermediates whose reactivity is centered around a positively charged carbon atom viz. carbenium ions (free or paired) and polarised ester molecules.

The subject has been subdivided into specific sections according to the chemical type of initiator(s) or the physical technique used to induce initiation. These include both classical methods as old as cationic polymerisation itself, and more modern ones, such as photo-, electro- and nuclear initiation, which offer some interesting new ways of exploring the birth of chain carriers. We have devoted our main effort to the discussion of initiation patterns with Brønsted and Lewis acids, which are the most documented and, to our taste, the most attractive ones. The remaining sections are perhaps less thorough, but nonetheless we have attempted to touch upon all the fundamental problems and achievements.

The interest of studying the sometimes very stimulating recent work (up to mid 1979) was coupled with the rediscovery of the classics. We found that much buried evidence for our mechanistic schemes could be dug out from near oblivion, a very pleasing exercise and a homage to some farsighted pioneers.

II. Fundamentals

Before analysing systematically the various types of initiation, we wish to introduce some basic notions and ideas concerning: the characteristics and properties of the various starting components which make up a system; the ways in which these chemicals an interact and how certain physical parameters can influence the course of initiation; our present knowledge or hypotheses on the different types of intermediates formed as a result of these interactions; and, finally, the advantages and drawbacks of the experimental techniques presently available for carrying out fundamental studies on these processes.

A. Monomers

Within the scope of this review we shall only consider those compounds possessing one or more alkenyl functions susceptible to activation by electrophilic attack. Included in this family is a vast array of monomers varying in basicity from ethylene, which is so resistant to protonation that the ethyl carbenium ion has hitherto eluded observations even under the most drastic conditions (see below), and which in fact is equally resistant to cationic polymerisation, to N-vinylcarbazole, whose susceptibility to this type of activation is so pronounced that it can be polymerised by almost any acidic initiator, however weak. We shall also deal with olefins which, because of steric hindrance, can only dimerise (e.g., 1,1-diphenylethylene) or cannot go beyond the stage of protonated or esterified monomeric species (e.g., 1,1-diphenylpropene). The interest of such model compounds is obvious: they allow clean and detailed studies to be conducted on the kinetics and mechanism of the initiation steps and on the properties of the resulting products which simulate the active species in cationic polymerisation. The achievements and shortcomings of the latter studies will be discussed below.

In order to construct a quantitative scale of the basicities of these compounds, one would need to know precisely the values of the thermodynamic parameters involved in the formation of the parent carbenium ions and in their solvation in the media currently used in cationic polymerisation. Unfortunately, we do not possess such values. Some interesting approaches to this problem have been discussed by Plesch[28,29] since the early fifties. He considered various types of initiators and proposed ways of obtaining criteria for the free energies of appropriate active species, but could not establish absolute conclusions because of the lack of pertinent data in the literature. Thus, the thermodynamics of initiation processes have not progressed far enough to allow quantitative predictions. However, despite the above limitations, the large body of qualitative evidence which has been gathered in the last few decades, together with elementary structural considerations, allows us to assess the relative ease of protonation of monomers, at least within a homologous series. It must be made clear however, that the citerion of high proton affinity does not necessarily imply that the species formed in the initiation process will be stable and thus build up to appreciable concentrations. In fact, as will be discussed when dealing with carbenium ions, secondary reactions (isomerisation, alkylation, ion-pair collapse, etc.) can drastically reduce the lifetime of chain carriers, to the detriment of the overall efficiency of initiation.

We have broadly classified monomers polymerisable by a cationic mechanism into four main categories, viz.: (i) aliphatic mono-olefins (of which isobutene is the most studied); (ii) aliphatic dienes (cyclopentadiene being the best-known member); (iii) aromatic olefins (with styrene as the most important); (iv) monomers in which the vinyl group is attached to an electrodonating atom (vinyl ethers, N-vinylcarbazole, etc.). Within each category, the ease of initiation for the most representative monomers follows approximately the order given below:

(i)   isobutene > butene-2 > cyclohexene > propene > butene-1;
(ii)   cyclopentadiene > pinenes > isoprene > butadiene > chloroprene;
(iii)   p-methoxystyrene > 1,1-diphenylethylene > α-methylstyrene > indene > styrene > p-chlorostyrene;
(iv)   N-vinylcarbazole > vinyl ethers ≃ vinyl sulphides > chlorovinyl ethers.

Among the four categories, the qualitative order of basicity is (iv) > (iii) > (ii) > (i). Of course, these criteria are only given as an indication of broad tendencies. Factors such as the nature of the catalyst used, the type of counterion and the physical conditions prevailing, are important in determining specific results concerning the course of initiation.

One of the soundest approaches to the relative facility with which a monomer can be converted into an active species by electrophilic attack, with respect to another monomer, is to study copolymerisation behaviours. The values of reactivity ratios have shed considerable light on this question, since they provide a good indication about the predominance of the more basic monomer in forming chain carriers at the expense of the less basic ones. Abundant literature is available on this topic with some studies standing out because of their thoroughness[30,31]. Also, interesting attempts have been made to correlate experimental results with semiempirical calculations[32,33].

A brief survey of the literature is sufficient to remind the reader that almost every conceivable vinylic monomer that can be polymerised cationically has been tried[2,3]. In this review we will confine our attention to those papers which deal with fundamental aspects of initiation, i.e., kinetic and mechanistic studies. Thus, references to studies reporting *tout-court* the successful cationic polymerisation of a new monomer will not be found here. For such an inventory we refer to previous books and reviews[2,3,13].

We have not commented on the criteria determining the actual polymerisability of a monomer by cationic initiation, since they are well known[18,34,35]. The present review deals with monomers belonging to the π-donor (olefins, dienes, vinyl aromatics) and π- and n-donor (vinyl ethers, vinyl amines, vinyl sulphides) families, but not with purely n-donor and δ-donor monomers.

B. Initiators

One can promote the initiation of cationic polymerisation by a variety of chemical and physical means. Since the behaviour of a system depends very heavily upon the type of catalysis employed, we have organised the present review according to the nature of the initiator rather than to any other parameter. An introductory discussion about the characteristics of the substances of techniques used to induce these processes follows.

1. Brønsted Acids

In this section we shall consider the state of protonic acids in the pure state and in solutions of three classes of solvents[324]: (i) amphiprotic-protogenic (mineral and carboxylic acids), (ii) aproticdipolar-protophobic (e.g., acetonitrile and nitromethane), and (iii) aprotic-inert (aliphatic and aromatic hydrocarbons and their haloderivatives). While classes (ii) and (iii) represent the only two families of solvents relevant to cationic polymerisation (with the possible exception of the polymerisation of N-vinylcarbazole, which can be carried out in certain dipolar-protophilic solvents), class (i) is interesting because it represents the interaction between two Brønsted acids, the initiator and the *solvent,* as a direct source of protonating species. Although the latter combination has not been used in cationic polymerisation, we will discuss its potentials and possible drawbacks.

Most undiluted pure protonic acids, from the very weak carboxylic acids to the very strong perchloric and trifluoromethanesulphonic acids, display a covalent structure with little self-dissociation into ions, but often considerable association into dimers or higher aggregates through hydrogen bonding. The specific situations will be outlined individually in the next chapter. Suffice it to say here that the strength of an acid cannot be gauged, even qualitatively, by the extent of its dissociation in the pure state. This strength only becomes manifest in the presence of a suitable base or proton acceptor, in which case appropriate values of such acidity parameters as Hammett's H_0 can be measured. In rather simplistic terms one could draw a parallel between the H_0 values and the relative ease with which a given monomer would be protonated if dissolved in the pure acids. Unfortunately, this protonation capacity is not a sufficient criterion to establish the potential usefulness of a Brønsted acid as initiator in cationic polymerisation, although it does represent an important parameter. Other factors, such as the solubility of the acid in the common solvents used, its inertness towards them, the relative stability of the ions formed in the protonation reaction against their collapse to the corresponding ester, etc., can play a major role in determining the real qualities of these possible initiators.

Mixtures of Brønsted acids, i.e., solutions of a strong acid into a weaker one usually give rise to ionogenic interactions. The behaviour of acetic acid as solvent for strong mineral acids is typical: the protonation reaction

$$HB + AcOH \rightleftharpoons (AcOH_2)^+ + B^-$$

leads to appreciable dissociation of HB. Initiation systems involving this type of combination have never been tried in cationic polymerisation. Yet, protonating ionised species such as $(AcOH_2)^+$ should be quite efficient. Of course, one would not use acetic acid as medium for the polymerisation because the esterification of the monomer would dominate over propagation. However, a suitable solvent could produce an interesting situation from the point of view of the potentials of studying the initiation and propagation processes. The possibility of following the fate of the initiator's positive ions conductimetrically, coupled with spectroscopic measurements of the active species arising from the protonation of the monomer would be one of the advantages of using such catalytic mixtures.

Solutions of carboxylic acids or anhydrides in strong mineral acids also give rise to the protonation of the weaker component, but often the reaction goes further and leads to the formation of acylium ions. Thus, in the case of acetic acid dissolved in such strong acids as sulphuric, perchloric and trifluoromethanesulphonic acids, dehydration of the protonated species $(AcOH_2)^+$ gives the corresponding acetyl ester and/or its ionised form. The same endproduct is obtained if acetic anhydride is used instead of its acid:

$$AcOH + HB \text{ (excess)} \rightarrow (AcOH_2)^+ + B^- \xrightarrow{-H_2O} AcB \text{ (or } Ac^+ B^-)$$

$$Ac_2O + HB \text{ (excess)} \rightarrow AcB \text{ (or } Ac^+ B^-) + AcOH$$

Acetyl salts and esters have been used as initiators, but not in the presence of excess acid (see next Chapter).

Considering the large number of publications dealing with the mechanism of cationic polymerisations promoted by Brønsted acids in various solvents, it is surprising to note that relatively little work has been carried out on the state and properties of these acids in solutions of (ii)- and (iii)-type solvents. The scattered reports originate mostly from laboratories not involved in polymerisation studies and, as will be seen in the next Chapter, are sometimes of limited value. A collection of available pK_a's is given in Table 1. They were calculated from potentiometric and/or conductimetric data. A word of caution is needed on these values, since the formulation of the dissociation equilibria is often erroneous, and the residual level of moisture in the solvents was not always minimised by appropriate drying techniques. Nevertheless, Table 1 is a useful guide to the relative strength of some typical protonic initiators in solvents of medium and high polarity. Unfortunately, no data can be found concerning solvents of low dielectric constant. The extent of dissociation of even very strong acids in ethylene chloride is so small that the covalent form of these solutes predominate. Similar conclusions can be drawn from unsystematic conductivity values in methylene chloride in some polymerisation studies[44-46]. The higher values of the dissociation constants in acetonitrile (Table 1) can be attributed to two possible causes: the much higher dielectric constant of this solvent and the possibility that at least the stronger acids might protonate it, despite its classification as protophobic. The relative amount of molecular species is however still very high even in this polar solvent. As for the state of the undissociated acids in various solvents, thorough studies are available for carboxylic acids in media covering a wide range of polarities, showing the extent of aggregation through hydrogen bonding and the structure of dimers and other polymeric forms. The situation is less documented and often controversial in the case of other, stronger acids, as will be discussed in the next Chapter.

Table 1. Selected pK_a values in different solvents

Acid \ Solvent	AcOH	(Ref.)	$(CH_2Cl)_2$	(Ref.)	CH_3NO_2	(Ref.)	CH_3CN	(Ref.)
CF_3SO_3H	4.7	36)	7.3	39)	3.0	37)	2.6	37)
$HClO_4$	4.9	36)	~3[a]	40)	2.0	37)	1.6	37)
HI	5.8	36)	7.9	39)				
HBr	5.6	36)	8.7	39)			5.5	42)
HSO_3F	6.1	36)			4.3	37)	3.4	37)
H_2SO_4	7.0	36)					7.3	42)
HCl	8.4	36)	10.8	39)			8.9	42)
CH_3SO_3H	8.6	37)			6.0	37)	8.4	37)
HNO_3	10.1	36)					8.9	42)
CF_3COOH	11.4	38)	~7[b]	41)			10.6	43)
CCl_3COOH	12.2	38)					12.7	43)
$CHCl_2COOH$							13.2	43)
$CH_2ClCOOH$							15.3	43)
CH_3COOH	12.8	38)					22.5	43)

[a] In CH_2Cl_2, for homoconjugated *ion pairs*
[b] Probably wrong equilibrium, homoconjugation ignored

Homoconjugation of an anion with its parent acid has been observed, characterised and quantified for almost all acids known in numerous non-aqueous solvents and in bulk. The formation of these ionic aggregates through hydrogen bonding is often nearly quantitative, i.e., virtually no free anions are present amongst the ionised species resulting from the dissociation of an acid or its interaction with a common-anion salt. Despite this well-documented state of affairs[324], which has emerged steadily during the last few decades, chemists interested in cationic polymerisation have almost ignored the evidence accumulating and seldom considered that, whenever anions are postulated as a result of the attack of a protonic acid upon a monomer, they will homoconjugate with the excess of free acid present and will not "survive" as simple species. More important still, because of this strong tendency to homoconjugation, studies of the effect of common anions, purportedly aimed at suppressing the contribution of free ions towards propagation, are in fact meaningless if conducted in the presence of free Brønsted acids. All that the addition of a common anion achieves is in fact the partial or total sequestration of the original acid to give the weaker homoconjugated form. Warnings to this effect were first given by Sigwalt at the Rouen meeting[47] in the specific context of perchloric acid, but these were largely ignored in later studies by various authors. We wish to emphasize this point in the hope that the situation will be rectified. For this purpose, abundant literature citations will be given in the appropriate sections of this monograph and the specific instances of the extent and stoicheiometry of homoconjugation available in the literature or from our own recent work will be expounded in detail. Essentially, the simplest interaction between an acid and its anion involves one unit of each species:

$$HB + B^- \rightleftharpoons HB_2^-,$$

the equilibrium lying heavily to the right-hand side in most situations pertinent to cationic polymerisation. More complex instances giving hydrogen-bonded aggregates of the
type $(HB)_2B^-$ and HB_3^{--} have also been reported. Heteroconjugation between an anion
and an acid is also frequent[43], and although this phenomenon is less relevant to the
present context, it should be borne in mind as a possible complicating factor when studying the effect of added salts in polymerisations initiated by protonic acids.

Equally misleading is the assumption that free protons can exist as a result of the
dissociation of an acid in the solvents commonly used in cationic polymerisation. Unfortunately ionogenic equilibria are often written with H^+ as the positive moiety formed.
Unless protonation of the solvent can occur, a rare situation in the media under discussion, the proton will interact with the parent acid to give the H_2B^+ species. Thus, hydrogen
halides give halonium ions, e.g., H_2Cl^+ and hydroxylic acids give oxonium ions, e.g.,
$H_2O^+ClO_3$.

In view of the two main points outlined above, the proper form of writing the equilibrium for the dissociation of a Brønsted acid in an undiluted form or in non-protophilic
media is in general the three-to-two reaction

$$3\,HB \rightleftharpoons H_2B^+ + HB_2^-,$$

going of course through the intermediate ion-pair aggregate.

High hopes were held that cationic polymerisations promoted by protonic acids,
and particularly by the stronger ones, would yield a simple mechanistic behaviour,
including rapid and complete acid consumption to give a corresponding concentration
of protonated active species. The results gathered in the last twenty years have deluded
those hopes, but new interesting phenomena were uncovered, including pseudocationic
polymerisation[8]. It is the purpose of Chap. III to analyse initiation studies involving
protonic acids varying in strength from the very weak acetic acid (capable of polymerising N-vinylcarbazole) to the extremely strong trifluoromethanesulphonic acid
(very active even with monomers of limited basicity). We shall discuss there the unexpected complications often encountered and the major factors responsible for this lack
of straightforwardness in the mechanisms yielding active species. We shall also attempt
to draw up a unifying rationale to explain the *ensemble* of behaviours.

2. Lewis Acids

According to well-established notions, Lewis acids and bases can be defined as follows:
an acid is a species which enters into a reaction by employing an empty orbital, while
a base uses a doubly occupied orbital. It must be underlined that it is not necessary
that the donor and acceptor orbitals be localised on a single atom or between two atoms.
They can in fact be multi-centered, even in a relatively localised representation, as in
the interactions involving delocalised electron systems such as that of benzene with
iodine. Also, despite frequent definitions to the contrary, the donor and acceptor orbitals involved in Lewis acid-base reactions need not be non-bonding in nature. The highest
occupied orbital of the donor molecule can be either bonding or non-bonding and the
same applies to the lowest unoccupied orbital of the acceptor. In these interactions all

degrees of electron donation are encountered, ranging from extremely weak intermolecular forces to the transfer of an electron from the donor to the acceptor.

One of the basic concepts implicit in the Lewis definitions is the *relative* nature of acidity and basicity. The following considerations place the accent on this important point[48]. Consider two Lewis acids A_1 and A_2 and suppose that A_1 possesses a high positive charge density on its acceptor atom and a high-energy lowest unoccupied orbital, while the opposite situation characterises A_2. At the same time consider a series of bases B_1, B_2, B_3, etc. for which the net charge density at the donor atom follows the order 1, 2, 3, but the energy of the highest occupied orbital follows the reverse order. Given the properties of A_1, it is clear that its interactions will be charge controlled and consequently its reactivity with the bases will follow the order $B_1 > B_2 > B_3$. As for A_2, the nature of its interactions with bases will be dominated by the size of the energy gap between the two relevant orbitals, i.e., the closer the two energy levels the larger the interaction between acid and base. Thus, the reactivity of A_2 will reflect the apparent

Table 2. Donicity (DN) of selected donor solvents, taken from Ref.[50]

Solvent	DN/kcal mol^{-1}
1,2-Dichloroethane	—
Benzene	0.1
Thionyl chloride	0.4
Acetyl chloride	0.7
Benzoyl chloride	2.3
Nitromethane	2.7
Nitrobenzene	4.4
Acetic anhydride	10.5
Phosphorus oxychloride	11.7
Benzonitrile	11.9
Acetonitrile	14.1
Sulfolane	14.8
Dioxane	14.8
Ethylene sulfite	15.3
Propionitrile	16.1
Ethylene carbonate	16.4
Methyl acetate	16.5
Acetone	17.0
Ethyl acetate	17.1
Water	18.0
Diethyl ether	19.2
Tetrahydrofuran	20.0
Tributyl phosphate	23.7
Dimethylformamide	26.6
Dimethyl sulfoxide	29.8
Pyridine	33.1
Hexamethylphosphoramide	38.8
Ethylamine	55.5
Ammonia	59.0
Triethylamine	61.0

basicity order $B_3 > B_2 > B_1$. Inversions of this type are rather common and constitute the basis for dividing ions into hard or soft acids and bases[49,54].

Various criteria have been followed in an attempt to establish quantitative scales of acidity and basicity. In order to account for solvation and ionic dissociation phenomena Gutmann[50] introduced a parameter called donor number, DN, which correlates the behaviour of a donor solvent towards a given solute with respect to the coordinating ability of a reference solvent towards the same solute. The basicity of a solvent can be related to the enthalpy of its reaction with a reference acid. Gutmann's DN scale is built on the equation

$$ - \Delta H_D = a \times DN_{D-SbCl_5} + b $$

where the enthalpy of the reaction between a dilute solution of the donor and a dilute solution of antimony pentachloride (reference acceptor) is correlated with the donicity DN through two constants, a and b, characteristic of the system chosen. Some experimental values of DN are given in Table 2 and refer to interactions measured in 1,2-dichloroethane. Satchell and Satchell[51] have used a similar approach to measure the relative strength of a series of bases. Taking the hydronium ion as reference acid, the pK of its equilibrium interaction with a base in dilute solution was measured and correlated with the Brønsted pK_a of the base.

Drago[52] has proposed a four-parameter equation designed to predict acid-base reaction enthalpies in the gas phase or in poorly solvating media:

$$ - \Delta H_{AB} = E_A E_B + C_A C_B. $$

Table 3. C and E parameters of selected acids and bases, taken from Ref.[52]

Acid Parameters			Base Parameters		
Acid	C_A	E_A	Base	C_B	E_B
Iodine	1.00	1.00	Pyridine	6.40	1.17
Iodine monochloride	0.830	5.10	Ammonia	3.46	1.36
Phenol	0.442	4.33	Methylamine	5.88	1.30
tert-Butyl alcohol	0.300	2.04	Trimethylamine	11.54	0.808
Trifluoroethanol	0.451	3.88	Ethylamine	6.02	1.37
Pyrrole	0.295	2.54	Triethylamine	11.09	0.991
Boron trifluoride (gas)	1.62	9.88	Acetonitrile	1.34	0.886
Trimethylboron	1.70	6.14	p-Dioxane	2.38	1.09
Trimethylaluminum	1.43	16.9	Tetrahydrofuran	4.27	0.978
Triethylaluminum	2.04	12.5	Dimethyl sulfoxide	2.85	1.34
Trimethylgallium	0.881	13.3	Dimethylformamide	2.48	1.23
Triethylgallium	0.593	12.6	Ethyl acetate	1.74	0.975
Trimethylindium	0.654	15.3	Acetone	2.33	0.987
Trimethyltin chloride	0.0296	5.76	Diethyl ether	3.25	0.963
Sulfur dioxide	0.808	0.920	Benzene	0.681	0.525
Antimony pentachloride	5.13	7.38	Mesitylene	2.19	0.574
Chloroform	0.159	3.02	Hexamethylphosphoramide	3.55	1.52

Here the acid A and the base B are each characterised by two independent parameters, E being a measure of their ability to participate in electrostatic bonding, and C a measure of their ability to participate in covalent bonding. Drago has optimised the calculation of these parameters and obtained a set of values for 33 acids and 48 bases (Table 3). Depending upon the parameter taken into consideration, two ranges of acidity are clearly distinguishable, which goes to show once again that one cannot draw an absolute scale of Lewis acidity, since the strength of an acid depends on the base used as reference. The above equation has been shown to be inadequate for strongly interacting systems such as cation-anion reactions. For these systems Marks and Drago[53] developed another correlation,

$$ - \Delta H_{AB} = [(D_A - D_B)^2 + O_A O_B]^{0.5} $$

where again two parameters were used to characterise each interacting species. Some of these values are shown in Table 4. For a detailed discussion of the merits and weaknesses of these various approaches towards acidity and basicity scales we refer to the excellent review recently published by Jensen[48].

Table 4. D and O parameters of various acids and bases, taken from Ref.[53]

Acid parameters			Base parameters		
Acid	D	O	Base	D	O
H^+	311.6	81.95	F^-	−42.6	94.47
Li^+	132.6	9.14	Cl^-	−16.0	63.57
Na^+	112.6	5.86	Br^-	−10.5	54.50
K^+	100.0	3.46	I^-	−3.6	47.06
Cu^+	160.1	27.65	O^{2-}	−176.4	100.33
Ag^+	158.0	18.87	S^{2-}	−135.0	62.06
Al^+	155.1	17.54	Se^{2-}	−122.9	9.77
CH_3^+	204.8	50.00	Te^{2-}	−100.4	21.51
$C_2H_5^+$	173.4	40.01	OH^-	−48.1	211.44
$n\text{-}C_3H_7^+$	164.0	41.33	CH_3^-	−58.6	152.65
$C_6H_5^+$	196.1	55.44	$C_2H_5^-$	−54.6	159.35
NO^+	140.0	35.41	CN^-	−30.1	136.48
Cl^+	251.8	52.73	NH_2^-	−75.9	97.76
Br^+	224.5	38.48	C_6H_5	−39.9	150.92
I^+	196.5	25.73	NO_2^-	−33.8	0.07
Mg^{2+}	253.7	17.75	H^-	−29.8	145.22
Ca^{2+}	213.2	12.45			
Fe^{2+}	270.0	22.49			
Hg^{2+}	294.5	6.87			

Pearson has classified acids and bases into two major groups which he called hard and soft[54]. He stressed that hard acids prefer to associate with hard bases and react more readily with them, while soft acids associate and react preferentially with soft bases. The general equation describing these interactions relates the equilibrium constant for adduct formation from an acid A and a base B to two parameters

$$\log K = S_A S_B + \sigma_A \sigma_B,$$

S representing a „strength factor" and σ a hard-soft factor. While it is often stated that the softness is a measure of the susceptibility to form covalent bonding and the strength is related to the ease of electrostatic bonding, we prefer the approach which describes soft acids as species possessing low-energy acceptor orbitals and soft bases as species characterised by high energy donor orbitals, hard acids and bases having opposite properties. Pearson's well known equation is in fact analogous to the correlations discussed previously in that it makes use of two independent parameters to quantify the acid-base interactions.

Particularly relevant to the present context is the fact that the olefinic double bond is considered as a soft base in Pearson's theory, while many Lewis acids used in cationic polymerisation (BF_3, BCl_3, $AlCl_3$, etc.) are classed as hard acids. Obviously, π-acceptors like chloranil or tetracyanoethylene are considered as soft acids. Thus, the interactions between Lewis acids and olefins must be considered as very weak in the context of the HSAB theory. This prediction is well substantiated by the tenuous character of the complexes observed in experimental studies (see Chap. IV). On the other hand, carbenium ions are usually placed at the borderline between hard and soft acids and are definitely softer than the Lewis acids mentioned above. Consequently, their interactions with olefins must be rather strong, which suggests that that propagation in cationic polymerisations promoted by Lewis acids should be faster than initiation. This is often verified both in systems giving direct initiation and those operating through cocatalysis.

These considerations are however quite superficial and we feel that it is premature to apply the HSAB principles quantitatively to the complex interactions which characterise most cationic polymerisations, in view of the lack of precise knowledge about the relative softness of different olefins and the relative hardness of the Lewis acids typically used as initiators.

Several studies of the relative strength of Lewis acids, particularly the metal halides used in Friedel-Crafts reactions, have been published[55–59]. These investigations involved the measurement of the magnitude of the association (complexing) of the acids with a standard base using infrared or NMR spectroscopy. Frequent inversions in the scales thus obtained are noticed from one paper to another, showing that the specific context in which they were established plays, as we have already stressed, a very important role. These criteria do not bear therefore any absolute significance and regrettably cannot be transposed to the realm of initiation of cationic polymerisation in an uncritical fashion. They should rather be considerd as useful primary guides of general interest, along with other previous considerations[60]. Many other factors intervene to determine the quality and usefulness of a given Lewis acid in a cationic polymerisation, and its *strength* must therefore be viewed in terms of the conditions peculiar to this field, as will be stressed in Chap. IV.

Like Brønsted acids, Lewis acids are known to homoconjugate and to heteroconjugate with anions derived from them. This point, completely neglected by researchers in the field of cationic polymerisation, is particularly relevant to direct initiation mechanisms and will be amply discussed in Chap. IV, where the evidence for the occurrence of such interactions is reviewed and our views on their role in provoking the premature stoppage of a polymerisation are presented.

3. Carbenium Salts and Related Initiators

Two major categories of carbenium salts have been used to induce the cationic polymerisation of vinylic monomers. The first comprises the so-called stable salts, i.e., chemical species which can be prepared and isolated in a pure form, manipulated even at room temperature, (provided precautions are taken to avoid their destruction by nucleophilic impurities such as water, alcohols, etc.), and generally handled in the same way as other catalysts. Triphenylmethyl- and cycloheptatrienyl- carbenium salts with various Brønsted and Lewis anions are by far the most commonly used initiators belonging to this first category. Their remarkable stability arises from the high degree of charge delocalisation available in the cation. Their systematic use as promoters of cationic polymerisation began just over ten years ago and has today become common practice among specialists. The reasons for this success are to be found in the relative simplicity of the mechanism of initiation, at least with the most basic monomers, and therefore in the possibility of arriving at values of the propagation rate constant from fairly straightforward kinetic studies coupled with complementary data or assumptions. The chemistry of the initiation step has been clearly demonstrated in some specific instances involving very reactive monomers: electrophilic addition of the carbenium ion of the catalyst to the monomer double bond seems to be the most typical event. The rate of this addition, relative to the rate of propagation, varies from system to system and gives rise to more or less complex kinetic patterns for the overall polymerisation process. Some authors have oversimplified this issue and their quantitative arguments will be criticised in Chap. V. Other groups have conducted more rigorous and refined experiments and thus obtained valuable new information about fundamental aspects of the elementary steps in cationic polymerisation. Many problems still subsist, but current and future research should succeed in solving them. The major intrinsic drawback, however is implicit in the stability of these initiators. It concerns the inevitable fact that a large number of monomers of limited nucleophilicity are not attacked by these salts and therefore cannot be polymerised: isobutene represents the best example of this situation, while styrene sits on the borderline between polymerisable and non-polymerisable monomers. The considerations sketched above will be discussed in detail in Chap. V, which is devoted to the literature on this topic.

The second category of carbenium salts includes much less stable entities which cannot be easily isolated. *t*-Butyl-, benzyl- and 1-phenylethyl carbenium salts with such anions as BF_4^-, ClO_4^-, etc., are typical examples of such highly reactive initiators. Polymerisations induced by these salts or esters have been studied by carrying out the initiator's synthesis in the presence of the monomer to minimise the rapid side reactions which tend to destroy the original structure of the promoter. Since some of these carbenium salts are stable at low temperature, better results are obviously obtained by conducting the polymerisations below the temperature at which decomposition reactions become important. The instability of these compounds is naturally accompanied by a higher electrophilicity, and suitable conditions can be found whereby even the less basic monomers can be polymerised. This wider applicability is however accompanied by an inevitable loss of precision in the evaluation of such parameters as the rate of initiation and the concentration and nature of the chain carriers. Thus, the work published on this topic is more interesting for its qualitative novelty and for some mechanistic implications than for reliable quantitative results. The discussion of these papers appears in Chap. V.

Acylium salts with appropriate anions (BF_4^-, SbF_6^-, $CF_3SO_3^-$, etc.) can be easily prepared and have been thoroughly characterised[61]. The cation in these species is a resonance hybrid between the two canonical structures

$$R\overset{+}{C}=O \quad \text{and} \quad RC\equiv O^+$$

and can be used to initiate cationic polymerisation. This field is beginning to develop, and the results obtained are very promising as will be seen in Chap. V. Also, some unionised acyl esters, such as acetyl perchlorate, have been recently employed as effective initiators. Their mode of addition to the monomer double bond seems to proceed in a manner similar to the esterification of olefins (see Chap. V).

Halonium ions have received a great deal of attention in the last decade and a whole book has already been devoted to these species[62]. The stability of dialkyl-halonium ions decreases in the order $R_2I^+ > R_2Br^+ > R_2Cl^+$ (the fluoronium homologues have never been observed), and iodonium and bromonium hexafluoroantimonates have been prepared and characterised in Olah's laboratory[63]. Under specific conditions these salts are active initiators of the cationic polymerisation of olefins, and this still little-explored field will be briefly analysed in Chap. IV.

4. Physical Means of Initiation

Active species can be generated in a cationic polymerisation by the use of appropriate physical techniques. Ionising radiation in the form of γ-rays or vacuum-uv photons has been used for a long time as a means of producing bare monomer cations formed by electron ejection. In fact, the first reliable values of propagation rate constants were obtained by this technique. More recent studies by pulse radiolysis have also produced some interesting information concerning the electronic absorption spectra of intermediates formed in the initial phases of the polymerisation. This field has been enriched by new ways of preparing bare cations, viz. electric field ionisation of the monomer and nuclear reactions generating extremely acidic cations. The achievement of the techniques which have by now become classical, and the potentials of the newer ones for improving the quality and understanding of initiation will be discussed in Chap. VI.

Electrolysis of monomer solutions in the presence of supporting electrolytes can produce cation radicals which will promote cationic polymerisation. Unfortunately, the contribution of this technique to a better understanding of the fundamentals of initiation and propagation has been disappointingly meagre, as will be shown in Chap. VII.

The photolytic excitation of charge-transfer complexes is another recent addition to the available physical expedients to promote cationic polymerisation. The cation radicals generated by the photolysis have been characterised in some systems. More recent still is the use of ultraviolet radiation to induce the photolysis of substances whose photoproducts are initiators of cationic polymerisation. These processes will be discussed in Chap. VIII.

5. Donor-Acceptor Complexes

The area of chemistry encompassing molecular association phenomena has developed into a well-established discipline and various exhaustive monographs describing its advances are available[64,65]. A detailed discussion of these interactions is beyond the scope of the present survey and only those aspects directly related to initiation processes in cationic polymerisation will be examined.

Donor-acceptor associations include Lewis acid-Lewis base compounds, already discussed in Sect. II-B-2, the pairing or collapse of ions, which will be outlined in Sect. II-J, and the formation of complexes varying in strength from very loose species, through π-complexes, to charge-transfer complexes. In the literature on cationic polymerisation, one encounters a pronounced tendency to invoke the formation of *complexes* in order to rationalise certain kinetic and mechanistic observations. The term complex has in fact become almost ubiquitous and we cannot help feeling that it is too often used as an easy way out whenever a given set of results cannot be otherwise rationalised. By this we do not wish to imply that complexation phenomena are absent from the very wide spectrum of interaction which characterise most cationic initiations. We simply want to warn against the excessive and unsubstantiated practice of postulating such intermediates. Throughout this review, the role of π-complexes in initiation mechanisms will be dealt with whenever direct or circumstantial evidence supporting their formation has been obtained. For example, olefins are known to associate weakly with Lewis acids to give π-complexes, some of which have been characterised spectroscopically or by other physico-chemical techniques. These interactions will be discussed in Chap. IV. Complex formation between alkenyl monomers and Brønsted acids have also been frequently invoked, but their existence is not always obvious to us, or at least their relevance to the initiation process. These points will be dealt with in Chap. III.

While π-complexes play, as we shall see, a rather subsidiary role in the sequence of events leading to initiation in typical cationic systems, the formation of charge-transfer complexes can be directly responsible for the synthesis of active species. A charge-transfer complex, in Mulliken's meaning of the term[66], arises from the interaction of a donor molecule having a low ionisation potential with an acceptor molecule possessing a high electron affinity. The resulting adduct in its ground state is a resonance hybrid between the two canonical forms:

$$A + D \rightleftharpoons AD \rightleftharpoons A^- D^+,$$

the contribution of the ionised species being small. Electronic excitation of these complexes is energetically easy, as shown by their typical absorption bands often lying in the visible region of the spectrum. The excited state hybrid possesses a major contribution from the ionised canonical form. Various instances of cationic polymerisations promoted by both ground or excited state charge-transfer complexes have been reported. The species responsible for initiation in these systems are most probably the cation radicals arising from the donor when electron transfer to the acceptor has taken place. These studies are discussed in Chaps. VIII and IX.

The effect of *added* electron acceptors and donors to cationic polymerisation systems or model systems simulating the latter have been the subject of some recent studies[67,68]. Although the addition of this extra component can yield some useful information concerning the possibility of enhancing the overall activity of the process, it also adds a further complicating element to already intricate situations. Thus, the results so far published cannot easily be explained in mechanistic terms for lack of accurate information, and only qualitative tendencies assessed.

C. Solvents

The types of solvents suitable for cationic polymerisation have already been defined in Sect. II-B-1. It is obvious that since they have to sustain the existence of acidic species, they must be protophobic or neutral. Any tendency of the medium to interact with the electrophilic initiator would jeopardize the chances of building up a reasonable concentration of active species. Even with formally acceptable solvents, care must be taken to ensure that no specific chemical interaction occurs, particularly when using strong acids as initiators. Examples of these detrimental situations are perchloric acid in nitromethane (strong hydrogen bonding) and acetonitrile (acid loses effectiveness probably because of protonation of the solvent), and trifluoromethanesulphonic acid in methylene chloride (slow chemical reaction decomposing the acid). Also, with aromatic solvents it must be kept in mind that complexes, however weak, are formed with both Brønsted and Lewis acids and these might decrease the strength of the catalyst. These aspects will be treated in detail in the appropriate sections below.

If the dielectric constant of the medium is taken as a measure of its ability to solvate and thus stabilize ionic or polarised species, one can broadly classify the solvents relevant to this context into three classes: non-polar solvents ($\epsilon = 2-6$), solvents of modest polarity ($\epsilon = 7-15$) and polar solvents ($\epsilon > 15$). Table 5 gives examples for each class with the value of the dielectric constant at room temperature. The polarity of the medium can drastically influence both the kinetics of initiation and the position of the equilibria among the various species formed therefrom. One often encounters limiting situations where a given monomer-catalyst system will yield no polymerisation in a

Table 5. Dielectric constants of solvents suitable for cationic polymerisation (taken from Ref. [69]).
The temperature is 20 °C unless otherwise stated

Solvent	ϵ	Solvent	ϵ
Pentane	1.84	1,2-Dichloroethane	
Hexane	1.89	(Ethylene chloride)	10.36 (25 °C)
Heptane	1.92	Butyric anhydride	12.9
Cyclohexane	2.02	Isocapronitrile	15.5 (22 °C)
Carbon tetrachloride	2.24	Valeronitrile	17.4 (21 °C)
p-Xylene	2.27	Propionic anhydride	18.3 (16 °C)
Benzene	2.28	Phenylacetonitrile	18.7 (27 °C)
Toluene	2.38 (25 °C)	Butyronitrile	20.3 (21 °C)
o-Xylene	2.57	Acetic anhydride	20.7 (19 °C)
Pentachloroethane	3.73	1-Nitropropane	23.24 (30 °C)
Chloroform	4.80	2-Nitropropane	25.52 (30 °C)
Chlorobenzene	5.62 (25 °C)	Propronitrile	27.2
Ethyl chloride	6.29	Nitroethane	28.06 (30 °C)
n-Pentyl chloride	6.6 (11 °C)	Nitrobenzene	34.38 (25 °C)
n-Butyl chloride	7.39	Nitromethane	35.87 (30 °C)
n-Propyl chloride	7.7	Acetronitrile	37.5
1,1,2,2-Tetrachloroethane	8.20		
Dichloromethane			
(Methylene chloride)	9.08		

non-polar solvent, but will function adequately if the dielectric constant of the medium
is raised. Moreover, it must be remembered that, since the dielectric constant of most
monomers considered here is very low (ϵ = 2—3 for olefins and for vinyl ethers), a high
monomer concentration will influence the polarity of the polymerisation medium accordingly. Thus, at the limit, bulk reactions, where the monomer is the medium, corre-
spond to non-polar conditions. A few examples are also known where high catalyst
concentrations demand that its polarity be taken into account (e.g., styrene-chloro-
acetic acids, see Chap. III).

Apart from the generalised solvating power of the medium, there exist more specific
interactions between solvent and other components of a polymerisation system which
must sometimes be taken into account. We have already mentioned examples of catalyst-
solvent interactions of a detrimental nature. In other situations the solvent can play a
cocatalytic role, or have peculiar beneficial properties, as in the case of liquid SO_2, a
compound of modest polarity (ϵ = 12.4 at 22 °C), but very conductive towards cationic
polymerisation and related reactions[70,71]. The active species can also give specific reactions with some solvents, as in the case of aromatic hydrocarbons which can suffer
Friedel-Crafts alkylations under certain conditions. These transfer and termination reactions are however outside the scope of the present work.

On the whole the choice of solvent is by no means a matter of routine, but requires
instead a good deal of judicious thinking along the lines discussed above. One last point
worth mentioning, although not related to the solvents' chemistry or polarity, is the
ease of purification of a given solvent. Considerable work has been carried out by the
best schools in the field to establish optimal procedures for the purification and drying

of solvents, and for some of them classical methods are now available[72]. Nevertheless some solvents are considerably more difficult to purify than others, e.g., nitrobenzene. It is for this reason, and because they are generally non-specific and possess an acceptable polarity, that methylene chloride and ethylene chloride are by far the most frequently employed solvents in cationic polymerisation.

D. Temperature

The fourth major parameter which defines a system after the monomer, the initiator(s) and the solvent, is the temperature at which the polymerisation is conducted. The effect of temperature upon the position of the propagation-depropagation equilibrium (ceiling temperature) is not directly relevant and too well-known to be discussed here. We are obviously more interested in discussing the specific role of temperature in the reactions leading to the *formation* of chain carriers. The following considerations are pertinent to the kinetics of such interactions and to the thermodynamics of the resulting equilibria.

In order to achieve the best yields of active species without the concurrent formation of unwanted products, side reactions between monomer and initiator(s) and of the carbenium ions produced should be minimised, without however taking measures which could drastically reduce the rate of normal initiation. Temperature can play an important role in this search for the optimum. Frequently, the activation energy of the unwanted reactions is higher that that of initiation. In those instances a lower temperature will give cleaner systems, but the extent of this decrease will depend upon the effect it has on the main reaction. Note that these considerations are not related to the classical problem of molecular weights in cationic polymerisation, but refer strictly to the process of formation and preservation of the chain carriers. The best operating temperature will therefore depend on the values of the activation energies of the possible alternative steps and is usually established after a series of orientative runs. This kinetic reasoning must however be accompanied by its thermodynamic counterpart, i.e., the effect of temperature upon the equilibria among the active species formed. It is well known that the dielectric constant of the medium increases as the temperature decreases. Thus, free ions wil be favored by cooling, on this simple principle. As for the temperature dependence of ionogenic (e.g., ester-ion pairs) and dissociation (ion pairs-free ions) equilibria, no general rule can be given and the situation will depend on the specific system.

E. Residual Impurities

The best laboratories engaged in cationic polymerisation have earned their reputation partly because of the painstaking care with which they design and set up their experiments. The reasons for demanding procedures stem from the very pronounced reactivity of both catalysts and carbenium ions. Any nucleophilic impurity can alter substantially the overall behaviour of a system and thus produce erroneous conclusions, unless of course the "impurity" is added in a controlled fashion for a specific mechanistic purpose. In some instances electrophilic impurities can also give rise to mystifying results, particularly if the real initiator is weak.

The following examples taken from the literature and from our own experience illustrate amply the observations given above. In a classical study on the polymerisation of N-vinylcarbazole by p-chloranil, Natsuume et al.[73,74] were able to show that an acidic impurity in the catalyst was in fact responsible for the polymerisation, chloranil itself being almost inactive as initiator. A concentration as low as 4×10^{-6} M of impurity (identified and isolated) gave a polymerisation rate much higher than that obtained with 2×10^{-2} M of carefully purified chloranil in the same conditions. This excellent piece of investigation proved that much of the speculations about the mechanism of the so-called charge-transfer polymerisation of N-vinylcarbazole by chloranil was based on fictitious arguments since the predominant phenomena observed by previous authors had in fact been caused by normal cationic polymerisation induced by the acidic impurity. Hamid et al.[75] discovered that an impurity in N-vinylcarbazole persisted even after several recrystallisations from hexane, producing appreciable cocatalysis when mercuric chloride was used as initiator. Although the level of these cocatalytic impurities could be reduced by different purifying procedures, they were held so tenaciously by the monomer that their concentration could not be reduced below a certain minimum value. Cardona and Schultz[46] recently reported that two alternative procedures for drying methylene chloride gave surprisingly different results when this solvent was used as medium in the polymerisation of styrene by trifluoromethanesulphonic acid. It is likely that of the two drying agents, phosphorus pentoxide and calcium hydride, the latter might have been responsible for introducing basic impurities into the solvent (see below). Problems with solvent purification are more frequent than with other reagents. The elimination of the last traces of impurities from dichloromethane in the delicate study of direct initiation by Lewis acids has been the subject of a careful investigation by one of us[72]. A complicating factor often arises from the change in quality from batch to batch. Recently in our laboratory, we suddenly encountered serious difficulties in reproducing previous results with a cationic system involving methyl chloride as solvent. After much tedious searching, it became apparent that a new batch of solvent (same manufacturer, same nominal specifications) contained a new nucleophilic impurity, not removed by our conventional purification technique. This should serve as a warning against adopting published purification procedures uncritically. The quality of chemicals varies in fact from one manufacturer to another and can evolve with time, not always for the better.

If the removal of specific impurities is fundamental to obtaining reliable results, equally important is the minimisation of the residual water concentration in a cationic system. This experimental aspect deserves some comments. Although the attainment of "dry conditions" should be a *sine qua non* to anyone working in cationic polymerisation, the actual moisture level acceptable in a given study can vary appreciably with such factors as the type of initiation, the nature and concentration of the catalyst, etc. Thus, bare-cation polymerisations initiated by γ-rays or field ionisation as well as studies on direct initiation by Lewis acids require the maximum degree of dryness attainable by present techniques ($\sim 10^{-6}$ M of residual water). On the other hand, certain systems involving Br∮nsted acids or Lewis acids with a cocatalyst (among which water itself) can tolerate as much as 10^{-4} M of residual water, but only if it has been proved that these relatively high moisture contents are not detrimental to the normal course of the reaction.

The above considerations must sound trite even to readers who are modestly familiar with this field, since they have been emphasised many times before. Yet, we feel that to date too many papers aspiring to tackle fundamental aspects are marred by inadequate purification or drying procedures, or by one weak link in the chain of experimental operations. This unfortunate situation has been a frequent source of frustration during the compilation of this review, because we encontered numerous potentially interesting studies, spoilt by some degree of negligence in their practical execution. If, for example, a system is set up where all reagents are properly purified and dried and the reactor adequately conditioned, but then the introduction of the 10^{-4} M catalyst solution is carried out with a hypodermic syringe, the amount of moisture reintroduced into the medium by the latter operation can easily be equal to the initiator concentration. Any later mechanistic discussion on the self dissociation of the acid used, based on conductivity measurements, is obviously futile, since the acid was mostly present in a hydrated (and therefore more dissociated) form. Moreover, the relative amounts of water added in each run will be different, thus endangering the reproducibility of the system. This type of manipulation should be confined to exploratory exercises, but ought to be replaced by more rigorous ones for in-depth investigations. Our own experience and the many instances of potentially good work spoiled by these experimental "simplifications", have confirmed our belief in the qualitative superiority of high-vacuum techniques for this kind of research.

F. The Interplay Among the Factors Governing Initiation

Assuming that a system is free from interfering impurities, the processes leading to the formation of active species are regulated by four major variables:
(i)   The proton affinity or nucleophilicity of the monomer and the stability of the ionic and polar species formed from it;
(ii) the acidity of the catalyst and the properties of the anion(s) formed after initiation;
(iii) the polarity of the medium and the occurrence of specific solvation phenomena;
(iv) the effect of temperature upon the overall process.
The rate (if any) at which chain carriers are formed, their lifetime and the relative abundance of various active forms, from polarised ester molecules to free carbenium ions, depends mostly on how these four variables are adjusted. All types of limiting situations are possible: very fast or extremely slow initiation reactions; relatively long-lived active species or very ephemeral ones; and predominance of free ions or almost exclusive formation of ester molecules. Usually, however, initiation processes display a behaviour which is intermediate between extreme cases. A few examples will illustrate the effect of each variable. The basicity of the monomer can alter the course of initiation in a given system.

Thus, trifluoroacetic acid is an extremely efficient promoter of the polymerisation of N-vinylcarbazole in dichloromethane even at concentrations around 10^{-5} M. In the same medium 2-vinylfuran and p-methoxystyrene are polymerised at reasonable rates if the acid concentration is raised above about 10^{-3} M. Styrene only gives the corresponding ester and oligomers unless the acid is used as "solvent" and the monomer dropped into the acid. Finally, isobutene only forms the ester even if it is added to the pure acid. In the same context we have already mentioned that stable carbenium salts are good initiators for the more basic monomers, poor ones for styrene and butadiene and totally inactive for isobutene. Also, perchloric acid at room temperature gives pseudocationic polymerisation with styrene in dichloromethane, fast cationic polymerisation with p-methoxystyrene, and only esterification with isobutene. The acidity of the catalyst is a very important parameter in determining the efficiency of initiation. Styrene is not polymerised by hydrogen halides (only the corresponding 1-phenylethyl halide is formed) in chlorinated hydrocarbons, but it is polymerised by sulphuric acid to limited yields (i.e., the polymeric ester is inactive), by perchloric acid to completion by a pseudocationic mechanism around room temperature, and very efficiently by trifluoromethanesulphonic acid through what is probably a true cationic process. Also, the rate of isobutene polymerisation in a given solvent at a given temperature and with a given cocatalyst increases with the strength of the Lewis acid used, e.g., $SnCl_4 < TiCl_4 < BF_3$. Examples of the effect of the dielectric constant of the medium upon the course of the initiation process are numerous. In general, an increase in polarity, all other parameters remaining constant, produces an increase in the rate of initiation and in the concentration of free ions relative to other active species, as in the case of the system styrene-trifluoroacetic acid at room temperature. Exceptions to this general behaviour are those polymerisations in which the solvent reduces the activity of the catalyst by some specific interaction, as in the system styrene-perchloric acid-nitrobenzene at room temperature. The effect of temperature is usually more complex: in some instances a decrease in temperature favours the formation of ionic species (styrene-perchloric acid-chlorinated alkanes), in others it simply reduces the overall rate of initiation (2-vinylfuran-trifluoroacetic acid-methylene chloride).

While it is fairly easy to understand the role of one of these parameters when all others are kept unchanged, the lack of quantitative correlations among them, based on sound theories, makes it extremely difficult to rationalise changes of behaviour when two or more parameters are altered at the same time. Hence our incapacity, mentioned in the introduction, to predict in a rational way the behaviour of new systems. This state of affairs has led researchers to attempt a simplification of the issue consisting in studying model systems: conductivity and spectroscopy of stable carbenium salts and other ionic compounds simulating the chain carriers, to obtain information on the equilibria between free ions and ion pairs; kinetics and thermodynamics of the interaction of non-polymerisable olefins with Lewis and Brønsted acids, to examine the details of initiation in a less complicated context; polarography of stable carbenium and oxonium ions and of Brønsted acid solutions, to establish quantitative criteria about the electrophilic character of these species; etc. While a good deal of valuable information has been gathered by these studies, its extrapolation to actual polymerisation systems is often unsafe, unless certain inherent limitations are recognised. For example, polarographic measurements are inevitably carried out in media of high ionic strength, a situation which can alter appreciably the state of an acid with respect to that encountered in the absence of added salts. Also, the dissociation constant of a stable carbenium salt in a given solvent can be different from that of a polymeric chain carrier in the same solvent, but in the presence of monomer.

Much work is still needed in this field if it is to attain the degree of maturity already reached in anionic polymerisation. The refinement of research on model systems, coupled with the development of such techniques as stop-flow methods, applied directly to real systems, should provide adequate tools for this quest.

G. Carbocations

Spectactular advances have been made in recent years in the chemistry of carbocations. The development of techniques which permit the preparation of appreciable concentrations of these entities in "superacid" media, has made their direct study possible. Thus, even though in the past strong evidence existed to suggest the almost inevitable presence of such intermediates in numerous reactions, their identification remained a stumbling block, except for some very stable members. Given the short lifetime, and therefore minimal concentration, of these species in electrophilic reactions, one had to resort to indirect proofs for their existence as transients, i.e., kinetic results, reaction mechanisms based on product distribution, electrical conductivity, etc. Today, one can prepare virtually any carbenium ion and many carbonium ions from a variety of precursors using suitable acidic environments, and carry out "at leisure" detailed studies on their structure and stability. Also, isomerisation to more stable conformations, charge migration and certain electrophilic reactions can be made to proceed smoothly, so that their course can be closely followed by various techniques, particularly ^{1}H and ^{13}C NMR spectroscopy. Thanks to these advances, many hypotheses concerning the chemistry of these intermediates have been confirmed or rejected with a high degree of confidence. The work of Olah and his school has been especially significant in this extraordinary progress. Their discovery of superacid media, their detailed and systematic study of the structure and properties of all families of carbocations, and their prolific output of articles, reviews[76-78] and books[79] have contributed substantially to the establishment of a new era in this field. Having referred to this exhaustive literature, where the reader will find ample information about the properties and chemistry of carbocations and carboxonium ions, we will limit ourselves here to a brief discussion of those aspects of the topic more directly relevant to cationic polymerisation.

1. Classification

Olah's definition of two main classes of carbocations is most appropriate. A *carbenium ion* is a "classical" entity which contains an electron-deficient trivalent carbon atom possessing sp^2 hybridisation and six electrons in the valence shell, e.g., CH_3^+, $(CH_3)_3C^+$, $C_6H_7^+$ (benzenium ion), etc. The three atoms bound to such as carbon atom tend to be coplanar with it. A *carbonium ion* is a "nonclassical" entity containing a penta- or tetra-coordinated carbon atom which posseses three single bonds and a two-electron, three-center bond, i.e., eight electrons in the valence shell, e.g., $[CH_5]^+$, $CH_3CH_2\overset{+}{C}H_3CH_3$, etc. This classification[77,80] is actually based on experimental observations and represents the culmination of decades of studies and speculations on the nature of carbocations. Although carbenium ions had gained undisputed status among chemists even before their systematic identification, a considerable amount of controversy still existed around the definition and structure of nonclassical carbocations. The work of Olah's school dissipated most of these uncertainties: many carbocations were inspected by NMR, IR and ESCA techniques, and decisive evidence was obtained to show that some of them were in fact fast-equilibrating mixtures of classical carbenium-ion isomers (e.g., dimethylisopropyl ions, cyclopentyl ions, and ethylenearenium ions), while others possessed nonclassical pentacoordinated carbon atoms (e.g., the norbornyl ion and the cyclopropylmethyl ion). The existence of alkonium ions such as CH_5^+ was proved by mass spectrometry in the gas phase, but also from the chemical behaviour of alkanes in superacid media.

Since our present interest is the reaction of π-bonded systems, the carbon-carbon double bond of alkenyl monomers, with electrophiles to give classical carbenium ions (or esters) as end products, we will concentrate on these entities. Whether nonclassical carbonium ions ("π-complexes") are involved in these reactions as high-energy transition states, is still uncertain, and this interesting aspect will be discussed specifically in later sections (see Chaps. III and IV).

2. Preparation and Characterisation

Convenient preparation of appreciable concentrations of stable carbenium ions is achieved by dissolving at low temperature the corresponding olefins, alcohols or halides in such powerful acidic mixtures as SbF_5-SO_2, $SbF_5-SO_2F_2$, FSO_3H-SbF_5, and $HF-SbF_5$[77,78]. The characterisation of these species by spectroscopic techniques has become today almost a matter of routine. Less stable ions, formed as initial products upon mixing, often undergo rearrangements or electrophilic reactions with the substrate, and these can be conveniently followed by NMR spectroscopy. Finally, particularly reactive ions, such as the ethenium and the benzylium ones, have so far eluded observation by this preparative technique.

If physical-organic chemistry as a whole has profited substantially from these studies, it is legitimate to ask whether the specific field of cationic polymerisation has been equally affected in terms of a better comprehension of at least some of its many controversial aspects. The answer to this question is both yes and no. Yes, because at last ions sought for a long time, such as $(CH_3)_3C^+$ and $Ph(CH_3)\overset{+}{C}H$, are now known to exist and have been characterised spectroscopically. No, because the specific conditions in which these carbenium ions have been prepared, i.e., under the stabilising environment of superacid media, are not those used in cationic polymerisation. Thus, the stability, reactions, and general behaviour of these species cannot be extrapolated to the much milder situations typical of a polymerisation. It is notorious that chain carriers have seldom been "seen" during a polymerisation, and this implies that either the concentration of carbenium ions is too low to be detected by spectroscopic techniques, or these particular species are in fact not present. In other words, although we are sure today of the position of the absorption maxima (ultraviolet-visible) for a large number of carbenium ions derived from monomers (Table 6), the actual search for these intermediates in a polymerisation system is not as straightforward as one would hope, due to their limited lifetime, and therefore low concentration, or in the limit, to their absence and to the presence of other active species, e.g. ester molecules.

Dorfman and collaborators have recently developped a very promising technique for the production of carbenium ions as transient species in halocarbon solvents[81], based on the dissociative ionisation of suitable precursors induced by pulse radiolysis of the solvent. While the extremely interesting kinetic results which this group is obtaining will be discussed in Sect. II-G-4, it is emphasised here that the fast time response of the apparatus used allows the characterisation of carbenium ions hitherto unobservable because of their excessive reactivity. The ultraviolet absorption spectrum and some reactions of the benzylium ion have been studied for the first time with this powerful tool. From the point of view of cationic polymerisation, the information obtained in this type of work is particularly relevant, since it deals with the identification and reactivity of carbenium ions formed in very low concentration in the *right* kind of medium. Cation radicals had already been prepared by pulse radiolysis involving nondissociative ionization (electron ejection or transfer), as will be discussed in Sect. II-K.

The use of stop-flow techniques to observe the formation of carbenium ions in actual polymerising systems was introduced by Pepper et al. about ten years ago[82], and is presently exploited by various research groups with increasingly fast equipment. These experiments consist essentially in mixing monomer and catalyst solutions in an appropriate flowing system coupled with a rapid detection apparatus which takes absorption spectra and can measure other physical parameters, such as the electrical conductivity of the reaction mixture. This technique is certainly the most appropriate for studying the rise and fate of ionic active species in cationic polymerisation and the few, but remarkable, results obtained so far will be reviewed in the various sections dealing with specific systems.

Simpler procedures are of course available for the preparation and characterisation of carbenium ions in solution, particularly for the more stable ones. Concentrated sulphuric acid was extensively used as protogenic medium before the superacid mixtures were shown to be superior, but many of the spectroscopic assignements in those earlier studies were later proved erroneous, particularly in the case of such reactive entities as the 1-phenylethylium ion[83]. Model "monomers" which cannot polymerise because of steric hindrance can generate fairly stable carbenium ions by interacting with Lewis or Brønsted acids in normal cationic polymerisation conditions. Thus, 1,1-diphenylethylene and its dimer, and 1,1-diphenylpropene give rise to typical visible absorption bands from which the concentration of the corresponding diphenylmethylium ions can be accurately calculated. As for carbenium ions capable of forming stable salts, their synthesis and characterisation is obviously easy.

The problem of identification of carbenium ions *during* a cationic polymerisation can be summarised on the basis of the above considerations. Ideally, the most rigorous proof for the existence of these active species in a given system would be their char-

Table 6. Absorption characteristics of carbenium ions related to the cationic polymerisation of aromatic olefins, including model compounds. Only the longer wavelength maximum of each spectrum is given

Ion	Solvent	λ_{max}/nm	ϵ/M^{-1}cm^{-1}	Ref.
$Ph\overset{+}{C}H_2$	$(CH_2Cl)_2$	363	–	[85]
$\sim\sim CH_2-\overset{+}{C}HPh$	H_2SO_4	318	$\sim 3 \times 10^3$	[86]
$\sim\sim CH_2-\overset{+}{C}HPh$	H_2SO_4	315	$\sim 10^4$	[83]
$\sim\sim CH_2-\overset{+}{C}HPh$	CH_2Cl_2	340	$\sim 10^4$ [a]	[82]
$\sim\sim CH_2-\overset{+}{C}HPh$	$(CH_2Cl)_2$	340	$\sim 10^4$ [a]	[87]
$Ph\overset{+}{C}(CH_3)_2$	H_2SO_4	327	$\sim 2 \times 10^3$	[86]
$Ph\overset{+}{C}(CH_3)_2$	HSO_3F-SbF_5	326	1.1×10^4	[88]
$Ph\overset{+}{C}(CH_3)_2$	H_2SO_4	327	1.5×10^4	[83]
$\sim\sim CH_2-\overset{+}{C}H$ — C$_6$H$_4$ — Cl	CH_2Cl_2	325	–	[82]
$\sim\sim CH_2-\overset{+}{C}H$ — C$_6$H$_4$ — OCH$_3$	$(CH_2Cl)_2$	380	2.8×10^4 [b]	[89]
$\sim\sim CH_2-\overset{+}{C}H$ — C$_6$H$_4$ — CH$_3$	$(CH_2Cl)_2$	332	10^4	[90]
$CH_3\overset{+}{C}(Ph)_2$	HSO_3F-SbF_5	422	3.7×10^4	[88]
$CH_3\overset{+}{C}(Ph)_2$	CH_2Cl_2	435	3.3×10^4	[91]
$CH_3\overset{+}{C}(Ph)_2$	CH_2Cl_2	436	3.5×10^4	[92]
$CH_3\overset{+}{C}(Ph)_2$	H_2SO_4	427	3.9×10^4	[97]
$C_2H_5-\overset{+}{C}(Ph)_2$	CH_2Cl_2	438	3.5×10^4	[92]
$C_2H_5-\overset{+}{C}(Ph)_2$	H_2SO_4	434	3.5×10^4	[92]
3-Isopropylindanyl	H_2SO_4	318	2.9×10^4	[93, 94]
3-Isopropylindanyl	CH_2Cl_2	322	2.9×10^4	[93, 94]
3-Methylindanyl	H_2SO_4	312	2.7×10^4	[93]
3-Methylindanyl	CH_2Cl_2	318	3.0×10^4	[93]
2,3-Dimethylindanyl	CH_2Cl_2	320	2.7×10^4	[93]

Table 6 (continued)

Ion	Solvent	λ_{max}/nm	$\epsilon/M^{-1}cm^{-1}$	Ref.
3-Phenylindanyl	H_2SO_4	412	3.2×10^4	93)
3-Phenylindanyl	H_2SO_4	412	3.7×10^4	97)
3-Phenylindanyl	CH_2Cl_2	414	3.2×10^4	93)
Acenaphthyl	H_2SO_4	560	5.4×10^4	95)
Acenaphthyl	CH_2Cl_2	570	–	95)
	CH_2Cl_2	~420	$>5 \times 10^3$	44)
	CH_2Cl_2	468	1.6×10^4	96)

^a Assumed similar to that of the corresponding anion.
^b Determined in H_2SO_4.

acteristic NMR spectrum. However, the concentrations needed for such a proof to be implemented are far too high compared with the typical populations encountered in polymerising solutions, except perhaps in some suitable model experiments. Electrical conductivity is too undiscriminating a parameter to be used alone, and is only interesting for dissociation studies when complementary techniques have already established the presence and nature of ionic chain carriers. There remain the measurements of electronic absorption spectra. Since the extinction coefficient of the bands given by carbenium ions possessing aromatic substituents are usually higher than 10^4, and the range of their appearence spans from about 310 nm to the visible, it is in principle quite possible to detect concentration of ionic species (free or paired) of less than 10^{-6} M if aromatic monomers are used. Of course, these measurements imply that the bands studied have been unequivocally assigned in previous experiments involving concurrent NMR and ultraviolet-visible spectroscopy on the ion in question prepared in appreciable concentrations in superacid media. Unfortunately, no study of this type is possible with aliphatic monomers, because the carbenium ions obtained from their protonation exhibit absorption bands below 210 nm, where catalyst and solvent interference is too serious for any search to be made. In fact, there exists no means of identifiying carbenium ions derived from aliphatic olefins during a cationic polymerisation.

In Table 6 we have collected the absorption maxima for a series of carbenium ions directly or indirectly related to aromatic monomers. Some of these data were actually obtained during polymerisation, while others are the result of specific studies in very acidic (protonating) media. Ultraviolet spectra of carbenium ions derived from dienic and cyclodienic monomers have not been reported yet, but polymethylated homologous ions give absorption maxima between 275 and 310 nm, with extinction coefficients[84] of about 10^4 1 mole^{-1} cm^{-1}. Consequently, given the relatively accessible region of these peaks, some work could be carried out to search for carbenium ions during the cationic polymerisation of such monomers as isoprene and cyclopentadiene, using ultraviolet absorption stop-flow systems. No study has however been reported in this field.

3. Thermodynamic Stability

The question of the relative stability of carbenium ions in the gas phase and in solution has been discussed in detail in various monographs devoted to the thermodynamics of formation of these species[98,99], and we will only deal with recent important contributions in this area.

In the gas phase, Lossing et al.[100,101] have published studies of free-radical ionisation potentials measured by energy-resolved electron beam bombardment. From these values and the corresponding C–H bond dissociation energies, the ionic heats of formation of a number of aliphatic species were computed and relative stabilities assessed. A different approach has been adopted by Beauchamp et al.[102,103] and Hehre et al.[104], who have used ion-cyclotron resonance techniques to obtain criteria of aliphatic and aromatic carbocation stabilities.

Of course, the traditional problem of the lack of precise knowledge of the heats of solvation for the passage of these ions into solution, makes the above criteria of stability less valuable to the condensed-phase chemist. A major breakthrough in this classical impasse has been achieved by Arnett and coworkers[105–107] who have recently carried out calorimetric measurements leading to reliable values of the enthalpy of ionisation of various alkyl, cycloalkyl and aryl halides in solution. These determinations owe their validity to the use of superacid conditions and the NMR verification that the ions expected were in fact formed in those media without the occurrence of secondary reactions. One of the most important conclusion of these studies is that on the whole the relative stabilities of carbenium ions are the same in the gas phase and in the solvents used. i.e., electrostatic solvation effects do not alter the order of stability. The importance of this new experimental approach is quite obvious and one can except in the near future considerable advances in the field of the thermodynamics of reactive carbenium ions in solution through the attainment of a precise knowledge of $\Delta G°$ values for their formation in various media.

The results of Arnett and Petro in methylene chloride[107] are directly relevant to cationic polymerisation. The following values of ΔH_i/kcal mole^{-1} for the ionisation of RCl by SbF$_5$ are a good illustration of the relative stability of typical carbenium ions involved in polymerisations as possible chain carriers or as initiators: $(CH_3)_2\overset{+}{C}H$, -7.5; $(CH_3)_3C^+$, -15.5; $Ph(CH_3)_2C^+$, -19.0; Ph_2ClC^+, -23.7; Ph_3C^+, -27.1. This series confirms an expected behaviour, but stands as the first experimental set of values directly related to the thermodynamics of initiation.

4. Reactivity and Kinetic Parameters

The reactions involving, or supposedly involving, carbenium ion intermediates cover a vast domain which embraces various branches of organic and physical-organic chemistry. Friedel-Crafts alkylations, $S_N 1$ substitutions, hydration and esterification of olefins, eliminations and rearrangements, fragmentations and, of course, cationic polymerisations, are a few important examples of the array of interactions requiring the participation of carbocations. Numerous books, reviews and essays have been devoted to these reactions and they need no detailed exposition here. Only those aspects which can shed some light on the complicated mechanisms of initiation in cationic polymerisation will be taken up in the appropriate contexts. It is interesting to note that despite the enormous amount of literature on these topics, little kinetic work had been done until recently on the reactions of carbocations, except with the more stable ones (see Chap. V). This was due to the difficulty of identifying these reactive transients in the actual context of a chemical interaction. Only overall rate laws were usually obtained, relating to the reactants-products transformation, and at best intelligent guesses were offered on the rate constants for the reactions of the carbenium ion intermediate. In the gas phase, ion molecule reactions have been studied closely for many years and the situation is completely different, since mass-spectrometric techniques allow rate constants to be measured with a fairly high degree of confidence[108]. Radiolytic techniques have also been successfully applied to these interactions[109].

We have already mentioned that Dorfman and collaborators have developed a versatile technique to observe short-lived carbenium ions in solution generated by dissociative pulse radiolysis. This novel approach to the characterisation of transient species has also allowed this school to measure the rate constants of many electrophilic reactions between carbenium ions (the benzylium ion in particular) and various nucleophiles. In the first paper of the series Jones and Dorfman[85] reported the rate constants of the benzylium ion reaction with methanol, ethanol, the bromide and the iodide ions in ethylene chloride at 24 °C. Values of about 5×10^{10} M^{-1} sec^{-1} were obtained for the halide ions and of around 10^8 M^{-1} sec^{-1} for the alcohols. Later studies[110] confirmed that the reaction of halide ions with benzylium, diphenylmethylium and triphenylmethylium ions is at the limit of diffusion control. Reaction rate constants of these three carbenium ions with amines and alcohols were also reported in the same paper. More recently, these studies have been extended to include cyclopropylphenylmethylium ion as electrophile, ammonia as nucleophile and methylene chloride and trichloroethane as solvents[111,112]. These results are extremely important because for the first time they reflect the reactivity of a (bare) carbenium ion in absolute terms as a function of its structure (stability and steric hindrance), the medium and the type of nucleophile. Interesting conclusions were drawn by Dorfman and his school about the effect of phenyl and cyclopropyl substituents upon the reactivity of carbenium ions. The major contribution is a steric one, reducing the rate constant appreciably when three phenyl groups are attached to the same carbon atom. Contrary to expectation there is no important effect on reactivity due to electronic factors, i.e., charge delocalisation both with phenyl and cyclopropyl substitutions. The very high values of the rate constants of the mono and disubstituted ions with ammonia and amines (all around 10^9 M^{-1} sec^{-1}) confirm earlier estimates by Williams et al.

Table 7. Rate constants for the reaction of benzyl and diphenylmethyl cations with various olefins in 1,2-dichloroethane at room temperature. Experimental uncertainty $< \pm 20\%$. Results kindly provided by Wang and Dorfman[114]

Cation	Olefin	$k/M^{-1}sec^{-1}$
$PhCH_2^+$	ethylene	$< 10^5$
$PhCH_2^+$	propylene	1.9×10^6
$PhCH_2^+$	isobutene	1.9×10^7
$PhCH_2^+$	1,3-butadiene	8.7×10^5
Ph_2CH^+	1,3-butadiene	$< 10^5$
Ph_2CH^+	3-methyl-1,3-butadiene	7.1×10^6
Ph_2CH^+	2,3-dimethyl-1,3-butadiene	2.7×10^7

concerning the reactivity of bare carbenium ions towards water and other strong nucleophiles[113].

Wang and Dorfman[114] have recently measured the rate constants for the reaction of benzilium and diphenylmethylium ions with various olefins in ethylene chloride at room temperature. Some of these values are reported in Table 7. Again, the very high reactivity of bare cations in solvents of medium polarity is underlined, even towards olefins of moderate nucleophilicity. It is not clear to us why these reactions should proceed so rapidly, particularly in the case of the diphenylmethylium ion, but it is obvious that the absence of counterion is the major factor affecting these rates. More details about the product of the interactions will certainly help in clarifying the matter. If the reaction is simply an addition of the original cation to the olefin, these systems would constitute excellent initiators with polymerisable monomers, given the high rate of formation of chain carriers.

H. Anions

Except in the initiating systems involving electron ejection from the monomer and formation of bare cations, whenever carbocations are the chain carriers in a polymerisation, there will be an equal amount of anions, A broad distinction can be established among counterions, namely those derived from Brønsted acids, which can form a covalent bond with the carbocation (e.g., ClO_4^-, HSO_4^-, CF_3COO^-, $CF_3SO_3^-$), and those derived from Lewis acids, which are composite species incapable of forming a covalent bond by direct union with the carbocation (e.g., $SnCl_5^-$, $SnCl_4OH^-$, $TiCl_5^-$, SbF_6^-, $AlBr_4^-$, BF_4^-). The simple forms written here are not necessarily the real ones in a given system, where homoconjugation might be important in both categories of anions, as already discussed. The implications of homoconjugation are mainly related to an effective decrease of the real concentration of initiator available, a phenomenon too often overlooked or ignored.

In general the role of the anions derived from Brønsted acids should be viewed in terms of their nucleophilicity, e.g. the facility with which they can recombine with the carbocation. Anions like Cl^- and CH_3COO^- have a pronounced tendency to form covalent bonds with carbon and therefore, except under very special circumstances such as extremely basic monomers or very polar media, the lifetime of ionic chain carriers will be very limited with such counterions. As we shall see, this situation is in fact much more general than commonly believed, and ester formation is quite important with practically all Brønsted acids.

The composite anions derived from Lewis acids possess intrinsically the potential of undergoing a neutralisation reaction with the carbocation:

$$\sim\sim CH_2 - \overset{|}{\underset{|}{C}}{}^+ + MtX_{n+1}^- \longrightarrow \sim\sim CH_2 - \overset{|}{\underset{|}{C}}X + MtX_n.$$

The importance of this displacement reaction will depend upon the relative electrophilicity of the Lewis acid and the carbenium ion.

Given the possible sources of neutralisation of the anions just discussed, it is surprising that almost no attempt has ever been made at following the fate of these species during a cationic polymerisation, except by measurements of electrical conductivity, which are however non-specific and incapable of detecting ion pairs. The only studies aimed at this purpose all relate to the spectroscopic monitoring of $SbCl_6^-$ in the polymerisation of heterocycles. Penczek and Kubisa used this technique to follow some anomalies in the polymerisation of 1,3-dioxolane by $Ph_3C^+SbCl_6^-$ [115]. Other authors have similarly observed a decrease in the concentration of $SbCl_6^-$ when this salt was mixed with various cyclic oxygen compounds [116]. Goethals [117] has recently found that the polymerisation of some cyclic sulphur monomers reaches a limited yield when triethyloxonium hexachloroantimonate is used as catalyst, and has followed the fate of the anion proving that termination is due to Cl^- transfer from $SbCl_6^-$ to the growing sulphonium ion. We quote these studies which are outside the scope of this review simply because we wish to illustrate the principle on which one can follow the changes in anion concentration during a polymerisation. No such experiment has ever been conducted to our knowledge with a system involving a vinylic monomer. Some authors have occasionally searched for halogen in the polymers to test the occurrence of displacement reactions with composite anions, but the results have often been indecisive. We think that more systematic work should be carried out on the general problem of the fate of counterions in the course of polymerisations. This would involve prior research on adequate analytical techniques to follow quantitatively the concentration of anions, particularly the possibilities of ultraviolet spectroscopy ($SbCl_6^-$ gives a band [115] at 272 nm in CH_2Cl_2, $\epsilon \simeq 10^4$). The importance of these studies would be twofold: it would give an alternative means of assessing the concentration of ionic species, most probably related to the chain carriers, and it would give direct evidence for the occurrence of any reaction consuming anions.

I. Covalent Species

The interaction of an acid with an alkenyl monomer can generate ionic chain carriers, but also covalent products with varying degrees of polarity. It has been shown that in certain systems these ester molecules can propagate the growth of a polymer chain, while in others they are inactive. Another source of covalent species in cationic polymerisation is the collapse (recombination) of the ionic pair or the X^- displacement from the anion to the carbocation discussed in the previous section.

The existence of active ester molecules (pseudocationic polymerisation) was recognised in 1964[9], but these new ideas were not easily accepted and a widespread controversy over their validity continued until recently. Today, many of the original opponents of the theory of pseudocationic polymerisation have explicitly[118] or implicitly[119] accepted it. While there is no doubt about the fact that propagating covalent species are less active than their ionic counterparts, their importance can be overwhelming if the specific conditions under which a polymerisation is carried out is highly unfavourable to the presence of the latter. The limit of this situation is encountered when no ions are allowed in a system and the ester molecules are not sufficiently polarised or activated to be able to propagate: in this case the polymerisation ceases. In conclusion, if the covalent species are inactive, their formation can be viewed as a termination reaction, since it implies catalyst consumption; if they are the only type of chain carrier present in a system, propagation will be relatively slow; if they are active but in the presence of ionic chain carriers, they will contribute in some proportion to the composite character of propagation.

The actual observation of ester molecules in a polymerising system involving alkenyl monomers should be a priority among the tasks of future investigations in this field. Extremely interesting results have been obtained in similar studies with heterocyclic monomers (see[25] and refs. therein), particularly with tetrahydrofuran.

Considerable progress has been made in recent years concerning the preparation, characterisation and study of the stability and chemistry of alkyl and aralkyl esters of strong acids (those of weaker acids have always been better known). Thus, after the pioneering work of Burton and Praill on perchlorates[120,121], these esters have been the object of several studies dealing with their preparation and stability[122−126], and have indeed been used as initiators for cationic polymerisation prepared in situ[127,128]. As for triflates, the alkyl derivatives are readily prepared and show remarkable stability[129−132]; some aralkyl esters have also been prepared, particularly the benzyl[133], the 2-phenylethyl[134], and the 1-phenylethyl[135]. The triflate anion is a facile leaving group[136] and triflates are effective alkylating agents to the point that they have been used with success as initiators for the cationic polymerisation of cyclic n-donor monomers[137,138], but not, to our knowledge, of alkenyl monomers. Chloro- and fluorosulphonates have also been prepared[130].

J. Equilibria Among Active Species

The ionic chain carriers produced in the initiation process are composed of species possessing different degrees of association in equilibrium among themselves. If one considers that covalent precursors or products are also present in these equilibria, whether or not they participate in the propagation, the number of intermediates to be reckoned with is further increased. Two types of equivalent sequences can be written, according to the nature of the anion, B^- representing the counterion derived from a Brønsted acid or MtX_{n+1}^- that were formed from a Lewis acid:

$$RB \overset{K_1}{\rightleftharpoons} R^+B^- \overset{K_2}{\rightleftharpoons} R^+//B^- \overset{K_3}{\rightleftharpoons} R^+ + B^-$$

$$RX + MtX_n \overset{K_1}{\rightleftharpoons} R^+MtX_{n+1}^- \overset{K_2}{\rightleftharpoons} R^+//MtX_{n+1}^- \overset{K_3}{\rightleftharpoons} R^+ + MtX_{n+1}^-$$

where R+ is the growing carbocation and // indicates ion pair separation by a third entity such as a solvent molecule.

These equilibria play a fundamental role in cationic polymerisation. The different species usually possess an increasing propagating capability as one moves from left to right, and therefore the relative abundance of each component determines the activity of a given system. These considerations are common to all ionic polymerisation and have in fact been the object of very accurate studies in anionic polymerisation, where the various species involved are more stable and the equilibria can be investigated spectroscopically and conductimetrically[139,142]. The situation is much more complicated in cationic polymerisation. The well-known difficulty in achieving systems where the active species are sufficiently stable to allow measurements of their relative concentration, has so far marred most attempts to determine reliable values of the equilibrium constants K_1 to K_3 and thus arriving at individual propagation rate constants. Only in the field of hererocyclic monomers, some progress has been made in recent years, precisely because the ions and esters present in those systems are considerably more stable than those derived from vinylic monomers[25] and refs. therein).

Notwithstanding these quantitative difficulties, a considerable amount of evidence has been gathered in recent studies to prove that free ions and ion pairs are present in equilibrium in most cationic polymerisations and that the former are usually much more reactive than the latter. The role of ester molecules has already been discussed in the preceding section. As for solvated ion pairs, they have been included in the above equilibria because they are known to exist in many similar non-aqueous systems (including anionic polymerisation), but no direct evidence has ever been produced for their formation in a cationic polymerisation.

The equilibria can assume different configurations depending upon the four basic factors determining the character of a cationic polymerisation, viz. the strength of the original acid used (or the weakness of the conjugate base formed from it, B^- or MtX_{n+1}^-), the nucleophilicity of the monomer (or the acidity of R^+), the polarity of the solvent, and the temperature. Shifting of all the equilibria in favour of free ions will be strongest with the strongest acids, the most basic monomers, the media of highest polarity and low temperatures, and vice versa (see Sect. II-F).

Recent studies showing that bimodality is a frequent occurrence in molecular weight distributions from cationic polymerisation, particularly when Brønsted acids are used as initiators, confirm the existence of at least two types of active species in these systems, but also pose some problems. It is known in fact that ion pairs and free ions exist in *fast* dynamic equilibria and therefore during the lifetime of a polymer chain the ionic carrier responsible for this growth will oscillate between the two states, free and paired, a large number of times. This implies that the degree of polymerisation (DP) of the polymer molecules generated by these species will be an average between two limiting values (a high one for free ions and a lower one for ion pairs) and no separation into a bimodal distribution can be expected. If the origin of bimodality cannot be ascribed to the existence of different types of ionic chain carriers in equilibrium among themselves, it seems logical to assume that the origin of this phenomenon is the existence of slow-equilibrating covalent and ionic species. In other words, the ester molecules can propagate a chain independently of the ionic species and gives rise to the lower DP portion of the polymer. This interpretation seems to be corroborated by the fact that when Lewis acids are used as catalysts and no active ester molecules can be formed, polymodality is never observed. These considerations will be dealt with in greater details within the context of the conclusions to Chap. III.

A word of caution must be finally given concerning the assumed general validity of the postulate that the looser the ionic association, the higher their propagating potential. A few exceptions to this rule have been reported in anionic polymerisation[140,141] and unusually high reactivity of ion pairs in certain specific situations has been discussed by Szwarc[142]. To our knowledge, no such reactivity inversion has been published concerning cationic systems.

K. Cation-Radicals

Many techniques of initiation in cationic polymerisation are based on the abstraction of an electron from the monomer with the consequent formation of its cation-radical. Two thorough monographs on these intermediates, one dealing with their general chemical behaviour[143] and the other with their formation and role in polymerisation processes[144], have been published recently. We will therefore limit our comments to a few essential points.

As in the case of carbenium ions, one must distinguish between stable cation-radical salts which can be prepared and characterised without excessive precautions, and which owe their stability to considerable delocalisation of the positive charge and the unpaired electron, and transient, reactive cation-radicals (often without counterion) which undergo rapid secondary reactions immediately after their formation. While recently a few instances have been reported of initiation by the first type of species (see Sect. V– E), the presence of the less stable entities is characteristic in such initiation processes as the activation of charge-transfer complexes, the anodic oxidation of suitable anions followed by attack on the monomer, pulsed or continuos radiolysis of olefins, and electric field ionisation of superdry monomers. The identification of the latter transients has been the object of considerable research and pulse radiolysis techniques coupled with fast systems of detection have permitted the recording of ultraviolet and visible spectra of these species and of entities formed as a result of their rapid ensuing reactions.

Although the formation of cation-radicals is certainly the first step in the reaction sequences resulting from the use of the initiation techniques enumerated above, the subsequent steps are less well established. Various pathways are available to this first transient: coupling to give dications, reaction with the monomer to give the carbenium ion or the dimer radical-cation, etc. Eventually, however the carbenium ions become the real active species, i.e., all these polymerisations involve a composite initiation mechanism where the cation-radicals are playing just a momentary role.

L. Experimental Techniques

In this section we will briefly review the most important experimental tools which have been used or especially developed to follow cationic polymerisations, to characterise their products, and to conduct experiments related to the understanding of its chemistry. Particular emphasis has been placed on those techniques which have a specific bearing upon the problems of initiation. We have already underlined the capital importance of purification and drying in this delicate field and our preference for high vacuum work, whenever circumstances permit it. Obviously, the quality of the results which a given technique can provide depend on the extent of care taken in preparing the reagents and conditioning the reactor. We will not give here a detailed description of each technique or apparatus, but rather an assessment of their potentials, advantages and drawbacks. To our knowledge this type of overview has not been given previously, except for the short chapter on experimental techniques written by Plesch for his second book in 1963[2].

1. Dilatometry

One of the oldest methods employed for following the course of polymerisations with half life longer than about 15 min is based upon the volume contraction which accompanies these processes. This technique can easily be adapted to high-vacuum manipulations and is quite reliable, provided accurate calibrations are carried out particularly when oligomers are present among the products. Apart from the limitation imposed by the initial dead time, dilatometry is also confined in scope, since it can only provide empirical kinetic relationships between the polymerisation rate and such variables as the concentrations of reactants, the temperature, the polarity of the solvent, etc. It is therefore more useful when used in conjunction with other tools devised to probe more mechanistic aspects of the process. High-vacuum equipment provided with a dilatometer, a cell for electronic spectroscopy and electrodes for measuring conductivity can be assembled without difficulties[145]. It is also possible to build recording dilatometers, which can help considerably when other determinations have to be made during the reactions.

2. Adiabatic Calorimetry

All polymerisations being exothermic, one can make use of this property to follow their kinetics by adiabatic calorimetry. Strictly speaking, this technique is not kinetically rigorous because of the temperature rise during the reaction. However, activation energies in cationic polymerisation are usually not high and if furthermore reasonably low monomer concentrations are used, overall temperature changes can be reduced to one degree or less, without loss of precision.

The first recording adiabatic calorimeter allowing high-vacuum handling of cationic polymerisations was constructed in Plesch's laboratory[146]. Its basic design has been conserved in later versions built by various groups, but interesting improvements and additions have considerably enlarged the scope and versatility of this type of apparatus. As with dilatometry, adiabatic calorimetry on its own only allows to obtain empirical rate laws including internal and external orders. It is however a much faster device, being able to handle half lives of a few seconds (the best calorimeters built to date have a time constant of less than 0.5 sec), i.e. relatively fast polymerisations. Further additions of reactants after the first phenomenology has been observed and recorded, are easily carried out in the better models by crushing more phials held in the medium. Another advantage of this technique over dilatometry is the fact that polymerisations can be followed even at very low temperatures, a fact very hard to accomplish with dilatometers.

In order to widen the potentiality of these instruments to cover complementary mechanistic aspects of cationic polymerisation, special versions have been carefully developped by Plesch's and Sigwalt's schools[147]. They permit the concurrent recording of electronic spectra and electrical conductivity on the same reacting solution which is being probed by the temperature sensor. Considerable successes have rewarded these painstaking modifications.

3. Stop-Flow Techniques

The combination of rapid mixing and fast detection systems allows cationic polymerisations to be followed on an even shorter time scale than with adiabatic calorimetry. Recent commercial stop-flow spectrophotometers have a dead time of about 15 msec, an improvement of more than one order of magnitude over previous home-made models. This implies that reactions with half lives of less than 100 msec can be analysed kinetically with a good degree of accuracy. High-vacuum techniques are not compatible with these instruments and all operations are therefore carried out in an inert atmosphere.

New fundamental studies can, and have begun to, be carried out with these devices. Typically, one can follow simultaneously the decay of monomer concentration and the appearence of new absorptions, namely those due to active species. Thus, the kinetics of both initiation and propagation can be obtained together with direct spectroscopic evidence of the presence of ionic chain carriers. The addition of a pair of electrodes downstream provides further scope to the technique, since measurements of electrical conductivity tuned to the optical density changes for the active species can in principle allow the determination of the relative proportion of free ions and ion pairs in the polymerisation medium. Rapid quenching experiments are of course also possible, by injecting a killing agent at a specified time after mixing, and this could provide further interesting information about the chemistry of initiation.

Given the versatility of this technique one can expect a remarkable surge of new studies in the near future. The few already published (see Chap. III) certainly testify to the enormous potential of further research.

4. Electrical Conductivity

It is essential to realise that this technique, used on its own, can only reveal the presence of free ions in the reaction medium, without any further indication about their origin or nature. Low values of conductivity, typical of the limited concentrations of free ions possible in non-aqueous media of limited polarity, can in fact arise from the presence of a variety of charged species, not necessarily related to the processes under investigation. Thus, interactions of impurities present in the medium or on the vessel's walls with an acidic initiator can give rise to a conductivity background (and changes in conductivity values) well above the often small signals one would expect from ionic chain carriers. Since it is dangerous to take conductivity readings at face value as evidence for the presence of free carbenium ions in a polymerisation system, it is advisable to couple this technique with a complementary one, capable of providing supporting information about the real nature of these ions, e.g., ultraviolet spectroscopy.

The above observations are obviously not intended as a rejection of this very useful technique. We shall have numerous opportunities to underline the important contributions made by its judicious exploitation. Perhaps the most interesting application of conductivity in cationic polymerisation relates to measurements in model systems such as the study of the self-ionisation of initiators, the interactions of Lewis acids with cocatalysts, and of course the extent of dissociation of stable carbenium salts.

We feel that, provided one is aware of the pitfalls which beset an uncritical usage of this technique, and therefore interprets its results with due caution, it is always beneficial to have a pair of electrodes in a polymerising solution. Some of the information obtained might in fact help in unravelling the mechanism under study.

5. Electronic Spectroscopy

The application of ultraviolet and visible spectroscopy to the identification and measurement of carbenium ions derived from aromatic and dienic monomer has already been discussed (see Sect. II-G-2). The use of this technique to monitor stable carbenium salts is also well known. We have finally stressed in a preceding section that the fate of certain anions could be followed spectrophotometrically during a cationic polymerisation. The limits of detection allowed by the values of the extinction coefficients of all these species and by the sensitivity of present-day instruments is 10^{-6} to 10^{-5} M.

One of the best illustrations of the potentials of this technique is the extensive work of Sauvet, Vairon and Sigwalt[148] on the dimerisation of 1,1-diphenylethylene. Other important applications include studies of initiation rates with stop-flow apparatus, of rates of carbenium ion consumption in systems involving stable salts as initiators, and hydride shift reactions leading to the formation of polyenylic carbocations. The use of this technique in conjunction with high-vacuum manipulations is not only possible, but in fact strongly recommended for obtaining precise and reliable results.

6. Nuclear Magnetic Resonance Spectroscopy

It is well known that even modern Fourier-transform instruments require concentrations of at least 10^{-3} M to give a reasonably resolved NMR spectrum. Also, the minimum time necessary for a spectrum to be taken or for one given signal to be scanned is of the order of a minute. These constraints rule out the possibility of identifying short-lived transients or any compound present in low concentrations formed during a cationic polymerisation. Thus, while carbenium ions have been characterised by this technique under suitable conditions, their concentration in a polymerising solution is always too low to permit detection by NMR spectroscopy. On the other hand, one can easily follow the disappearance of the monomer and/or the formation of the polymer and of any other product (such as an ester) formed in appreciable quantities, an operation which is extremely convenient, not only from the point of view of the kinetics of polymerisation, but also because it can lead to mechanistic information concerning the nature and the structure of various products.

The application of this technique to the determination of the polymer structure and tacticity is now almost universally practiced and needs no further discussion.

The filling and sealing-off of NMR tubes on a high-vacuum line is a straightforward operation, which permits a rigorous way of following cationic polymerisations and related reactions.

7. Infrared Spectroscopy

Little has and can be done to profit from this technique in terms of gaining information concerning the initiation processes in cationic polymerisation. One could in principle use infrared spectroscopy to follow the rate of monomer consumption through the decrease of some typical vinylic band, but other techniques are superior and easier to handle. As for looking at products, polymer or other, *during* the reaction, this method is rather limited in scope, given the difficulties of characterisation from a band-crowded spectrum derived from such complex mixtures. Specific interactions, on the other hand, are easily and profitably followed and examined by infrared spectroscopy. Thus, for example, one can analyse hydrogen bonding phenomena and complex formation in model ractions, and extract important information about the mechanism of initiation. While gas-phase spectra can be taken with cells filled under high-vacuum conditions, the same procedure for liquid-phase work presents some technical problems.

8. Electron-Spin Resonance Spectroscopy

As already mentioned, cation-radicals are difficult to characterise in solution, unless they are derived from polynuclear hydrocarbons or similar stabilising systems. Only fast systems such as pulse-radiolysis coupled with spectrophotometry have permitted the identification of the cation-radicals of some common vinyl monomers. No report has been published to our knowledge on the detection of these species by electron-spin resonance spectroscopy *during* a cationic polymerisation. In systems where such intermediates have been postulated, ESR signals have occasionally been obtained in glassy matrixes, where cation radicals are trapped and therefore incapable of displaying their usual high reactivity. This technique is therefore a poor one for studying fundamental processes in cationic polymerisation, although it has sometimes been used to justify certain postulated initiation mechanisms.

9. Polarography

Plesch and collaborators have recently introduced this technique for the study of carbenium and oxonium ions[149] and of Brφnsted acid dissociation equilibria[150]. The ultimate aim of this work is of course to gain a better understanding of the mechanism of initiation in cationic polymerisation. For the moment however it is difficult to assess the real potential of polarography in that context and we feel that, although it certainly possesses a high sensitivity, the conditions required for meaningful measurements are still too distant from those employed in an actual polymerisation, in particular the high ionic strength of the medium. Its use is therefore still doubtful and one can only hope that more research will bridge the gap.

10. Gel Permeation Chromatography

The molecular-weight distribution of a polymer can sometimes offer interesting indications of the mechanisms by which its chains have been generated and built up. Of the various methods available for determining the DP distribution, gel permeation chromatography is today the most practical and reliable. As we have already mentioned, this technique has provided stimulating results concerning the bimodal character of some polymers obtained with Brønsted acids and olefins (p. 34).

11. Other Specific Chemical and Physical Methods

Depending upon the system under investigation, various specific tricks have been contrived for elucidating certain aspects of the mechanism of a cationic polymerisation. Since these ideas and methods are scattered through the literature, it is impossible to refer to all of them briefly and systematically. We will give just a few examples in this section and discuss the other important cases in the appropriate contexts throughout this review.

The use of labelled molecules to clarify a reaction mechanism has often been applied to initiation of cationic polymerisation. The most successful results have been obtained by Russell et al. in the context of water cocatalysis in the polymerisation of isobutene by stannic chloride.

Saegusa et al. have recently developed a method for killing a cationic polymerisation with phenoxide ions in order to "count" the number of active chains present in certain systems[151].

Cheradame et al.[93] have looked into the formation of titanium tetrachloride-indene addition products to prove the occurence of direct initiation.

One of the best studies of the relationship between polymer structure, end groups and molecular-weight distribution, and the mechanism of its formation is still an old classical study by Brown and Matheson[152] on the system styrene-trichloroacetic acid.

The determination of residual moisture has always been a difficult problem. Penczek and Slomkowski[153] have developed an ingenious method based on the reaction of the triphenylmethylium ion with water.

Plesch's and Sigwalt's schools have established a sound tradition in building specific devices for high vacuum handling of precursor solutions and cationic polymerisations. Some of these are highly sophisticated, others simple but extremely useful, such as Rutherford's mid-point method for determining the amount of substance contained in a vacuum-filled breakable phial[154].

III. Initiation by Brønsted Acids and Iodine

Although the reasons for the inclusion of iodine in this chapter will become more obvious as the discussion develops, we wish to anticipate that the formation of diiodides and hydrogen iodide in the interaction of iodine with olefins is a well-established feature which determines a close similarity between these systems and those involving Brønsted acids.

The reactions between protonic acids or iodine and olefins has been extensively studied under a wide variety of conditions. Four major types of investigations dealing with the kinetics and mechanism of these processes are of interest in the present context:

— the hydration of olefins and the determination of the rate constants of "protonation" in aqueous media;
— the modes of addition of acids onto the double bond in non-aqueous media;
— the kinetics of protonation of olefins;
— the polymerisation of olefins, with particular emphasis on the phenomenology of initiation.

While the last topic is the focal point of this chapter, the other three represent an essential complement to its understanding, particularly in the light of much important work recently published.

A. The Hydration of Olefins

In an excellent monograph published more than ten years ago, de la Mare and Bolton[155] discussed the various aspects of electrophilic addition to unsaturated systems. The hydration of olefins was postulated to occur via the rate determining reaction

$$
\begin{array}{c}
{}^+H\cdots OH_2 \\
\vdots \\
RCH\!\!=\!\!CH_2 \;\; \rightarrow \;\; R\overset{+}{C}H\!-\!CH_3 + H_2O \\
a \qquad\qquad\qquad b
\end{array}
$$

involving an undefined complex a. The following collapse of the carbenium ion b with the nucleophile OH^- was thought to be very fast. These concepts have been reassessed as a result of more recent work and Nowlan and Tidwell[156] have formulated a unified mechanism based on general acid catalysis proceeding through an $A_{SE}2$ transition state. The rate-determining step is therefore the protonation of the olefin

$$H_3O^+ + \overset{\backslash}{\underset{/}{C}}=\overset{/}{\underset{\backslash}{C}} \rightarrow H-\overset{|}{\underset{|}{C}}-\overset{|}{\underset{|}{C}}^+ + H_2O$$

followed by its rapid reaction with water to give the corresponding alcohol and regenerate the hydrated proton. Tidwell's school has conducted extensive research on a large number of alkenes and applied successfully Brown and Okamoto's concepts[157] related to the electronic effect of substituents in resonance with a positive charge. Schubert and Keeffe[158] had already obtained excellent correlations between σ^+ and the rate constant of hydration for a series of substituted styrenes, thus showing the positive character of the transition state of the rate-determining step. Tidwell's group[159,160] extended these considerations to non-aromatic olefins and expanded considerably the range of values of the hydration rate constant, to cover about 18 powers of ten at 25 °C. The linear relation between $\log k_2$ ($= k_{obs}/H_3O^+$) and σ^+ was most satisfactory, while the parameter σ_p, which fails to take into account direct resonance, proved inadequate.

In conclusion, overwhelming evidence points today to the general acid catalysis of these reactions and to the formation of a short-lived intermediate with the characteristics of a "symmetrical" open ion. This behaviour includes an enormous range of nucleophilicity of the substrate, from ethylene to 1,1-diethoxyethylene. Thus, the postulation of a π-complex precursor in the mechanism of acid-catalysed hydration of olefins is now unjustified, and the second order rate constants experimentally obtained are in fact a reflection of the protonation reaction of the hydronium ion onto the double bond.

B. The Kinetics and Mechanisms of the Addition of Acids to Olefins

The topic of electrophilic addition of Brønsted acids to the carbon-carbon double bond has been surveyed by various authors up to a few years ago[155,161,162]. Since these reviews were published, some interesting work has appeared in the literature and we will concentrate on the latter studies in relation to previous theories and generalisations.

As pointed out by Fahey[161], three types of limiting mechanisms can be formulated for these reactions. In the gas phase, molecular syn-addition through a polarised transition state is the dominant feature of the process, which follows in general the Markownikov rule:

$$\overset{\backslash}{\underset{/}{C}}=\overset{/}{\underset{\backslash}{C}} + HB \rightleftharpoons \left[\begin{array}{c} H \cdots B^{\delta-} \\ \vdots \quad \vdots \\ -\overset{|}{C}{=}\overset{|}{C}{}^{\delta+} \\ / \quad \backslash \end{array} \right]^{\neq} \rightleftharpoons \begin{array}{c} H \quad B \\ | \quad | \\ -\overset{|}{C}-\overset{|}{C}- \\ | \quad | \end{array} .$$

The reversibility of this reaction is well illustrated by studies on the pyrolysis of esters and halides[163,164]. In solution, two types of mechanism govern the course of the addition, viz., the Ad_E2 process involving a carbenium ion intermediate and giving both syn- and anti-addition,

$$
\begin{array}{c}
\diagdown \diagup \\
C=C \\
\diagup \diagdown
\end{array}
+ HB \xrightarrow{\text{slow}}
\begin{array}{c}
\underset{|}{\overset{H}{\underset{|}{}}} \\
-\overset{|}{C}-\overset{+}{\underset{|}{C}} \\
\end{array}
+ B^-
\xrightarrow{\text{fast}}
\begin{array}{c}
\overset{H}{\underset{|}{}}\ \overset{B}{\underset{|}{}} \\
-\overset{|}{C}-\overset{|}{C}- \\
\end{array} ,
$$

and the Ad_E3 mechanism in which the simultaneous addition of "H" and "B" onto the double bond takes place from two distinct molecules of acid (again syn- or anti-), e.g.,

$$
\begin{array}{c}
\diagdown \diagup \\
C=C \\
\diagup \diagdown
\end{array}
+ 2\,HB \xrightarrow{\text{slow}}
\left[
\begin{array}{c}
\delta^- \ B \cdots H \ \overset{\delta^-}{B} \cdots H^{\delta^+} \\
\diagdown \ \vdots \quad \vdots \diagup \\
C \cdots C \\
\diagup \quad \delta^+ \diagdown \\
(syn)
\end{array}
\right]^{\neq}
\xrightarrow{\text{fast}}
\begin{array}{c}
\overset{H}{\underset{|}{}}\ \overset{B}{\underset{|}{}} \\
-\overset{|}{C}-\overset{|}{C}- \\
\end{array}
+ HB.
$$

The latter two mechanisms are most relevant to cationic polymerisation and deserve a close inspection through the evidence published in favour of each one in the recent literature.

Most of the detailed kinetic and mechanistic studies have been conducted with HCl, HBr, CH_3COOH and CF_3COOH. Pocker et al.[165,166] have studied the addition of HCl and DCl to various substituted butenes and cyclopentenes in nitromethane. In this solvent hydrogen chloride exist predominantly as undissociated monomeric HCl. The kinetic law governing these additions was third order overall, with $R = k_3$ [olefin] $[HCl]^2$. Given the lack of effect of added chloride ions on the rate constant for addition and other mechanistic considerations, it was concluded that the initial step was the interaction of two molecules of undissociated acid with one of olefin. The question of the intermediate activated species leading to the chloride is of particular interest: Pocker et al. preferred to postulate the formation of the ion pair R^+, HCl_2^-, but did not rule out the possibility of synchronous addition of proton and anion by two molecules of acid. In fact, the Ad_E3 mechanism had not yet received much attention at the time, but it seems now clear that the systems studied by Pocker and collaborators behaved according to that pathway.

Fahey's school has produced a series of very interesting papers dealing with the addition of HCl to various olefins in acetic acid. Kinetic studies including isotope effects and the role of added chloride ions were coupled with detailed product analysis. Styrene and ter-butylethylene[167] were found to react with molecular HCl (or its ion-pair form) in a typical Ad_E2 fashion:

$$
\begin{array}{c}
\diagdown \diagup \\
C=C \\
\diagup \diagdown
\end{array}
+ HCl \xrightarrow{\text{slow}}
\left[
\begin{array}{c}
Cl^- \quad H \\
\diagdown \ \overset{+}{} \diagup \\
C-C- \\
\diagup \quad \diagdown \\
AcOH
\end{array}
\right] ,
$$

this rate determining step being followed by fast collapse of the ion pair to give the corresponding chloride or by the fast addition of AcOH onto the carbenium ion followed by elimination of HCl (acetate formation). Cyclohexene on the other hand was shown to react predominantly via anti Ad_E3 pathways[168,169] involving the two transition states

$$
\left[\begin{array}{c} \overset{H\quad Cl^-}{\underset{Cl^\delta}{\overset{\delta^+}{C}}{=}C} \end{array} \right]^{\neq} \quad and \quad \left[\begin{array}{c} \overset{H\quad Cl^-}{\underset{AcOH}{C}{=}C} \end{array} \right]^{\neq} ,
$$

although a second order component was also found in these reactions (Ad_E2). Thus the rate law contained three terms:

$$
R = k_2 \text{ [olefin] [HCl]} + k_3 \text{ [olefin] [HCl] [Cl}^-\text{]} + k_3' \text{ [olefin] [AcOH] [HCl]}
$$

A very similar behaviour was encountered with 1,2-dimethylcyclohexene[170]. In this study, Fahey and McPherson investigated various solvents apart from AcOH and found that in methanol, a stronger ionising medium, the Ad_E3 mechanism prevailed, in acetyl chloride and CH_2Cl_2 the Ad_E2 route was more favoured, and in pentane the addition involved two molecules of undissociated HCl, i.e. a different type of Ad_E3 intermediate.

The addition of hydrogen bromide to olefins in acetic acid has been the subject of two recent studies[171,172] in which both the kinetics and the stereochemistry of the reaction were investigated. The predominance of Ad_E3 mechanisms were proved, syn and anti additions taking place, the latter being favoured. Different types of transition states were postulated, depending on the olefins involved in each study, but they all involved simultaneous attack by HBr at one carbon and a second moiety (N = Br^-, HBr or AcOH) at the other:

$$
\left[\begin{array}{c} H{\cdots}Br^{\delta-} \\ \overset{\delta^+}{-C}{\cdots}C{-} \\ N \end{array} \right]^{\neq} \quad \text{(anti } Ad_E3 \text{ transition state)}
$$

Fahey and collaborators have repeatedly stated that an analysis of older work on these addition reactions points to the frequent occurrence of the Ad_E3 mechanism.

As pointed out by Pasto et al.[172] the Ad_E3 mechanism does not necessarily imply a termolecular collision in the transition state. One can in fact envisage the formation of an acid-olefin complex producing the necessary positive polarisation on the carbon atom prior to the rate-determining step consisting in the formation of the activated complex with the in-coming nucleophile. Such complexes have been identified in various investigations[173,174].

The occurrence of an Ad_E3 mechanism has also been shown in the reaction of hydrogen iodide with cyclohexene in acetic acid[175]. Moreover this mechanism seems to apply equally well to the addition of HCl to acetylenic systems[176,177].

The esterification of the olefinic double bond with acetic acid is usually accomplished in the presence of catalytic amounts of a strong acid ($HClO_4$, H_2SO_4, HSO_3CF_3).

Coussemant et al.[178,179)] have studied the influence of substituents on the rate of addition of AcOH to styrenes and established that the good correlations obtained with H_0 and σ indicate the classical behaviour of this family of compounds, in terms of general acid catalysis and substituent effect. Unfortunately, however, no mechanistic interpretation was given to the course of these reactions. The addition reaction of carboxylic acids to cyclopentene and cyclohexene in the presence of various salts (perchlorates, acetates, methanesulphonates) and catalysed by strong acids was studied by Guenzet et al.[180–183)]. They showed that while the ionic strength has no effect on the rates, specific solvation and common-ion effects are important. The general mechanism involves the protonation of AcOH followed by the rate-determining addition of $AcOH_2^+$ to the double bond, the rate being proportional to the first power of both the olefin and the acid concentration. More recently, Roberts[184)] has studied the addition of AcOH to cyclic and bicyclic olefins in the presence of trifluoromethanesulphonic acid. This fine piece of work showed that $AcOH_2^+$ could not be the proton source for the olefin and instead a mechanism was proposed involving the attack of the stronger acid in its ion-pair form to give a π-complex with the olefin, which in turn gives the σ-bonded state (protonation). The rate-determining step was the "solvation" of the latter ion pair by acetic acid followed by the rapid collapse of this solvent-separated species to give the corresponding acetates in a non-stereospecific process. The transition state can here be viewed as a complex entity resembling those postulated in Ad_E3 mechanisms, with the concomitant participation of both acids around the nucleophilic center. Gandini and Prieto[185)] have studied the kinetics and mechanism of the addition of both acetic and chloroacetic acids to N-vinylcarbazole in various solvents. Owing to the very high nucleophilicity of this olefin, these reactions proceed at appreciable rates without the assistance of a catalyst. Strong evidence was obtained in favour of an Ad_E3 mechanism requiring two molecules of acid in the activated complex.

Pasto and Gadberry[902)] recently investigated the addition of acetic acid to 2-butene in the presence of various acidic catalysts. With DCl, DBr and CH_3SO_3D, the lack of diastereomerisation, positional isomerisation and hydrogen-deuterium exchange clearly indicated the occurrence of an Ad_E3 addition mechanism "inconsistent with the reversible formation of intermediate tight ion pairs or free carbenium ions". With the much stronger catalyst triflic acid all evidence pointed to an Ad_E2 pathway involving carbenium ion intermediates.

Trifluoroacetic acid is a strong enough acid to add onto common olefins without the need of strong-acid catalysis. Peterson et al.[186,187)] have studied the effect of remote substituents on the rate of addition of this acid on various types of olefins in the absence of solvent. A good correlation of the inductive effect of the substituent upon the rate of addition as a function of its distance from the double bond was obtained, except for 5-halo-1-pentenes which showed deviations due to substituent participation in the formation of the transition state (cyclic halonium ions). A similar phenomenon was observed by Mason and Norman[188)] in the esterification of 5-phenyl-1-pentene by trifluoroacetic acid, as part of a wider study. Anchimeric assistance by the phenyl group with formation of a bicyclic carbenium ion intermediate was proved by product analysis and ring-substituent effects. Brown and Liu[189)] have studied the addition of this acid to norbornene and other bicyclic olefins with the purpose of proving that the intermediate carbocation is in fact the classical unsymmetrical species. This investigation was entirely devoted to the solution of the controversy about the classical or non-classical nature of that ion and did not touch the kinetic aspects of the reaction. Much more revelant to the present context are two studies concerning the kinetics of addition of trifluoroacetic acid to $(+)$-(R)-limonene and isobutene respectively. In the first Roberts[190] has shown that the Ad_E3 mechanism applies in solvents benzene, 1,2-dichloroethane and probably nitromethane, where synchronous attack by two monomeric acid molecules produces the transition state

$$
\left[
\begin{array}{c}
H-OCOCF_3 \\
\vdots \\
{>}C{\cdots}C{<} \\
\vdots \\
HOCOCF_3
\end{array}
\right]^{\neq}.
$$

In cyclohexane the reaction was also first order in olefin and second order in acid, but the predominantly dimeric state of the acid in this solvent suggested that the one molecule of the aggregate was responsible for the attack on the olefin. Of course the transition state might again involve the concerted participation of two acid molecules arising from the opening of the dimer. In the second study, Latrémouille and Eastham[191] also obtained third order kinetics for the reaction of trifluoroacetic acid with isobutene in 1,2-dichloroethane. Given the low value of the dimerisation constant in this solvent[192], it seems obvious that here too the Ad_E3 mechanism should hold, a point which was in fact implicitly touched on by the authors in the discussion of their results.

The above summary of esterification studies delineates the present state of the theories concerning the mechanism of these reactions. While the Ad_E2 mechanism seems to be favoured by stronger acids in ionising media, it becomes evident that the alternative Ad_E3 route has gained a very firm status and appears to be the rule rather than the exception. Thus, the formation of open carbenium ions as intermediates in these interactions is by no means a general feature – a fact which bears important implications for the mechanism of initiation by Brønsted acids. Of course all the reactions discussed above are to be seen as too "weak" with respect to a polymerisation system. This simply means that on one hand the rate of acid attack on the olefin is too low to give appreciable concentrations of active intermediates – may they be carbenium ions or species polarised by an acid and a nucleophile – and on the other hand the rate at which these activated complexes collapse to give the ester is too high to allow their accumulation. In other words the lifetime of potentially active species is minimal in these conditions, as testified by the absence of polymerisation of styrene in such systems as HCl–AcOH[167] and AcOH–H$_2$SO$_4$[178] and of isobutene in pure trifluoroacetic acid[191].

Many other reports of the addition of Brønsted acids to olefins are to be found in the literature, but their interest is less relevant to this section than to that in which the possibilities of polymerisation will be discussed.

An excellent compilation of numerical values of the rate constants for these reactions, including the overall order observed, has been published in Bolton's monograph[162].

C. The Addition of Iodine to Olefins

It is now well established that iodine reacts with olefins to give diiodides. The process is complex due to the apparent occurrence of both a radical and a polar contribution to the mechanism. Fraenkel and Bartlett[193] studied the addition of iodine to styrene and showed that radical inhibitors reduced considerably the rate of reaction, but did not inhibit it completely. Sumrell et al.[194] found that the addition of iodine to propene, butenes and 1-pentene is rapid, but the diiodides formed tend to decompose if iodine is present. More recently, Ayres et al.[195] investigated the kinetics and the mechanism of the reaction of iodine with pentenes and proposed a sequence of steps involving the attack of iodine atoms on the complex formed by the olefin with molecular iodine. They also characterised the diiodides by NMR spectroscopy, underlining that these are stable in the absence of iodine. With traces of iodine they generate hydrogen iodide and tend to polymerise. It is interesting to note that the solvent has practically no effect on the rate of the addition reaction with cyclohexene[196], an observation which sustains the atom-complex interpretation. Also, the fact that water can be used effectively as a medium for the process[194] rules out the possibility that this involves such reactive intermediates as iodonium ions. Other addition reactions studied include those with acenaphthylene[197], in which the diiodide formed generates HI in chlorinated hydrocarbons, and with vinyl ethers[198,199].

In conclusion, the relevant points about this reaction, which obviously needs to be studied further to be understood more clearly in terms of mechanism, are the fact that the addition onto the double bond is rather fast and reaches high equilibrium conversions and the ease with which the diiodides generate hydrogen iodide in the presence of iodine. These points are the basic prerequisites in understanding the cationic initiation of certain monomers, as will be discussed in Sect. III-E-13.

D. The Kinetics of Protonation of Olefins

Despite the vast amount of work which has recently been done on the protonation of olefins in the context of the search and characterisation of carbenium ions, very little information is available on the kinetics of this proton transfer process in non-aqueous media. A study of this nature requires the use of an analytical system capable of detecting quantitatively small concentrations of carbenium ions and possessing an adequate time response. Two types of investigation have been carried out, namely those using polymerisable monomers and those using olefins which cannot be polymerised. We will not analyse here reactions in such basic solvents as methanol, although an excellent study has been conducted in this medium[200], because they are outside our scope.

It is only in the last few years that researchers have tackled this difficult problem and the rapidity of these interactions have imposed the use of stop-flow techniques. The system hydrogen chloride-anhydroretinol-1,2-dichloroethane at room temperature was examined spectroscopically by Bulgrin and Lookhart[201] with a fairly fast stop-flow apparatus. The acid concentrations were around 10^{-2} M and those of anhydroretinol about ten times smaller. In the first second or so, after the mixing of the reactants the following processes gave rise to the carbocation precursors:

$$A + HCl \underset{\text{fast}}{\overset{}{\rightleftharpoons}} C \xrightarrow{\text{slow}} AHCl, \quad A=$$

where C is an "intermediate of unknown structure" and AHCl a species immediately preceding the protonated anhydroretinol (A). In the second phase of the process (1 to 10 sec after mixing) ionisation took place, characterised by a maximum at 620 nm,

$$AHCl + HCl \xrightarrow{k_2} AH^+HCl_2^-,$$

followed by cyclisation and polymerisation reactions. The equilibrium constant K had values ranging from 39 M^{-1} at 30 °C to 250 M^{-1} at 10 °C; k_1 went from 1,500 sec^{-1} to 260 sec^{-1}; and k_2 from 21.1 to 15.7 M^{-1} sec^{-1}. This study is the first we know of the protonation of an olefin in an aprotic solvent. Of course the olefin is highly nucleophilic because of its six conjugated double bonds and this is the reason why a relatively weak acid can protonate it without rapid collapse of the ensuing ion pair.

Lorimer and Pepper[202] have recently reported a study of the polymerisation of styrene by perchloric acid in methylene chloride at low temperature. Although the stop-flow apparatus they used was not very fast, some rough ideas on the kinetics of initiation could be obtained. Assuming that the initial increase in absorption at 340 nm was due to the protonation of styrene, unhampered by side reactions (esterification), the approximate rate law observed at −80 °C,

$$R = k_3 \, [\text{styrene}] \, [\text{HClO}_4]^2,$$

gives a value for the third-order rate constant of about 100 M^{-2} sec^{-1}. These calculations were carried out by us on the basis of the authors' kinetic relationships and published graphs. The value of the protonation rate constant thus estimated is almost certainly too low since the presence of an excess of styrene over the acid certainly did not favour the formation of ionic species.

Kunitake and Takarabe[87,203] measured rates of protonation in dichloethane at room temperature using a spectroscopic stop-flow technique with a dead time of about 20 msec. Trifluoromethanesulphonic acid was the proton donor. 1,1-diphenylethylene gave a simple kinetic pattern, $[CF_3SO_3H] = (1-2) \times 10^{-3}$ M, [olefin] = $(0.7-3) \times 10^{-4}$ M, 30 °C, viz.

$$d[C^+]/dt = k_2 \, [\text{olefin}] \, [CF_3SO_3H]$$

indicating that the protonation process involved the interaction of one molecule of each reactant. The second-order rate constant was 3,000 to 4,000 M^{-1} sec^{-1}. Given the nature of the ion followed, whose absorption spectrum is well characterised both in terms of λ_{max} and extinction coefficient, there is little doubt about the reliability of this study. In fact, it is the first thorough examination of the kinetics of initiation in a cationic "polymerisation" promoted by Brønsted acids. A similar investigation with styrene as nucleophile, carried out in typical polymerisation conditions — $[CF_3SO_3H] = (3-100) \times 10^{-3}$ M, [styrene] = $(3-30) \times 10^{-2}$ M, 30 °C — gave a kinetic pattern close to the simple one obtained with 1,1-diphenylethylene, but not as clear-cut, probably because of incomplete protonation particularly at the higher styrene concentrations. However, a second-order rate constant of 15 to 30 M^{-1} sec^{-1} could be estimated for this process. The decrease of two orders of magnitude of this constant with repect to that obtained with 1,1-diphenylethylene in the same conditions reflects the increase in basicity produced by a second phenyl substituent on the α-carbon atom.

Sawamoto and Higashimura[204,205] recently used a similar apparatus to follow the protonation of p-methoxystyrene by HSO_3CH_3 and by HSO_3CF_3 in 1,2-dichloroethane at room temperature. These authors assumed bimolecular protonation, without actually carrying out a detailed kinetic study of the system, and reported a second-order rate constant of 0.6 and about $5 \times 10^4 M^{-1} s^{-1}$ respectively, an increase reflecting the large difference in strength between the two acids. These results must however await kinetic confirmation.

It is surprising to note the paucity of research in this domain, considering its paramount importance in cationic polymerisation. One can only hope that the example given by the excellent work of Kunitake and Takarabe will stimulate other authors to take up similar studies.

E. The Polymerisation of Alkenyl Monomers

This section is devoted to a systematic analysis of publications concerned in a direct or indirect way with the mechanism of initiation in the cationic polymerisation of alkenyl monomers by Brønsted acids. Included in it are many examples of failures, i.e., systems in which no polymerisation was observed. Such experiments are important because the lack of production of active species can give considerable information about the alternative processes taking place when the two reactants are mixed. We have also included all the information we could gather about the state and properties of the initiators in the media relevant to the general context of this review.

1. Hydrogen Halides

On the whole hydrogen halides are poor initiators of cationic polymerisation, except in the case of very nucleophilic olefins. The well-known order of acidity HI > HBr > HCl > HF applies equally well in water and in many non-aqueous solvents, but of course the extent of dissociation of these acids in the latter media (particularly the aprotic ones) is much lower than in the former. Thus, in 1,2-dichloroethane the pK's are 7.9, 8.7 and 10.8 for HI, HBr and HCl respectively[39]. One would expect these values to be even higher in such non-polar solvents as aromatic and aliphatic hydrocarbons or carbon tetrachloride. In more polar media the reverse is true; thus, in acetonitrile pK(HBr) = 5.5 and pK(HCl) = 8.9[42] and in nitromethane pK(HCl) = 6.22[165]. Other data are available for a variety of polar solvents[206,207]. Since all the evidence indicates that these acids do not show any tendency to molecular association in these media, it cam be concluded that in a cationic polymerisation system involving one of these initiators the predominant species will be the monomeric molecular acid. The weak dissociation must be viewed in two alternative schemes, one not involving the solvent (i.e., when the latter cannot "accept" a proton),

$$3 \; HCl \; \rightleftharpoons H_2Cl^+ + HCl_2^- \;\; [208],$$

and the other when the solvent has some nucleophilic capacity, e.g.

$$CH_3NO_2 + 2 \; HCl \rightleftharpoons CH_3NO_2H^+ + HCl_2^- \;\; [165].$$

The homo- and heteroconjugation of halide anions with hydrogen halides is well known[209] and virtually complete in chlorinated hydrocarbons and nitromethane[210-212] as shown by NMR spectroscopy. Thus, the presence of halide ions in a system containing one of these acids in such solvents, and most probably in less polar ones, produces the sequestration of an equivalent amount of acid. Higher homoconjugated aggregates have recently been reported in the case of HCl in propylene carbonate[207].

a) Polymerisation of Aliphatic Monomers

Olefins and dienes are not polymerised by these acids. The addition products[213,214] such as *ter*-butyl halides, are formed at different rates depending on the acid used and the reaction conditions (see Sect. III-B). However, propylene gives a certain amount of hexyl halides with HBr[215] and HCl[216].

This behaviour can be rationalised in terms of a very fast collapse of the activated species formed in the attack of the acid molecule(s) onto the double bond, and/or by considering that the transition state is itself inactive towards propagation (tight ion pair or polarised Ad_E3 intermediate). The dimerisation of propylene could arise from some specific interaction, and is not to be considered as the first step in a cationic polyaddition because other oligomers have never been detected and the ceiling temperature of propylene is quite high.

Recently, Sangalov et al.[217] have reported that complexes of HCl with copper sulphate are effective heterogeneous catalysts for the oligomerisation of isobutene at room temperature. These authors claimed that the strength of the acid in considerably increased in the complex due to the "loosening" of the H—Cl bond. It seems to us that the heterogeneous nature of the catalyst must play an important role in favouring the formation of active species in this system, but obviously more work is necessary before a reasonable explanation can be given of this interesting behaviour.

Vinyl ethers readily polymerise under the influence of dry hydrogen halides. This is understandable, given the high nucleophilicity of these monomers. It is regrettable that no kinetic or mechanistic study has ever been carried out on these potentially interesting systems.

b) Polymerisation of Aromatic Monomers

It has been reported that styrene can be polymerised by HCl and HBr[299, 218], but that the balance between polymerisation and simple additions is a delicate one which can be easily swung to either side by a change of solvent. The original work of Pepper and Sommerfield showed that uncontrollable factors tended to marr the reproducibility of these systems and unfortunately this study was not pursued[300]. Tsuda found[219] that dry HCl in fairly high concentrations in methylene chloride gave low yields of polystyrene, $\overline{DP}$ = 1,000 at −78 °C and about 500 at −21 °C. This author however did not search for 1-phenylethyl chloride among the products. In sulphur dioxide a high concentration of HBr produced some polystyrene[220], but HCl failed to polymerise this monomer[221]. Finally, Giusti and Andruzzi[222] showed that in 1,2-dichloroethane styrene is not polymerised by HI, even when high acid concentrations are used. The only product of this reaction was the 1-phenylethyl iodide, presumably at room temperature.

One can conclude that the 1-phenylethyl halide is by far the preferred product of these interactions, but that suitable conditions of polarity and temperature can create the conditions required for some polymerisation to take place. Pocker et al.[223] have demonstrated that 1-phenylethyl chloride exchanges chlorine atoms with HCl in nitromethane, i.e., that carbenium ion pairs can be generated in this system. It seems likely that the low yields of polymers observed in specific conditions as described above resulted from the solvation of the halide by excess acid, a contingency which restricted active species to a short period. From the few scattered results published it is nevertheless difficult to derive a sound interpretation of such apparent anomalies as the fact that the strongest of all hydrogen halides, HI, failed to induce polymerisation.

The situation is quite different if the complex HCl · CuSO₄ is used as catalyst[217]. Just as in the case of isobutene, this initiator showed a higher acidity than HCl alone towards styrene as demonstrated by its high activity in benzene suspension at room temperature: 90% yield in 30 minutes, $\overline{DP}$ = 27.

The interaction of HCl and HBr with indene in methylene chloride[224] and with acenaphthylene in pentane, methylene chloride and acetic acid[225], as well as the reaction of HI with the latter monomer in methylene chloride[226] only yielded the addition products, no polymerisation being detected. However, according to older reports the dimers of both these monomers can be obtained with HCl[227]. Whatever the reason for this discrepancy, it is obvious that indene and acenaphthylene fall in the same category as styrene in that they do not offer the best conditions for the formation of active species in reacting with hydrogen halides. It would be interesting to try these interactions in more polar solvents.

Moving towards more nucleophilic monomers one finds a brief but instructive study by Andruzzi et al.[228] on the reaction between *trans*-anethole and hydrogen iodide. In 1,2-dichloroethane ($\epsilon = 10$) a 0.5 M solution of this monomer was polymerised quantitatively with acid concentrations ranging between 2×10^{-3} and 0.25 M, although the DP's were low. In carbon tetrachloride and toluene ($\epsilon = 2.2-2.4$) no polymerisation was observed but the acid was totally consumed to give the addition product. All these reactions were conducted at 25 °C. One witnesses here an excellent example of the type of limiting situations described in the previous chapter. A change in polarity of the medium from $\epsilon = 10$ to $\epsilon = 2$ brings about a radical change in the behaviour of the system. In the absence of supporting evidence, it is difficult to comment on the nature of the chain carriers formed when polymerisation took place. They could be ion pairs formed from the iodide in the more polar solvent, or ester molecules solvated by excess acid becoming active in a favourable medium.

N-vinylindole[229] and N-vinylcarbazole[230] are easily polymerised by HCl, HBr and HI. The latter monomer can even be polymerised by HF and by gaseous HCl passing over the solid olefin[231]. It is in fact so basic that a 10^{-6} M solution of HCl in benzene at 31 °C is sufficient to produce its (slow) polymerisation[232]. It is a pity that the only extensive paper on these systems[230] leaves a lot to be desired both in the quality of the experimental work and in the depth of the discussion, and no mechanistic conclusions can actually be drawn about this investigation.

c) Conclusions

Although no thorough study has been published on the mechanism of initiation and propagation for cationic polymerisations promoted by hydrogen halides, a few general conclusions can be drawn. First, it is clear that, except for the most nucleophilic monomers, the simple addition across the double bond tends to predominate over the formation of chain carriers. Thus, most of the common olefins are not polymerised at all or give polymers only under special conditions. Second, whenever polymerisation does occur, inclusion of halogens in the polymers seems to be a general feature. These halogen atoms must be present as end groups, a fact which suggests that the active species are probably ion pairs or polarised ester molecules which can easily lose their propagating capacity by forming a stable carbon-halogen bond. Finally, we would like to suggest that fundamental information about the way these acids operate could probably be obtained from a study of the kinetics and mechanism of the polymerisation of *p*-methoxystyrene by one of these acids. This monomer is certainly basic enough to polymerise and its carbenium ion has been characterised unequivocally by ultraviolet spectroscopy, i.e. it could be searched for during the polymerisation. Many basic points could be approached in such a study and this specific area of cationic polymerisation would greatly benefit from a thorough investigation of this type.

Hydrogen halides have been frequently used as cocatalysts in cationic polymerisations involving Lewis acids. These systems will be discussed in Chap. IV.

2. Acetic Acid and Homologues

Unsubstituted carboxylic acids are very weak acids. The autoprotolysis constant of acetic acid is around 10^{-13} [38] and its extent of dissociation in non-aqueous solvents extremely low as indicated by the value of pK = 22 in acetonitrile[43]. In the pure state acetic acid exists essentially in the form of cyclic dimer and only a small fraction is present as open hydrogen-bonded oligomers. In non-aqueous media the extent of dimerisation is still important, but decreases with increasing polarity of the solvent[233]. Homoconjugation of acetic acid with its anion is well documented[234] and very strong, the $pK_{HB_2^-}$[43] in acetonitrile being -3.9. Moreover, the acetate anion conjugates readily with many carboxylic acids[43].

The only monomer which is sensitive to such weak acids is N-vinylcarbazole. Conflicting reports had been published to this effect[44,235,236], but recently Gandini and Prieto[237] have studied this system in detail and established beyond doubt that acetic acid and some of its homologues do polymerise this very nucleophilic monomer. The situation can be summarised as follows. The reaction between N-vinylcarbazole and acetic acid in solvents of low dielectric constant (hexane, benzene, CCl_4) only yields the corresponding acetate, probably through an Ad_E3 mechanism. If methylene or ethylenen chloride are used and the acid concentration is higher than about 0.5 M the esterification is accompanied by some polymerisation giving oligomeric acetates. If the monomer is mixed with glacial acetic acid a fast polymerisation occurrs and the products are a mixture of oligomeric esters and unsaturated oligomers. Finally, when gaseous acetic acid is placed in contact with the solid monomer in a high-vacuum system polymerisation starts immediately and again oligomeric esters are produced together with low-moleculer weight poly(vinylcarbazole) with terminal double bonds. Kinetic and mechanistic studies of this system suggest that no ionic chain carriers are formed and that propagation is due to the solvation of ester molecules by excess acid followed by the reaction of this activated species with monomer, or to the interaction between two ester molecules with release of a molecule of acid. It seems likely that both processes are operative, the first being more efficient at the beginning of the reaction and the second taking over when most of the acid has been consumed. The pseudocationic nature of this polymerisation is confirmed by its insensitivity to added water.

A similar phenomenology was observed when acetic acid was replaced by propionic, acrylic and methacrylic acids[237]. However, with these initiators there was less polymerisation than with AcOH, and more esterification.

3. Chloroacetic Acid

$CH_2ClCOOH$ is somewhat stronger than acetic acid but remains within the family of weak acids. Its pK in acetic acid must be close to 12.5[38] and in 1,2-dichloroethane[39] it should be around 17. Chloroacetic acid is appreciably dimerised in non polar solvents[238,239] and its tendency towards homoconjugation with its anion is very pronounced[43] ($pK_{HB_2^-} = -2.94$ in acetonitrile).

Brown and Matheson[240] found that this acid was too weak to induce the poly-merisation of styrene, but that it could activate α-methylstyrene. This is yet another typical example of limiting situation where the critical parameter is the nucleophilicity of the monomer. That particular system will be discussed in the next section together with the bulk of Brown and Matheson's work.

Gandini and Prieto[241] have studied in detail the polymerisation of N-vinylcarba-zole with chloroacetic acid in 1,2-dichloroethane. It was found essential to work under high-vacuum conditions and with highly purified chemicals in order to achieve repro-ducible results. The investigation on the actual polymerisation was preceded by a specific study of the kinetics and mechanism of esterification using excess acid and very low monomer concentrations. Evidence for an Ad_E3 mechanism was obtained. Polymerisations were followed dilatometrically and the extent of free acid in the system was also determined throughout these reactions. On the basis of these results and of other mechanistic tests, including the effect of added strong nucleophiles, it was concluded that two types of active species are formed in this system. One is for-med immediately after mixing monomer and acid and is sensitive to the addition of methanol. Its concentration decreases steadily as the polymerisation proceeds and as the acid is consumed by esterification. This chain carrier is either an ester molecule activated by solvation of free acid (this activation being quenched by methanol which displaces the acid molecules which are coordinated to the ester), or an ion pair which slowly collapses to form the ester as the acid is being consumed. The absence of an increase in conductivity in the polymerising solution with respect to the conductivity of the acid solution on its own excludes the presence of free ions in this system. The other active species is insensitive to the presence of methanol and builds up slowly as the esterification proceeds. Obviously the ester itself is a chain carrier, but much less active than the polarised species. Polymerisations by the ester alone were in fact very slow.

Concomitant propagation by ionic and non-ionic active species is not new. Per-chloric acid and styrene give rise to a similar situation as will be seen later in this chap-ter.

4. Di- and Trichloroacetic Acids

These two acids, though stronger than AcOH and $CH_2ClCOOH$, are still weak acids ($CCl_3COOH > CHCl_2COOH$) with all the same characteristics: they display a signif-icant tendency to associate in non polar solvents[238,239,242,245], and they possess a pronounced affinity for their anions. Thus the pK of trichloroacetic acid in acetic acid is 12.2 and in acetonitrile dichloroacetic acid has pK = 13.2 and trichloroacetic acid 12.65[43], and the $pK_{HB_2^-}$ in the latter solvent is -2.81 and -2.48 respectively[43].

No recent work has been published on the use of these acids in cationic poly-merisation. We will review some of the older work because it certainly deserves a reassessment in the light of present-day knowledge.

Alkenes do not polymerise with these acids and only the corresponding esters are formed, even in the presence of excess acid. Arenes and some dienes are activated by these catalysts and the polymerisation rates depend of course on the basicity of the specific monomer. Brown and Matheson[243] have given a qualitative picture of this situation and have classified the polymerisable monomers in two families: those which give polymers with one ester group *per* chain, and those which do not, i.e. which give at least a portion of product without acid fragments. Gandini and Prieto[241] have recently found that N-vinylcarbazole polymerises rapidly with traces of these acids with the formation of products belonging to the second family.

a) Styrene and α-Methylstyrene

To our knowledge, the classical papers of Brown and Matheson[152,240,243–247] represent the first thorough kinetic and mechanistic investigation of the polymerisation of styrene by Brønsted acids. The considerable amount of evidence which they gathered has already been summed up in Plesch's second book[248] and we will therefore limit ourselves to express some fresh views on these interesting systems.

Among the general conclusions which these authors reached in their discussion, two are of particular relevance to us, namely: the formation of polar acid-monomer complexes of various stoichiometry, and the ionic nature of the active species involved in all propagations. A reexamination of the ensemble of the results raises some serious questions as to the validity of these two assumptions.

The existence of complexes was inferred both by the kinetics of the polymerisation and by the rates of acid consumption durring the reactions. The determination of the amount of acid consumed was carried out by shaking an aliquot of the polymerising solution with water and by titrating the acid extracted. It was argued that both the acid bound to the polymer and that complexed with the monomer would remain in the organic phase and would therefore not be titrated. However, since the formation of these complexes was postulated as leading to equilibria, it seems unlikely to us that upon the removal of the free acid into the aqueous phase the complexes would not revert to the original compounds thus allowing all the acid they contained to pass to the aqueous layer. One of the reasons behind the apparent necessity of postulating the existence of such complexes was the observation that the acid consumed (titrated) was more than that incorporated into the polymer chains in terms of one acid residue *per* macromolecule. However, a close inspection of the results reveals the possible origin of this discrepancy. In the preliminary check carried out by Brown and Matheson to ascertain that all the monomer consumed corresponded to the amount of polymer formed (Table 1, Ref.[244]) the result obtained is in fact misleading, because while it appears to show the correct equivalence between the two quantities, the amount of acid incorporated in the polymer as end groups was not taken into account. Since the average DP of these polystyrenes was less than 15, there is a difference of at least 10% between the monomer consumed and the polymer formed, i.e. something is missing among the products. We feel that the monomeric ester could have been distilled away together with the the monomer during the isolation of the polymer. Alternatively, the monomeric ester could have been hydrolysed more easily than the oligomeric esters upon treating with water. In both cases the amount of acid titrated would seem higher than that incorporated in the polymer, and this alternative explanation would eliminate the necessity of advocating "water-resistant" complexes. In conclusion, the lack of attention towards the monomeric ester caused a great deal of complication. Both the high initial uptake of acid by these systems and the high order in acid observed in all cases are compatible with a simple esterification reaction between styrene and the acid.

As for the ionic nature of the chain carriers, it seems to us that the arguments given to support such conclusions are weak. In view of the recent discovery of pseudocationic polymerisation, it is necessary to reanalyse all the data available in order to assess the evidence in favour of one or the other type of chain carrier. We have done so and wish to propose the following comments. Many aspects of the phenomenology of these polymerisations favour the presence of non-ionic chain carriers. First of all the role of water. Small amounts of added water were reported to accelerate the reaction, probably by increasing the extent of acid dissociation, in the absence of solvent. But the most interesting results were obtained in nitromethane (Table 7, Ref.[244]). For water concentration up to the initial acid concentration the polymerisation rate was practically the same as in dry conditions, and only when the water concentration was raised well above that of the acid did the rate decrease, and then only by about a factor of two. Such insensitivity towards excess moisture is not compatible with carbenium ions being the active species. Secondly, the fact that in all systems involving styrene each polymer molecule contained one acid residue as end group, and at the same time transfer was important, was rationalised by Brown and Matheson in terms of a reaction

$$M_n^+ C^- + MC \longrightarrow M_n C + M^+ C^-$$

in which the transfer agent is the acid-monomer complex, and a termination reaction consisting in the collapse of the ion pair to give the macroester

$$M_n^+ C^- \longrightarrow M_n C.$$

Since these complexes were not identified and are in fact rather ill-defined, it seems more logical to us to think in terms of active species consisting of ester molecules solvated by an acid molecule. These chain carriers could lose activity by the departure of the polarising acid molecule towards a monomer molecule, a reaction which would in effect play the role of monomer transfer, viz.:

$$E \cdot HB + M \longrightarrow E + HMB.$$

Thirdly, the different external order in acid depending on the medium can be better rationalised in terms of rate-determining esterification (initiation) than by the complicated sets of complexation equilibria invoked by the authors. It is well known that in the addition of acids to olefins the order with respect to acid increases as one goes from a solvent of high polarity to one of lower dielectric constant, and this is precisely what was observed in these polymerisations.

In conclusion, we are inclined to think that polarised ester molecules are responsible for the propagation in the polymerisation of styrene by di- and trichloroacetic acid. However, we realise that only a new study oriented specifically at understanding the role of the esters and at trying to ascertain the presence or absence of acid-monomer complexes would provide an adequate answer to this interesting problem. The same considerations apply to the polymerisation of α-methylstyrene by the three chloroacetic acids, despite the small differences in behaviour encountered by Brown and Matheson with this monomer[240,247].

b) Cyclopentadiene

In another classical investigation, Wassermann and coworkers[249−252] studied the system cyclopentadiene-trichloroacetic acid for several years and were able to disentangle its intricate behaviour through a series of brilliant and meticulous experiments. The identification of the dimeric ester and the study of its polymerisation to give polyconjugated products, a reaction enhanced by the presence of such promoters as mono- and trichloroacetic acids, demonstrate the possibility of propagation by ester-ester interaction assisted by acid.

While there is no doubt that as the degree of conjugation in the polymer increases, stable polyenic carbenium ions are formed in the acidic medium, the nature of the chain carriers involved in the formation of the dimeric ester and in the early stages of its polymerisation is worth discussing. The authors invoked the formation of carbenium ions from the first step of the process[250], but no proof of their presence, or of the presence of the dimer cation and dication was offered. An alternative mechanism based on activated ester molecules would be more plausible to us in view of the high reactivity of the cyclopentadienyl cation and the low likelyhood that it would be formed in such mildly acidic conditions as those employed in these experiments.

We have refrained from dicussing the details of this work, because it has already been adequately reviewed in Plesch's second book[253].

c) 1,1-Diphenylethylene and Homologues

The mechanism of dimerisation of these compounds has been the subject of many recent studies, but the pioneering work of Evans and collaborators with trichloroacetic acid as catalyst[254−257] is to be considered as the starting point of a long series of interesting investigations. These authors took a number of diaryl ethylenes and studied the kinetics of their dimerisation and of the reverse reaction, the thermodynamics of the equilibria involved and the electronic spectra of the intermediate species. Benzene was normally used as solvent, but nitroethane was also tried with 1,1-diphenylethylene. This work has already been discussed by Bywater[2], but we would like to add the following remarks concerning the nature of the active species. Important qualitative and quantitative differences exist between the behaviour of 1,1-diphenylethylene and 1,1-di-p-methoxyphenylethylene, i.e., the least and the most nucleophilic of the substances studied. Spectroscopically, while the reacting solutions of 1,1-diphenylethylene did not exhibit the characteristic absorption at 430 nm due to the corresponding carbenium ion, even when very high acid and olefin concentrations were used, the solutions containing trichloroacetic acid and 1,1-di-p-methoxyphenylethylene always gave rise to the carbenium ion absorption at 490 nm. In fact the authors could calculate from these spectra the stationary concentration of intermediates. Kinetically, while the dimerisation of 1,1-diphenylethylene was second order in olefin, the same process was first order with 1,1-di-*p*-methoxyphenylethylene. Also, while the activation energy for these reactions decreased smoothly from 9.1 to 5.2 kcal mole^{-1} when passing from 1,1-diphenylethylene to more nucleophilic homologues, it rose again for the most nucleophilic, 1,1-di-*p*-methoxyphenylethylene, to 9.4 kcal mole^{-1}.

These basic differences are strong indication of a change in mechanism between the two processes. Our interpretation is that 1,1-diphenylethylene dimerises via a pseudocationic mechanism where the active species are the trichloroacetate ester molecules (absence of spectrum) and the rate-determining step is the dimerisation itself (second order in olefin). On the other hand, 1,1-di-*p*-methoxyphenylethylene dimerises under the action of a carbenium ion intermediate (typical spectrum), and the rate-determining step is the protonation reaction (first order in olefin and higher activation energy).

Once again we encounter a change in mechanism brought about by the sole change of monomer basicity, all other conditions being the same. This confirms our views on the sensitivity of a given system to an apparently slight modification of conditions.

The appeerance of a long wavelength band (600–700 nm) in these systems is neither due to the formation of acid-olefin π-complexes[256] nor to cation radicals derived from the diarylethylene[258], but rather to minor side reactions producing fluorenyl-type carbenium ions, as shown by Sauvet[259].

d) Conclusions

We consider that the three studies examined above present considerable evidence in favour of the existence of non-ionic chain carriers, but that more work ought to be done on those systems before a definite conclusion can be drawn. It is clear however that the phenomenon of pseudocationic polymerisation is more frequent than usually supposed, and conversely, carbenium ion intermediate are not as ubiquitous in cationic polymerisation as previously believed.

We would also like to draw the reader's attention to the thoroughness and high quality of those three investigations. Despite the fact that they are all about twenty years old, they remain examples of work to be conducted in cationic polymerisation. One would welcome today a similar high level of approach, while one reads so many papers in this field marred by technically doubtful set-ups, ill-conceived experiments, incomplete results, and debased by unsubstantiated conclusions. This situation is the more shameful considering the wealth of experience accumulated and the vast array of sophisticated instruments presently available.

5. Chloroformic Acid

Sakota et al.[260] have reported a curious but interesting experiment in which styrene, trichloroacetyl chloride and water were mixed in equimolar amounts in sealed tubes. The oligomeric product was fractionated and each fraction analysed and characterised. It was concluded that the process had generated a series of oligostyrenes from DP = 1 to DP ~ 15, each molecule containing a terminal chloroformate group. Chloroform was also found among the liquid products together with unreacted trichloroacetyl chloride. These results were interpreted according to the following reaction scheme:

$$Cl_3COCl + H_2O \rightleftharpoons CHCl_3 + ClCOOH$$

$$ClCOOH + CH_2{=}CHPh \rightarrow CH_3CH(Ph)OCOCl$$

followed by polymerisation promoted by 1-phenylethyl chloroformate. The authors actually referred to the cationic polymerisation of styrene promoted by the proton of chloroformic acid. In fact if one considers that the reaction between the initial acid chloride and water did not go to completion, it follows that the polymerisation took place in the presence of excess water, a situation highly unfavourable to the formation of the 1-phenylethylium cation. The presence of ester end groups in all oligomers despite the very low DP of the product can be interpreted by the transfer mechanism we proposed in the system styrene-trichloroacetic acid (see Sect. III-E-4-a) or by considering that propagation is very slow. This system represents for us an excellent example of pseudocationic polymerisation.

6. Trifluoroacetic Acid

The strongest among carboxylic acids, trifluoroacetic acid is nevertheless weaker than most mineral acids. Its pK in acetic acid is 11.4[38], while in acetonitrile the acid is slightly more dissociated, pK = 10.6[43]. Bolza and Treloar[41] have reported a study of the conductivity of this acid in 1,2-dichloroethane and have derived a value of pK = 7. They did not write the dissociation equilibria in the correct manner, ignoring homoconjugation and postulating the existence of solvated protons in a solvent incapable of such a task. Judging from the scale of pK's established by Bos and Dahmen[39] in this solvent and the values in other media reported above, it seems more likely that the value of pK should be close to 12. Solutions of trifluoroacetic acid in non-aqueous aprotic solvents contain mixtures of monomers, cyclic dimers and open dimers[192,242,190,261,262]. The dimerisation constants at 25 °C are 192, 149, 2.6 and 1.5 M^{-1} in cyclohexane, CCl_4, benzene and 1,2-dichloroethane, respectively[192]. Homoconjugation of trifluoroacetic acid with its anion is extremely pronounced. Various studies prove that the trifluoroacetate anion can conjugate with one and even two molecules of acid[263-268], and in acetonitrile $pK_{HB_2^-} = -3.88$[43]. The acid heteroconjugates strongly with many other carboxylic anions[43], and vice versa.

Alkenes do not polymerise in the presence of trifluoroacetic acid, but only give the corresponding esters. Isobutene can be bubbled through the pure acid without any noticeable polymerisation[191].

No studies have been reported on the action of this acid on dienes, but in view of the effectiveness of trichloroacetic acid on cyclopentadiene[249-253], one would expect an even easier catalysis by the stronger trifluoro derivative. Furan polymerises in the presence of fairly high concentration of trifluoacetic acid[269,270] giving rise to poly-conjugated products. The structure of these polymers has been analysed in detail[270], but the mechanism of initiation and the nature of the chain carriers which produce them remain unexplored, particularly in the early stages of the reaction.

Aromatic and heteroaromatic olefins can be polymerised by this acid, but regrettably the number of studies of these systems is still very limited. In fact, only styrene and the alkenyl furans have received some serious attention as discussed below. N-vinylcarbazole is extremely sensitive to this initiator; concentrations as low as 10^{-5} M are sufficient to produce a rapid polymerisation in chlorinated alkanes[241].

a) Styrene and Homologues

The first report on these systems is the famous study of Throssell et al. on the peculiar behaviour of mixtures of styrene and trifluoroacetic acid alone or in different solvents[271]. The essence of those experiments was to show that while in a medium of low polarity ($\epsilon \leqslant 5$) addition of the acid to the monomer to give 1-phenylethyl trifluoroacetate was the predominant feature (if the acid was dripped into undiluted styrene the same product was obtained, with some oligomeric esters), the yield and molecular weight of polymer increased progressively as the dielectric constant of the medium was increased. When styrene was dripped into the pure acid a fast polymerisation was observed. The interpretation of these results was based on the formation of the 1-phenylethylium cation (ion pair) and the relative rate of its collapse to the ester as a function of the polarity of the medium. In solvents possessing good solvating power the active species had a finite lifetime during which propagation could take place before the collapse; in solvents of low polarity the collapse was virtually immediate and no polymerisation occured. While we agree that in the pure acid the protonation of styrene could give a relatively long-lived carbenium ion (very fast polymerisation, high DP), we think that the situation is pobably different in diluted media where the solvated ester is more likely to be the active species. It is a pity that Throssell and coworkers did not carry out longer experiments to see if the yields increased, however slowly, after the first burst of polymerisation, particularly in the media of low polarity: it would have been interesting to see if the ester molecules could combine to give more polymer in the presence of excess acid, as in the case of the cyclopentadiene dimeric esters. Gandini and Plesch[8] have noticed that in methylene chloride trifluoroacetic acid gives a slow polymerisation and low DP's, the reaction being insensitive to water and have proposed a pseudocationic mechanism for this system.

    Nikolayev and collaborators[272,273] have restudied the polymerisation of styrene by trifluoroacetic acid and $H(CF_2)_6COOH$. Limiting yields were often obtained, coinciding with the total consumption of the acid, and polymerisation could be reinitiated at this point by addition of further acid. It seems obvious from the ensemble of their results that although the ester is inactive in the absence of free acid, the active species are probably ester molecules solvated by the acid, and not carbenium ions. The dielectric constant in these reactions was in fact low and it is difficult to envisage that the 1-phenylethylium cation could "survive" in these media.

    Recently, Sawamoto et al.[274] have restudied this system in 1,2-dichloroethane, nitrobenzene and mixtures of 1,2-dichloroethane and benzene, at 50 °C. They noticed again that esterification of styrene accompanied the polymerisation, but limiting yields were not detected. In fact, even in the media of lower polarity the polymerisation proceeded slowly but steadily over periods of days. The authors argued in favour of ionic chain carriers because the monomeric trifluoroacetate failed to promote the polymerisation of styrene. They also used alternative kinetic treatments to prove that only that involving the formation of protonated styrene was compatible with the experimental results. These arguments do not really prove the point claimed, because the active species could well be ester molecules polarised by free acid, and because the kinetic schemes proposed implied bimolecular initiation, a reaction not proved experimentally. In fact, in solvents such as chlorinated hydrocarbons and benzene, there is ample evidence to show that the addition of trifluoroacetic acid to olefins requires two molecules of acid[190,191]. Also, the high orders in acid found by Nikolayev et al.[273] show that the initiation in these systems is far from kinetically simple.

p-Methoxystyrene is readily polymerised by trifluoroacetic acid in methylene chloride and the initiator concentrations needed for a reasonable rate are much lower than those used for styrene. Gandini and Plesch[8,275] have briefly studied this system with the specific aim of establishing the nature of the chain carriers. The absence of electrical conductivity and of absorption in the region 340–800 nm during the polymerisation suggests that either the chain carriers are not ionic, or they must be present in concentrations lower than 10^{-6} M. With those maximum concentrations, one can easily calculate the minimum value of k_p for the ionic propagation from the initial rates of polymerisation. This turns out to be unreasonably high compared with published values. It is interesting to note that the trifluoroacetate of 1-p-methoxyphenyl-ethyl has been synthesised and characterised by Bourne et al.[276], but it was noticed that the pure compound was unstable and gave polymeric materials. It seems that once again one is witnessing a phenomenon of ester-ester propagation with expulsion of acid.

The problem of the nature of the active species in the polymerisation of styrene and p-methoxystyrene by trifluoroacetic acid is still unsettled and although we favour the pseudocationic mechanism (except in pure acid), more work is needed to reach a clear answer. Stop-flow studies would undoubtedly help in this context.

Some puzzling experiments were recently reported by Nikolayev and coworkers[277–279]. They found that the addition of small amounts of nickel-or copper(II)-acetate to the system styrene-trifluoroacetic acid produced a dramatic increase in both the rate of polymerisation and the DP of the products. Critical values of the ratio salt-to-acid were observed, above which the system began to slow down again. These values were surprisingly low, namely 0.007 for the nickel salt and 0.003 for the cupric one. Similar results were obtained with manganese(III) acetylacetonate[279]. The interpretation of these phenomena is difficult, but given the drastic change in behaviour that the addition of these metal salts brings about, one is obviously witnessing a new mechanism of initiation, i.e. the formation of much more active chain carriers. First of all, the salts added must rapidly be converted into the corresponding trifluoroacetates. The problem is thus to rationalise how trifluoroacetic acid can become a much more powerful initiator in the presence of small amounts of metal trifluoroacetates, and why, when these amounts are increased, the actviity of the initiating entity decreases again. It seems that ionic complexes involving the metal, the acid and styrene are formed[277] and eventually this complex interaction leads to the formation of carbenium ions which cannot easily collapse to give the ester and are thus very active chain carriers. An increase in the concentration of the metal salt probably produces a corresponding increase in ion pairs with a decrease in the activity of the propagating species. Whatever the intimate mechanism of these processes, it is certain that the complexing properties of the cations added are at the origin of these peculiar phenomena, never observed before in cationic polymerisation.

b) Alkenylfurans

Gandini and collaborators[280-283] have carried out a detailed investigation of the polymerisation of four alkenylfurans by trifluoroacetic acid in methylene chloride. Important side reactions accompanied the normal vinylic propagation and introduced serious complications for the study of the more fundamental aspects of these processes. Only indirect evidence could therefore be obtained concerning the initiation mechanism and the nature of the chain carriers. Alkenylfurans are considerably more nucleophilic than styrene and rather resemble p-methoxystyrene in their behaviour towards Brønsted acids. The kinetics of these polymerisations showed that initiation requires two molecules of trifluoroacetic acid:

$$M + 2\,HB \rightarrow P_1^*, \text{ fast.}$$

It was also deduced that a deactivating reaction competes kinetically with propagation to give inactive ester species which can be reactivated by free acid:

$$P_n^* + M \Big\langle \begin{array}{l} \nearrow P_{n+1}^* \\ \searrow P_nB + HMB \end{array}$$

$$HMB + HB \rightarrow P_1^*$$

Water was found to decrease the rates of polymerisation, but not to inhibit the process. Moreover, the four corresponding carbinols were successfully polymerised by trifluoroacetic anhydride. These reactions went through a dehydration stage, and since polymerisation occurred even with about 50 times more carbinol than anhydride it was again proved that excess of water did not impede the polymerisation of these monomers by trifluoroacetic acid. It was concluded that two types of active species coexist in these systems, one sensitive to moisture (carbenium ions?) and one insensitive to it (ester?).

Addition of common-anion salts produced a considerable decrease of the rates of polymerisation, but this phenomenon was *not* due to the shifting of free ions-ion pairs equilibria, but to the homoconjugation of the trifluoroacetate anions added with the acid and the corresponding decrease in the strength of the initiator. Stable homoconjugated salts were isolated and characterised analytically and by IR and NMR spectroscopy[268].

7. Phosphoric Acid

Apart from scattered reports merely mentioning the activity of this acid as initiator for the cationic polymerisation of various monomers[220,235], the only interesting studies published to date are the old investigations of the system phosphoric acid – propene. Fontana's review of this work[284] is particularly interesting today in the light of recent advances, because the possibility that the active species might be esters or "modified esters" is specifically mentioned. These observations are perfectly in tune with modern views on the pseudocationic nature of certain polymerisations. Butenes seem to give a similar behaviour with this catalyst[284].

8. Nitric Acid

This acid has never been used in a systematic study in cationic polymerisation. This is not surprising, given the low stability of pure HNO_3. Homoconjugation of this acid with its anion has been reported[285].

Tazuke et al.[286] investigated briefly the role of this initiator in the polymerisation of N-vinylcarbazole in the context of a wider study involving metal salts as promoters. They found that aqueous nitric acid in dioxane was able to induce the polymerisation of this very sensitive monomer. Concentrations of 10^{-3} to 10^{-4} M gave 50% conversion in several hours. No useful speculation can be offered about the meaning of these results considering the very odd medium used and the well-known idiosyncrasy of the monomer.

The pK of this acid is 8.9 in acetonitrile[42] and 10.1 in acetic acid[36].

9. Sulphuric Acid

The strength of sulphuric acid is usually considered as intermediate between those of HCl and HBr. Thus, the following pK values have been determined or interpolated: 7.0 in acetic acid[36], 7.25 in acetonitrile[42] and about 9.5 in 1,2-dichloroethane. This acid exists essentially as monomer in non-aqueous solvents, but has a strong affinity for its anion as shown in homoconjugation studies in nitromethane[287], acetonitrile ($pK_{HB_2^-} = -3$[42]) and in propylene carbonate[207].

Despite its fame as the first initiator used in cationic polymerisation, sulphuric acid has not frequently been used for fundamental studies in this field. Its low vapour pressure hinders proper purification procedures even in high vacuum, and its limited solubility in the solvents currently used for these investigations is another serious drawback. While alkenes are only oligomerised by this acid, aromatic monomers can give high polymers under suitable conditions. Our present discussion will be limited to the only system investigated in depth, i.e. the polymerisation of styrene.

Tsuda[219] and Asami and Tokura[220] reported the polymerisation of styrene by sulphuric acid in methylene chloride and sulphur dioxide respectively, but did not carry out any extensive kinetic or mechanistic study of these systems. It was Pepper and his collaborators who undertook the task of looking into this process methodically[288–291]. Although previously reviewed[292], this work is worth rediscussing.

On the basis of the kinetic results obtained, an overall mechanism was proposed which included two fundamental assumptions:
– fast and complete initiation involving the total consumption of the acid present and the formation of an equal amount of active species (ion pairs);

— non-stationary character of the polymerisation due to a termination reaction consisting of the unimolecular collapse of the active ion pair to give polystyryl bisulphate.

The kinetic scheme associated with these postulates predicts a progressive decrease of
the rate of polymerisation and the attainment of a limiting asymptotic yield of polymer
which will depend upon the acid concentration used and upon the ratio of the rate
constants of propagation and termination. These phenomena were in fact observed
and moreover a linear relationship was obtained between $\ln (M_0/M_{00})$ and the initial
acid concentration, as expected. Finally, from a graphical fitting of the experimental
points a propagation rate constant was derived at 25 °C, $k_p = 7.6 \, M^{-1} \, sec^{-1}$.

Several weak points have been found in this scheme since it was first postulated.
Gandini and Plesch[8,275)] found that electronic spectra taken less than one minute
after mixing the reagents in vacuo did not show any absorption above 320 nm, indicating that the concentration of 1-phenylethyl carbenium ions was smaller than $10^{-6} \, M$,
i.e. nearly four orders of magnitude lower than the acid concentration used. This observation ruled out either complete initiation to give carbenium ions, or the ionic nature
of the chain carriers. The abnormally low value of the propagation rate constant was
also placed in a different perspective, because if the concentration of ionic active
species was in fact much lower than the initial acid concentration, would rise corre-
spondingly; and if the active species were non-ionic one would expect a low k_p. Some
years later, Peniche and Gandini[293)] repeated Hayes and Pepper's experiments at room
temperature, but carried out all operation under high vacuum. They observed good
reproducibility, but considerable disagreement with the original work, both in the shape
of the time-conversion curves and in the values of the asymptotic yields. In particular,
it was found impossible to fit a series of curves reflecting polymerisations at the same
temperature and the same initial monomer concentration but variable acid concentration with the same value of the supposed termination rate constant using Burton and
Pepper's kinetic scheme.

It now seems clear that the behaviour of this polymerisation is much less simple
than originally postulated. Peniche and Gandini[293)] carried out some mechanistic tests
to try and understand the real nature of the initiation process. To solutions which had
reached the asymptotic yield they added small amounts of trichloro- or trifluoroacetic
acid (the actual resulting concentrations of these acids were well below those necessary to produce appreciable polymerisation of styrene): the polymerisation suddenly
started again and the yield reached a value approximately double that previously attained. They also showed that the solutions which have arrived at the asymtotic yield
are sufficiently acid to ionise Ph_3COH and to induce the rapid polymerisation of
N-vinylcarbazole. All these observations suggested that the initial polymerisation ceases
when all the acid has been consumed to give bisulphate esters. These end groups are not
polar enough to induce the propagation with styrene but their acidity is sufficient to
ionise triphenyl carbinol and polymerise a very basic monomer. Moreover, they can be
reactivated for the polymerisation of styrene by the addition of acidic substances capable of solvating them. On the basis of their results, Peniche and Gandini proposed a
different mechanism for this system. The active species are formed by the polarisation
of a molecule of 1-phenylethyl bisulphate by excess sulphuric acid. As the reaction proceeds the number of chain carriers (and/or their activity) decreases because the free acid

is being consumed by the monomer progressively to form ester. The polymerisation ceases when all the acid has been used up in this reaction. This situation is particularly complex and difficult to verify quantitatively because the various rate constants are not known. As to the intimate nature of the chain carriers, nothing precise can be said, but in view of the sensitivity of this system to water they could be ion pairs at least at the beginning of the polymerisation when the acid concentration is still high and therefore the specific solvation of the ester could induce its ionisation.

One last point is worth mentioning. As already noticed by Hayes and Pepper, Peniche and Gandini observed that the polymerisation of styrene proceeded very slowly after the apparent attainment of the asymptotic yield. If the solutions were left for several days, complete conversions were attained. This simply means that even the neutral ester is active but its propagating capacity is extremely low.

Several authors have employed Burton and Pepper's scheme to calculate propagation rate constants for systems displaying the rather frequent behaviour of fast initiation-limited yields. Ykeda et al.[294] and Yamamoto et al.[295] have applied this principle to the polymerisation of styrene derivatives by sulphuric acid, and Kohjiya et al. to the system cyclopentadiene-perchloric acid[296]. We wish to caution about the uncritical usage of this procedure, since, as has been shown above, it can lead to gross underestimates. In fact, it should only be used where there is sufficient mechanistic evidence to justify its application.

10. Methanesulphonic Acid

The strength of this acid lies between those of sulphuric acid and hydrogen chloride. pK values of 8.6, 8.4 and 6.0 have been obtained in acetic acid, acetonitrile and nitromethane, respectively[37]. By interpolation, a pK of about 10 can be assumed in 1,2-dichloroethane[39]. No evidence of dimerisation has been reported in these solvents. Like all acids considered before, methanesulphonic acid readily homoconjugates with its anion, $pK_{HB_2^-} = -3.5$ in acetonitrile[298].

Methanesulphonic can be purified and handled in vacuo more easily than sulphuric acid, and its solubility in the solvents typically used in cationic polymerisation is adequate for most studies. Despite these positive characteristics and the fact that it can be purchased in an extremely pure form, no study had been reported until very recently on its use as initiator.

Masuda et al.[297] reported a very brief study of the polymerisation of styrene by methanesulphonic acid in methylene chloride, nitrobenzene and benzene at 0 °C. Internal first orders were obtained with the first two solvents, but the investigation was not pursued and nothing can therefore be said about these systems except that they seem interesting and would be worth studying further.

The only other monomer which has been investigated with this acid is p-methoxystyrene. Sawamoto and Higashimura[89,119,204] studied the initiation and propagation processes in various solvents at 30 °C, using a spectroscopic stop-flow apparatus. In the first paper[89] of this series the monomer carbenium ion was characterised spectroscopically and both the rate of its appearance and that of the monomer consumption were then studied simultaneously. From these measurements (two experiments only) the value of $k_p = 2 \times 10^4$ M^{-1} sec^{-1} was calculated for polymerisations in 1,2-dichloroethane. In the second paper[119] this study was extended to media of lower polarity by mixing the original solvent with increasing quantities of carbon tetrachloride. Paradoxically, it was found that the apparent value of k_p *increased* as the dielectric constant of the medium *decreased*. In fact, starting from a mixture containing 60% of CCl$_4$, no carbenium ions were detected spectroscopically and yet the polymerisation proceeded normally. The authors concluded that an "invisible" species must be formed which is responsible for a progressively higher share of the propagation process as the dielectric constant of the medium is lowered. It is hard to avoid the conclusion that Higashimura from his own experimental evidence was rediscovering the very pseudocationic mechanism which he had consistently been rejecting since its original formulation by Gandini and Plesch[127] fifteen years earlier.

It is obvious that the values of k_p reported by Sawamoto and Higashimura, including that in pure 1,2-dichloroethane, are too high. The amount by which they should be reduced is not known, since the proportion of ester among the active species in the various solvent mixtures cannot be calculated.

In the third article of the series[204] the authors set out to determine the kinetic initiation parameters and the lifetime of the ionic chain carriers. The values of k_i were computed from the measured rates of carbenium ion formation assuming bimolecular initiation. This assumption is unacceptable a priori since the interaction between Brønsted acids and olefins in solvents like the one used in this work has been shown to involve kinetic patterns which are almost always more complicated than a simple first order in each reactant (see Sect. III-B). As for the calculation of the mean lifetime of the active species based on the expression

$$\tau = [P^*]_{max}/R_i,$$

we fail to see the kinetic meaning of such treatment. In fact, the lifetime of a transient species can be calculated under steady-state conditions: if $R_i = R_t$, the lifetime is the ratio of the concentration of the transient to either rates (cf. free radical polymerisation); if R_i = finite and $R_t = 0$, the lifetime of the transient approaches infinity (see living polymerisations). In the absence of steady state, various possibilities arise, but the lifetime of the transients will vary with time during the reaction. No study of possible termination reactions was carried out in this work, so that the overall kinetic behaviour of the system is not known. Thus, the value of the ratio between the maximum concentration of active species, attained well after mixing[89], and the *initial* rate of carbenium ion formation, measured at time zero, is internally and externally inconsistent. Unfortunately therefore none of the parameters calculated in this third paper are acceptable.

The potentials of methanesulphonic acid as initiator have by no means been exhausted. It seems a very interesting initiator and should be exploited in the future. The fact that it forms active ester molecules with *p*-methoxystyrene, a very basic monomer, suggests that the polymerisation of styrene with this acid is very probably pseudocationic.

11. Chloro- and Fluorosulphonic Acids

These two acids, which are stronger than sulphuric acid but weaker than trifluoro-methanesulphonic acid, have not been used in any detailed fundamental study in cationic polymerisation. Alkenes give the corresponding esters, but do not polymerise with these initiators[130,303]. Aromatic olefins, on the contrary, are readily polymerised.

Pepper has briefly reported on their activity vis-à-vis styrene, but has never published a full account of his investigation[299,300]. Asami and Tokura[220] used $ClSO_3H$ in sulphur dioxide for the polymerisation of the same monomer and found it to be a very efficient initiator. Recently, Masuda et al.[297] published a concise study of the polymerisation of styrene by both acids in various solvents. As expected, the chloro derivative was found to be a weaker promoter than the fluorosulphonic acid. Both systems gave asymptotic yields indicating that fast initiation was followed by an important termination reaction. Incorporation of chlorine into the polystyrenes prepared with chlorosulphonic acid (but absence of sulphur) suggested that this acid decomposes in the polymerisation process (hydrolysis?). On the whole, the limited amount of evidence obtained in this study does not allow any mechanistic conclusion.

We feel that these acids are not particularly suited for fundamental studies, particularly because of their readiness to hydrolyse. However, for those interested in the possible use of fluorosulphonic acid as catalyst we refer to a recent contribution by Goypiron et al.[301] concerning the state of its molecules in the undiluted liquid. This acid homoconjugates readily with its anion[302].

12. Trifluoromethanesulphonic (Triflic) Acid

Perfluoroalkanesulphonic acids are the strongest acids known. The first member of this series has received considerable attention in the last decade owing to its exceptional acidity coupled with a remarkable chemical stability (cf. perchloric and halosulphonic acids) and useful properties of its anion. The following pK values have been published: 4.7 and 5.1 in $AcOH$[36,37], 2,6 in acetonitrile[37], 3.0 in nitromethane[37], 2.2 in propylene carbonate[207], and 7.3 in 1,2-dichloroethane[39]. Kolthoff and Chantooni have found evidence of dimerisation of this acid in acetonitrile[304]; with the recent pK value in this solvent the dimerisation constant $(HB)_2/HB^2$ is about 10^3 at room temperature. Chmelir et al.[305] have postulated the existence of the double equilibrium

$$2\,HB \underset{}{\overset{K_1}{\rightleftharpoons}} (HB)_2 \underset{}{\overset{K_2}{\rightleftharpoons}} H_2B^+ + B^-$$

in dichloromethane, and from conductivity measurements obtained $K_1 > K_2$ and $K_1K_2 = 9 \times 10^{-8}M^{-1}$. Kabir-ud-din and Plesch[150] calculated a value of $K_1K_2 = 7.2 \times 10^{-3}$ in CH_2Cl_2 containing 0.1 M of base electrolyte. These values are most probably wrong because the above treatment did not take into account the very strong homoconjugation of this acid in CH_2Cl_2[135,306]. The polymerisation of alkenes with this acid (commonly known as triflic acid) has not been reported, but it is known that ethene and propene react with it to give the corresponding esters[130]. Styrene and other aromatic monomers are known to polymerise efficiently under the action of this initiator. These reactions are analysed below.

a) Styrene and Homologues

The first study of the cationic polymerisation of styrene by triflic acid was undertaken before 1970 by Mathias and Plesch[307]. They found that very low catalyst concentrations ($10^{-4}-10^{-5}$ M) were needed to induce a fairly rapid polymerisation in dichloromethane. These reactions were followed calorimetrically, conductimetrically and spectrophotometrically. The major obstacle precluding a systematic study was the poor reproducibility of the results. This problem was finally traced to the aging of the acid solutions contained in sealed phials. It was concluded that a slow reaction between the catalyst and the solvent was the cause of this aging phenomenon:

$$CF_3SO_3H + CH_2Cl_2 \longrightarrow CF_3SO_3CH_2Cl + HCl,$$

but possible complications arising from impurities in the solvent or on the glass surface of the phials could not be excluded. It must be pointed out that other authors have noticed that triflic acid reacts with acetonitrile[304] and with 1,2-dichloroethane[203].

Despite this drawback, the general features of the system could be assessed. Polymerisations were accompanied by an increase in conductivity and, towards the end, by the appearance of a yellow tint. A second addition of styrene provoked a sudden decrease both in conductivity and in colour intensity, but during the second polymerisation they increased again. Although these phenomena recalled a similar behaviour in the system styrene-perchloric acid, a major difference distinguished between these two systems, namely the relatively high conductivity *during* the polymerisation with triflic acid. However, in view of the problems encountered with the catalyst solutions, the authors could not draw any reliable mechanistic conclusions from their study.

A few years later, Chmelir presented a communication on this system at the Rouen Meeting[308]. Since then this author has published several papers with Schulz and Cardona on the subject[46,305,309−312]. There is a good deal of repetition in these publications, and numerous contradictions are scattered throughout this work. The history and evolution of this study is worth a close analysis.

The first report by Chmelir[308] can be summarised as follows. The system consisted of styrene (0.05−0.15 M), triflic acid ($6 \times 10^{-5} - 5 \times 10^{-4}$ M) and methylene chloride, T = −52 to 25 °C.

(i)   Solutions of triflic acid in CH_2Cl_2 showed appreciable conductivity κ, directly proportional to the acid concentration. This behaviour was taken as evidence for the equilibria

$$2\,TfOH \rightleftharpoons (TfOH)_2 \rightleftharpoons TfOH_2^+ + TfO^-, \quad Tf = CF_3SO_2;$$

no intermediate ion pairs were considered, nor the possible homoconjugation of the anion with excess acid.

(ii)   Polymerisations were quite fast, but the time-conversion curves had curious shapes, not compatible with a simple kinetic treatment. No induction period was observed, notwithstanding the incorrect terminology of the author describing the curves as S-shaped.

(iii)   Moisture did not affect the process sensibly, except when the water concentration was more than ten times that of the acid. Then, the polymerisation rates were reduced.

(iv)   Solutions were colourless both *during* and *after* the polymerisations.

(v)   A complex reaction scheme was proposed involving initiation by both the neutral dimeric acid and the protonated acid, and a reaction between monomer and monomeric acid giving an unspecified inactive species. The algebraic treatment of this scheme led to an equation which was supposedly verified experimentally (see discussion below).

(vi)   It was finally stated that the formation of active species from two molecules of acid was "supposed to be a determining-step in the mechanism of polymerisation" (sic).

Other points related to transfer reactions and molecular weights are not relevant to the present context. A year later, Cardona and Chmelir reported some new points[309], namely:

(vii)   100-fold dilutions of "concentrated" acid solutions produced a change in conductivity (first a rise, then a slow decrease) which took several minutes to reach a stable value. This phenomenon was explained in terms of slow reequilibration of free ions to dimer molecules (again, no ion pairs were mentioned).

(viii) In a polymerisation, the conductivity suddenly increased upon mixing the reactants, then levelled off. With a second addition of monomer at the end of the first polymerisation, κ decreased abruptly and then rose again slowly to reach a new steady value, higher than that attained at the end of the first reaction. This cycle could be repeated several times. The authors wrongly established a parallel between this behaviour and that observed by Gandini and Plesch with styrene-perchloric acid[45].

At the same meeting, Chmelir gave a second communication[310] and new aspects were analysed:

(ix)   Different time-conversion curves were obtained by adding a concentrated acid solution to the monomer solution and by adding the monomer to the diluted and "equilibrated" acid solution. This was taken as further proof of point (vii).

(x)   The overall kinetic equation was modified by the addition of a new term accounting for yet another equilibrium

$$TfOH_2^+ + (TfOH)_2 \rightleftharpoons (TfOH)_2 \cdot TfOH_2^+$$

which produced a third species capable of initiation.

(xi)   The "initial rate of polymerisation", defined as the rate observed immediately after achievement of the various equilibria (?), goes through a maximum for a given value of the monomer concentration (at constant acid concentration) and decreases for values of M above and below this critical concentration.

In an "expanded version" of his Madrid paper, Chmelir[311] reiterated some of the previous points but did not present any new material. At the Akron meeting, Chmelir[312] revised point (i) by writing the equilibria as

$$2\,TfOH \;\rightleftharpoons\; TfOH_2^+\,TfO^- \;\rightleftharpoons\; TfOH_2^+ + TfO^-$$

(dimeric acid not mentioned). He also changed point (iv) which now became:

(xii) Solutions turned yellow after polymerisation. An absorption band at 415 nm was recorded and its variation in optical density reported to follow the same type of cycles as the conductivity (see point viii).

An additional point was made in this paper:

(xiii) The dispersity of the polymer molecular weight increased considerably as the reaction temperature was decreased. This was interpreted as evidence for the plurality of active species below $-15\ °C$, and for only one type of chain carrier above that temperature ($\overline{DP}_w/\overline{DP}_n = 2$). Oddly, the temperature at which most of the kinetic measurements had been made in previous study was $-15\ °C$, so it is difficult to know how many types of chain carriers are present at this borderline temperature.

Meanwhile, Cardona and Schulz[46] published some disturbing results concerning the differences in the rate of polymerisation and in the conductivity of acid solutions with two batches of methylene chloride purified by different procedures. CH_2Cl_2 dried only with P_2O_5 gave higher conductivities and much higher polymerisation rates than the solvent purified by treatment with P_2O_5 and CaH_2. The interpretation given by the authors about the origin of this discrepancies was based on the presence of a "cocatalytic impurity" in the solvent treated only with P_2O_5, i.e. the solvent used for all previous studies of Chmelir and Cardona's.

The last paper of this series of seven, purported to give the complete picture on the system[305]. For the first time in four years some experimental details were reported. Although the actual polymerisation apparatus has not yet been described, some revealing details were finally published about purification and handling of materials. Reagent solutions were injected in the polymerisation vessel with hypodermic syringes against a stream of dry nitrogen. It is not clear *how* the acid solutions were prepared, but they were renewed before each set of experiments. Apparently, the polymer was isolated without removal of the acid (solvent evaporation). In this paper the situation is as follows. Point (i) was retained, no mention of point (xi) being made. A new dichotomy was introduced by the fact that while the conductivity measurements of the acid solutions were carried out with a solvent purified by both P_2O_5 and CaH_2 and were therefore much lower than those reported in the first paper[308], the kinetics of polymerisation were investigated with a solvent purified only with P_2O_5, thus reflecting a situation where much higher rates are obtained[46]. Point (iii) was changed to "water addition decreased the rate of polymerisation". Point (iv) was retained despite the fact that in the same paper visible absorption bands (415 nm) were reported and discussed. Point (v) was expanded to include point (x). Point (vi) was reasserted verbatim. Point (vii) was expanded by more experimental evidence but now the time scale for the attainment of the new "ionic equilibria" was of the order of 30 min, no explanation being given of the discrepancy with previous results[309]. Point (xii) was reinforced by an actual spectrum of the "*polymerising solution*" exhibiting an absorption at 415 nm. Point (xiii) was not mentioned again.

We have devoted so much space to this group of seven papers to clearly underline their weaknesses with respect both to the often contradictory information and to the mechanisms proposed. For example, the problem of the selfionisation of the acid, so basic to the mechanism postulated thereafter, is alternatively treated by different equilibria[305,312] and moreover the actual values of conductivity are changed from the first paper to the last without a clear explanation of the origin of the discrepancy. On the specific problem of solvent purification we feel that the treatment with calcium hydride might have introduced some basic impurities of hydride or hydroxide carried over in the distillation. This would explain the lower conductivity and the lower polymerisation rates, since the real acid concentration would be appreciably lower than the nominal one, given the fact that the amount of acid used was very small (conc. $1-4 \times 10^{-4}$ M). Also, the use of syringes for the addition of the acid solution must inevitably have introduced a non-negligible amount of moisture. We feel that the amount of water present in these systems must have been at least of the same order of magnitude as the acid concentration. Thus, the conductivity could well have originated from the hydration of the acid and not from its self-ionisation. We also find difficult to understand the occurrence of ionic reequilibration which take first a few minutes, then about half an hour to establish upon a simple dilution. No proof is given that the conductivity of the polymerising solutions is due to the presence of ionic (free ions) active species. These values are quite high, which would suggest a considerable concentration of free carbenium ions propagating in a medium of modest polarity. This list of doubts and inconsistencies could continue, but the point we wish to make is that when one looks at these seven papers closely, one realises that the useful conclusions which can be extracted from them are just a few qualitative features of the system, not much more than already observed by Mathias and Plesch, who felt that their results were too meagre to be published.

Higashimura's school has also studied the polymerisation of styrene by triflic acid in various solvents. In the first paper of this investigation it was shown that with [M] = 1 M, [TfOH] $\sim 10^{-4}$ M at 0 °C, polymerisations were first order, in methylene chloride, benzene and nitrobenzene, and did not show the odd behaviour reported by Chmelir[297]. Also, the first order rate constants in CH_2Cl_2 and benzene were found to be directly proportional to the initial acid concentration. The effect of water was also investigated. First, it was established that about 3×10^{-4} M of water were present in normal polymerisation systems (i.e. a concentration comparable to that of the acid). Then water was purposely added and its effect on the rate of polymerisation studied. Considering the moisture already present, it was found that water up to twice to three times the amount of catalyst did not alter appreciably, but larger quantities sensibly reduced the rate.

In their second paper[313] these authors investigated in detail the role of the dielectric constant of the medium. Their ingenious experiments proved that the apparent dependence of the conversion upon the initial monomer concentration was solely due to changes in the solution polarity and that for a given acid concentration the time-conversion curves relative to a series of polymerisations with different monomer concentrations could be reduced to a single line simply by adjusting the dielectric constant to a fixed value, with additions of benzene or CCl_4. Thus, Chmelir kinetic scheme originated from an artifact and all polymerisations were shown to follow the simple law

$$-d[M]/dt = k_2 [TfOH]_o [M],$$

i.e., a stationary state situation. Very interesting was also the observation that the second-order rate constant k_2 at 0 °C was sensitive to the dielectric constant of the medium in the range of 7 to 9 units. In that small interval of polarity k_2 went from 2 to 20 M^{-1} sec^{-1}, while below $\epsilon = 7$ and above $\epsilon = 10$ the changes in k_2 were quite modest. This steep variation of rate was accompanied by a corresponding qualitative change in the DP distribution of the products. In the region of low ϵ the polymers possessed unimodal distribution centered around a low value of DP; in the region corresponding to the sharp increase of k_2 the polymers displayed a bimodal distribution; finally above $\epsilon = 10$ the products were again unimodal, but with a fairly high DP. All these observations indicate that the active species suddenly gain a considerable amount of propagating capacity probably through a change in their intimate structure, within a critical range of polarity. Yet, the overall kinetic behaviour does not change qualitatively, i.e. the polymerisation conserves its simple stationary-state character. Two questions come to mind at this point:
— is k_2 the propagation rate constant, i.e. has all the acid been consumed to give active species?
— what is the difference between the structure of the two chain carriers?
The answer to the first question cannot be given because no tests were carried out to check the presence of free acid in the polymerising solutions (see styrene-perchloric acid in next section). Moreover, the first-order dependence on the acid concentration was verified over only a relatively narrow range of acid concentrations[297]. Kunitake and Takarabe[87] have in fact shown that with about ten times more acid the polymerisation is extremely fast and apparently only involves a few percent of the total acid added. The latter conclusion was based on a comparison of the concentrations of carbenium ions formed and of acid used, but of course it could well be that the rest of the acid had reacted to give the ester; in other words no direct demonstration was given of the presence of *free* acid in those experiments. Thus, the value for the propagation rate constant calculated by these authors (3×10^5 M^{-1} sec^{-1} at 30 °C) could well be too high for the same reasons already discussed in the case of the system p-methoxystyrene-CH_3SO_3H (see Sect. III-E-10).

As for the second question, no direct evidence as to the nature of the chain carriers was offered by Higashimura's group[313]. Their experiments with added $Bu_4N^+CF_3SO_3^-$ do not prove the existence of free ions since homoconjugation of the anion with the acid was the real reason for the decrease in rates. Souverain[135] has shown conclusively that such interaction is very strong in methylene chloride. The work of Kunitake and Takarabe[87] has certainly proved that with high acid concentration, i.e. above 10^{-3} M, 1-phenylethyl carbenium ions are formed, although it is impossible to comment on their aggregation state. The extrapolation of these observations to experiments conducted with ten times less initiator are dangerous, before a better overall knowledge of these systems is acquired. In particular, the role of residual moisture must be better understood. As we have said, all of Higashimura's work was in fact conducted with the acid hydrate, since the water concentration was higher than the acid concentration, in CH_2Cl_2. On the contrary, Kunitake and Takarabe's experiments most probably involved unhydrated acid as the principal initiating species, given their higher concentrations. This difference ought to be considered seriously.

Higashimura's group has published a third paper in this context[314]. Apart from showing that p-methylstyrene polymerises much more readily than styrene in the same experimental conditions, and that p-chlorostyrene does the opposite, as expected, this investigation does not add anything important to the understanding of the fundamental questions discussed above. The latter can also be said about other investigations by the same school concerning the selective dimerisation of styrenes by triflic acid in various solvents[315,316], and the role of transfer reactions in the system styrene-triflic acid[317].

Cardona-Sutterlin has recently reopened the dossier concerning the system triflic acid-styrene with two papers dealing again with the kinetics and mechanism of this polymerisation[318,319]. Her results and interpretations are in open contrast with previous work carried out by Chmelir and herself in the same laboratory. Without explaining why or how this was achieved, the author affirms that the polymerisation rates are now not affected by the type of purification procedure used for the solvent, viz. treatment with P_2O_5 or with P_2O_5 and CaH_2, in marked contrast with her previous work with Schulz[46]. The course of polymerisations carried out between -15 and $-60\,°C$ are now said to follow a first-order pattern and a criticism is given of the previous complex treatment (see discussion of Chmelir's work above). The variation of the first-order constant with monomer concentration is also presented and despite the elegant explanation already given by Higashimura's school, a rather confused discussion is presented as to the origin of this phenomenon. We consider the arguments about conductivity during the polymerisation and nature of the active species to be obscure, mainly because the author is wrong in considering the changes in conductivity (particularly as conversion increases) as a direct reflection of corresponding fluctuations in the concentration of "polystyril ions". Obviously, by far the major contribution to this electrical phenomenon is due to the formation of phenylindanyl cations, as already recognised in previous publications from the same laboratory, and *not* to propagating species in the form of free ions. An interesting information which can be retained from these two papers comes from the systematic observation of external orders higher than unity in triflic acid. Of course this does not imply, as claimed by the author in her discussion, that the active species are triple ions (we fail to understand the connection between third order dependence in acid and triple ions as chain carriers), but simply that the initiation process must involve more complex interactions than a straightforward one-to-one reaction between monomer and catalyst. This is in open conflict with Higashimura's findings[297], and we have no explanation for the discrepancy.

Chmelir and Schulz[320] have already defended their original work against the new information and interpretation given by Cardona-Sutterlin, who has given her answer to their comments[321]. Unfortunately, however, this debate adds nothing to the understanding of the processes in question. In particular, we found it surprising that Chmelir and Schulz completely ignored the fundamental contribution of Higashimura's group to the unravelling of this system.

An extremely important observation has recently been reported by Souverain in his thesis carried out in Sigwalt's laboratory[135]. 1-phenylethyl triflate was synthesised in methylene chloride from the parent carbinol and triflic anhydride. This ester was shown to be stable at room temperature and characterised by NMR spectroscopy. The implications of this experiment are all too obvious to be discussed at length, but it must be emphasized that future investigations with that compound as initiator in suitable media could provide very valuable information concerning the mechanism of formation and the nature of the active species in the polymerisation of styrene by triflic acid.

Within the last few years that system has become one of the most studied in the field of Brønsted-acid initiated cationic polymerisations. Both initiation and propagation reactions have received close attention and the nature of the active species has been discussed from various points of view. However, many conflicting pieces of evidence still need to be reconciled and therefore the understanding of these processes is far from complete.

b) 1,1-Diphenylethylene

The mechanism of the dimerisation of this olefin with triflic acid has been the object of several recent investigations[135, 148, 203, 306]. Since this is one of the best-understood systems in cationic polymerisation we will give below a summary of the most salient features which characterise it.

We have already analysed the work of Kunitake and Takarabe[203] in Sect. III-D. With spectroscopic quantities of olefin, and acid concentrations about 100 times higher, the protonation reaction was found to be bimolecular

$$CPh_2{=}CH_2 + CF_3SO_3H \xrightarrow[\quad 30° \quad]{CH_2Cl_2} CH_3{-}\overset{+}{C}Ph_2 \ CF_3SO_3^-.$$

However, the limited range of acid concentrations used in this work do not allow an unequivocal conclusion about the molecularity of the protonating agent in the above reaction.

LeBorgne[306] and Souverain[135] have in fact found that the interaction of these two reactants in approximately equal concentrations in CH_2Cl_2 leads to the formation of an amount of 1,1-diphenylethylium ions corresponding to only one third to one fourth of the amount of acid used. The interpretation of these results involves the necessity of free acid to provide adequate solvation for the carbenium ion and to account for the homoconjugation of the anion formed in the reaction.

The fast initiation step is followed by an equally fast propagation reaction. While the rate constant of the former has been measured by Kunitake and Takarabe[203], the dimerisation kinetics have not been measured for this particular system. The following slow reactions consist of the proton transfer between the dimeric cation and the monomer to give alternatively the unsaturated dimer or the indanylic one. Finally, the protonated dimer can isomerise to a more stable configuration due to the direct interaction of the two phenyl groups through space polarisation effects[148,322]. Thus:

$$3 \ (4) \ CF_3SO_3H + CH_2{=}CPh_2 \ \rightarrow \ CH_3{-}\overset{+}{C}Ph_2/(nCF_3SO_3H)/H(CF_3SO_3)_2^-$$

$$HM^+HB_2^- + M \rightleftharpoons HD^+HB_2^-$$

$$HD^+HB_2^- + M \begin{array}{c} \nearrow D_{unsat} + HM^+HB_2^- \\ \searrow D_{ind} + HM^+HB_2^- \end{array}$$

$$HD^+ \rightleftharpoons X^+ \text{ (stabilised dimer).}$$

All the equilibrium constants were determined in this scheme thanks to a very detailed spectroscopic study initiated by Sauvet et al.[148] and completed by LeBorgne[306].

c) 3-Substituted Indenes

Two interesting studies concerning the protonation of 3-substituted indenes which cannot polymerise because of steric hyndrance, have been conducted in recent years using triflic acid as electrophile. Hung et al.[94] have followed the formation of the carbenium ion derived from 3-isopropylindene by ultraviolet and ^{13}C NMR spectroscopy. The characterisation of this ion was achieved and it was claimed that each molecule of triflic acid generated one carbenium ion, i.e., 1:1 stoicheiometry.

Similar results were obtained by Cheradame et al.[93] with other substituted indenes, and again 1:1 protonation was advocated[323]. One of these systems was restudied recently by Souverain[135] who showed that protonation was fast but incomplete, since approximately two molecules of acid were needed to protonate each molecule of olefin. This result was interpreted in terms of specific solvation of the cation by excess acid and/or homoconjugation of the triflate anion. The origin of the discrepancy between Hung's and Souverain's results is being investigated in Sigwalt's laboratory where all this work was carried out.

d) p-Methoxystyrene

Sawamoto and Higashimura[205,891] have continued their investigation of the stop-flow spectroscopic behaviour of the polymerisation of p-methoxystyrene and added triflic acid to the series of catalysts previously tried. Again, they observed the typical absorption around 380 nm, attributed to the protonated monomer and calculated an initiation rate constant at 30 °C in ethylene chloride (assuming a bimolecular interaction; orders not determined): $k_i = 5 \times 10^4 \, M^{-1} \, s^{-1}$. This value is expectedly much higher than that obtained by Kunitake and Takarabe[87] for styrene under similar conditions, given the much higher basicity of the monomer under study. It is also higher by a factor of ten than that for the protonation of 1,1-diphenylethylene[203]; this would indicate that p-methoxystyrene is more nucleophilic than the latter olefin, but the comparison must await confirmation of the assumed initiation stoicheiometry.

Sawamoto and Higashimura also determined values of the propagation rate constant by measuring at the same time the amount of monomer consumed and the integrated amount of active species. This was done in media of decreasing dielectric constant and it was found that k_p apparently increased as the polarity decreased. This anomaly, already encountered with methanesulphonic acid, was again explained by invoking the presence of "invisible" active species. This is another "rediscovery" of pseudocationic polymerisation (cf. Sect. 10 above) fifteen years after the event. The figure of about $1 \times 10^5 \, M^{-1} \, s^{-1}$ for the propagation rate constant at 30 °C in ethylene chloride is probably close to the true value if, as the authors affirm, most of the acid had been used up to give carbenium ions.

It is interesting that pseudocationic polymerisation can occur even with the strongest acid known and one of the most basic monomers available, if the solvent has a relatively low polarity. The presence of propagating ester molecules is therefore much more widespread than originally thought, and must be considered a totally general phenomenon in the cationic polymerisation initiated by Brønsted acids.

13. Perchloric Acid

The high strength of perchloric acid has been known for more than a century, and numerous studies have been devoted to its properties, stability and behaviour in various solvents[40,325,326]. The following values of pK have been determined: 5.1 in acetic acid[37], 1.57 in acetonitrile[37] and 2.0 nitromethane[37]. Specific interactions have been detected spectroscopically with the latter solvent[325] which explain the high degree of dissociation. A hydrogen bond is in fact established between the NO group of the solvent and the OH of the acid. Very probably a similar type of complexation occurs in acetonitrile. Dissociation of perchloric acid in chlorinated hydrocarbons is minimal[40,45]; its pK in 1,2-dichloroethane can be estimated at about 7.5[39]. Coutagne has concluded that this acid is not dimerised in dichloromethane[40], and Pavia has found no indications of such an association in chloroform and CCl_4[325]. Homoconjugation of the perchlorate anion with its acid readily occurs in dichloromethane[327] and chloroform[328], and the biperchlorate anion has been characterised by Karelin et al.[329]. Safe methods have been described to prepare anhydrous $HClO_4$ solutions in inert solvents[330,331].

The use of anhydrous perchloric acid as a catalyst for cationic polymerisation dates from 1960, when Tauber and Eastham[332] published a study of the interaction of this catalyst with 2-butene in 1,2-dichloroethane. The complex behaviour of this system did not permit any definite conclusion about the reactions involved, but the possibility of initial formation of iso-butyl perchlorate was seriously considered. Today the formation of alkyl perchlorates is well documented as discussed in Sect. II-I, except for tertiary ones. Plesch and Westermann[333] noticed that isobutene did not polymerise in CH_2Cl_2 when perchloric acid was mixed with it, but only gave t-butyl perchlorate.

Also in 1960, Asami and Tokura showed that 60% perchloric acid was a very effective catalyst for the polymerisation of styrene in sulphur dioxide[220]. But it was the now classical study of Pepper and Reilly[334] on the polymerisation of styrene by the anhydrous acid in various solvents that initiated a long series of investigations on the role of this promoter in the cationic polymerisation of aromatic olefins and more recently of cyclopentadiene.

a) Styrene

Pepper and Reilly's original investigation[334] underlined the "docile" behaviour of the polymerisation of styrene by perchloric acid in chlorinated aliphatic hydrocarbons. Reactions were first order internally and insensitive to moisture up to $[H_2O] \sim 10\,[HClO_4]$; no termination reaction could be detected, since a second addition of monomer at the end of a polymerisation produced a new polymerisation with the same rate as the first; the first-order rate constant was directly proportional to the initial acid concentration over a wide range of values; the dielectric constant of the medium, changed by mixing different proportions of CH_2ClCH_2Cl and CCl_4, affected the rate of polymerisation in an expected fashion, i.e., the rate constant decreased smoothly as ϵ was lowered; the activation energy for the second-order rate constant increased with decreasing dielectric constant.

On the basis of the above kinetic observations a mechanism was proposed which involved fast and complete initiation followed by standard bimolecular propagation (transfer reactions will be ignored in this context):

$$CH_2=CHPh + HClO_4 \;\rightarrow\; P_1^+$$

$$P_n^+ + M \;\rightarrow\; P_{n+1}^+.$$

Thus, if $k_i \gg k_p$, the experimentally determined second-order rate constant is the propagation rate constant k_p. At 25 °C in 1,2-dichloroethane ($\epsilon = 9.7$) $k_p = 17.0\,M^{-1}\,sec^{-1}$ and $E_p = 8.3$ kcal mole^{-1}.

The nature of the active species was also investigated spectroscopically. An absorption peak at 415 nm was erroneously attributed to the presence of 1-phenylethylium ions *during* the polymerisation. Pepper and Reilly proposed therefore that the chain carriers were ionic, and more specifically ion pairs[334,335].

Some disturbing features of this investigations, namely the low values of the propagation rate constant and the lack of sensitivity of the polymerisation towards moisture, prompted Gandini and Plesch to carry out more work on this system. These studies[8,45,127,330,336] led to the discovery of pseudocationic polymerisation. While the kinetic features in methylene chloride were an excellent replica of those described by Pepper and Reilly, the lack of absorption above 310 nm, confirmed later by other authors and in the original laboratory[86,83], and of increase in conductivity *during* the polymerisation were taken as evidence for the absence of ionic chain carriers. An experiment involving the in situ preparation of 1-phenylethyl perchlorate gave polymerisation[127] and was therefore taken as strong supporting evidence for the non-ionic nature of the propagating species.

The proposal of the theory of pseudocationic polymerisation provoked a great deal of controversy, but since Gandini and Plesch[45] already discussed the various objections put forward, we refer the readers to that rebuttal and refrain from repeating all the arguments against or in favour of these new ideas. We will briefly comment on a more recent paper by Hamann et al.[337] concerning the mechanism of this polymerisation in carbon tetrachloride. These authors rejected the pseudocationic interpretation on the basis of the lack of signals due to the ester in the NMR spectrum of the polymerising solutions. However, it is clear from the acid concentrations they used that the detection of oligostyryl perchlorates, i.e. the resonance of the sole proton which is likely to appear at a characterisitic field value ($\sim$CH$_2$–C$\underline{\text{H}}$Ph–OClO$_3$), would have been impossible with the resolution they disposed of, and in presence of the broad signals due to the accumulating polymer. Also, we find it surprising that they did not follow the fate of the acid proton at low fields to verify their alternative mechanism involving the formation of a π-complex between HClO$_4$ and styrene. This search would have probably shown the absence of acid during the polymerisation, as already proved by Worsfold and Bywater[86].

We find in the fifteen years following the discovery of speudocationic polymerisation a considerable amount of evidence from the study of numerous systems showing that polarised ester molecules do intervene in propagation reactions as more or less active species. It is also evident that in many instances the esters alone, i.e. in the absence of a suitable polarising micro-environment, are not capable of propagating at a measurable rate. These points have been emphasised all along the present chapter.

The system styrene-perchloric acid is however more complicated than originally thought. Gandini and Plesch[336] reported in 1965 that under "special" conditions the true cationic polymerisation of styrene had been observed. The presence of ionic species resulted in a very fast polymerisation even at temperatures as low as −90 °C. These observations were complemented in a later paper[45], but never studied systematically. It was in Pepper's laboratory[338,339] that this investigation was continued and led to the use of stop-flow techniques to attempt a complete rationalisation of the very complex features displayed by this system at low temperature[17,82,118,340,341]. Lorimer and Pepper[20?] have given a detailed account of these studies and their difficult interpretation. Essentially, three distinct situations are now recognised, depending on the temperature range at which the reactions are carried out:
- from room temperature down to about −25 °C the behaviour is essentially pseudo-cationic, although a minor contribution from ionic carriers has been detected in highly purified systems[342];
- between −30 and −80 °C two stages are clearly distinguishable in the conversion-time curves, a first rapid and non-stationary consumption of monomer (characterised by the presence of polystyryl ions detected spectroscopically) and a slower stationary polymerisation leading to complete conversion through the action of ester carriers;
- below −80 °C the fast non-stationary polymerisation by ionic active species attains a limiting yield, probably because the ester is too inefficient at these temperatures.

The direct identification of the ionic chain carriers during the flash polymerisations and the rate-suppressing effect of common-anion salts used in concentrations well below those of the acid (i.e., in the absence of serious homoconjugation problems), strongly suggest that free ions are also present in these initial "bursts" of monomer consumption. The fate of the ionic chain carriers is not altogether clear, but they probably collapse into the corresponding ester after their ephemeral but vigorous existence.

The complicated kinetic pattern obtained for the non-stationary phase of these polymerisations could not be entirely mastered by Pepper and Lorimer. Unfortunately their stop-flow apparatus was not very fast and we feel that with a more modern version of this technique, such as the instruments used by other authors in the last few years[87,89], more accurate and informative results could be gathered. However, tentative values of the propagation rate constants for ion pairs and free ions were calculated: $2,000 \, M^{-1} sec^{-1}$ and $30,000 \, M^{-1} sec^{-1}$, respectively at $-80 \, °C$. The corresponding activation energy was approximately 4 and 1 kcal mole^{-1}, respectively. The kinetics of initiation were analysed by us in Sect. III-D on the bases of Pepper and Lorimer's data, but must await experimental confirmation, particularly in view of the uncertainty about the molecularity of the reaction.

In conclusion, this very prolific system is now well understood in its pseudocationic aspect, but needs further probing before the intricacies of the fast non-stationary polymerisation at low temperature are fully unravelled.

It is interesting that $HClO_4$ is incapable to polymerise styrene in nitromethane, nitrobenzene and acetonitrile[334,335]. The reasons for this inefficiency must be found in the specific interactions of these solvents with the acid. These media sequestrate the acid by forming strong hydrogen-bonded complexes and the nucleophilicity of styrene is not high enough to remove a proton from them.

Kucera[343,344] has recently speculated about the mechanistic features of the system perchloric acid-styrene and has concluded that "peripheral solvated ion pairs" (sic) are in fact the dominant chain carriers, in contrast to the pseudocationic theory. His arguments are very hazy and unconvincing. In particular, this author seems to forget that however tight his proposed ion pairs might be, they should exhibit the typical spectrum of the 1-phenylethyl carbenium ion, which is not oberserved during pseudocationic polymerisation. This important fact, coupled with the absence of free acid in the system leaves little choice but to postulate that the ester molecules are responsible for propagation. Kucera's alternative proposal would also lead to much more complex kinetics than those observed. Note finally that his ionisation equilibria for the acid are wrongly formulated since they do not take into consideration homoconjugation nor the improbability that a proton can exist as such in methylene chloride.

Apart from the original papers by Gandini and Plesch, the arguments supporting the theory of pseudocationic polymerisation have been expounded by Plesch[9,10,14,15] and by Dunn[345].

b) Other Aromatic Monomers

p-Chlorostyrene behaves qualitatively like styrene in that it gives pseudocationic polymerisation at room temperature[346], and a contribution from true-cationic polymerisation at $-80 \, °C$ as testified by the presence of carbenium ion during the reaction[82,300]. Quantitatively, it is less reactive than its parent compound, as expected from the electron-withdrawing effect of the chlorine atom in the para position of the ring; these effects in copolymerisation are yet another confirmation of pseudocationic polymerisation at room temperature.

2,4,6-Trimethylstyrene has also been studied at $-80 \, °C$ and evidence for the presence of the corresponding carbenium ions obtained by ultraviolet spectroscopy[82,300]. Both α-methylstyrene and p-methoxystyrene were studied at $-80 \, °C$ by stop-flow spectrophotometry. They were found to polymerise so rapidly that the spectra taken were in fact those of species present at the end of the polymerisation[82,300]. The polymerisation of p-methoxystyrene by perchloric acid in methylene chloride at room temperature is also extremely fast and accompanied by both an increase in conductivity and colour formation[8,275], i.e. it bears all the symptoms of a true-cationic process. These observations are compatible with the very pronounced nucleophilicity of the latter monomer which would favour the formation of carbenium ions with such a strong acid as perchloric acid.

Acenaphthylene on the other hand polymerises with $HClO_4$ in CH_2Cl_2 with no detectable presence of carbenium ions. It seems that this system is another example of a pseudocationic polymerisation[8,275]. Finally, in the case of N-vinylcarbazole it is difficult to establish the character of its polymerisation by perchloric acid in methylene chloride, since its very high rate even with less than 10^{-5} M of acid did not permit a serious search of possible carbenium ions formed *during* the process[275]. The fact that this monomer can be polymerised by dilute aqueous solutions of perchloric acid in such nucleophilic solvents as ethyl acetate and acetone[347] is hardly compatible with a cationic mechanism. The nature of the chain carriers in this bizarre system is totally unclear.

c) Cyclopentadiene

The only study of the polymerisation of a dienic monomer by perchloric acid is a brief report by Kohjiya et al.[296]. Reactions were carried out in toluene and methylene chloride at $-78\,°C$ with acid concentrations ranging from 2×10^{-3} to 1.4×10^{-2} M. Asymptotic yields were observed and the polymers showed an appreciable amount of branching and crosslinking. The authors' treatment of the results was optimistically based on Pepper and Burton's fast-initiation unimolecular-termination kinetics. In fact, the evidence collected was extremely limited and any conclusion other than the obvious occurrence of a termination reaction seems totally unwarranted.

14. Iodine

Solutions of iodine in chlorinated hydrocarbons can be characterised by the existence of the double equilibrium[348]

$$I_4 \rightleftharpoons 2\,I_2 \rightleftharpoons I^+ + I_3^-$$

where the species I_2 predominates and the ionic species are more abundant than the tetraatomic molecular form. To our knowledge, no study of these equilibria has been published in the solvents used in cationic polymerisation. The tendency of iodine to give charge-transfer complexes with aromatic and olefinic compounds is well known and documented.

Alkenes do not polymerise in the presence of iodine, but give instead the corresponding diiodides (see Sect. III-C). Aromatic olefins and vinyl ethers on the other hand are susceptible to this initiator and readily give polymers in suitable media. Our present discussion of these processes will concentrate on work published after Plesch's second book, where previous investigations are thoroughly reviewed.

a) Aromatic Monomers

In 1961 Okamura and collaborators[349] proposed a method for determining propagation rate constants based on the general assumption that the total amount of initator used in a polymerisation was distributed among three species (in the early stages of the reaction): free initiator, initiator complexed to the monomer, and active species. Measurements of polymerisation rates and of the amount of complexation (or of free initiator present) would thus lead to the value of k_p through suitable algebraic manipulation of the experimental data. The first application of this method involved the system iodine-p-methoxystyrene-chlorinated hydrocarbons. In a later publication[350] the same authors studied the polymerisation of styrene from the same viewpoint. The values of k_p thus obtained were surprisingly low for the proposed ionic chain carrier. Moreover they showed a definite trend with both the iodine and the monomer concentration. The method was in fact erroneously applied because it neglected the iodine being bound to the polymer and it assumed stationary-state conditions which were not achieved in either system.

A few years later Giusti and Andruzzi[351-353] carried out an excellent fundamental investigation on the system styrene-iodine 1,2-dichloroethane, using rigorous high-vacuum techniques to follow the kinetics of polymerisation and the electrical conductivity of the solutions. Some ingenious mechanistic ideas were also successfully applied to complete the study. Their experiments led to the following conclusions:

— styrene reacts with iodine to give the corresponding diiodide which in turn expels hydrogen iodide in an equilibrium reaction;
— hydrogen iodide adds rapidly onto styrene to give 1-phenylethyl-iodide; the latter can act as a chain carrier for the polymerisation because it is activated by polarising solvation of free iodine;
— the above sequence of interactions is the cause of the acceleration periods observed in all polymerisations, which can be eliminated by the addition of HI or 1-phenylethyl iodide to the initial styrene-I_2 mixtures; note that HI alone does not cause any polymerisation;
— water deliberately added only had a detrimental effect on the rate of polymerisation if it was mixed with HI and iodine before styrene was introduced (hydration of the acid reduces the initiation rate); no effect was observed on the other hand if an equimolar mixture of water and iodine was added to styrene containing 1-phenylethyl iodide;
— the conductivity during the polymerisation was due to the dissociation of iodine; its increase when the conversion has attained high values is caused by HI regeneration; the acid in the absence of styrene reacts with iodine to give HI_2^+ and I_3^-.

The authors concluded that the polymerisation must be pseudocationic, the active species being both 1-phenylethyl iodide and styrene diiodide polarises by the specific solvation of iodine molecules.

Similar work was carried out by Giusti's group[197,354] on acenaphthylene and the same mechanistic conclusions were reached. The lack of effect of high electric fields applied to the polymerising solutions was an additional element in favour of pseudocationic propagation.

In a third study, the same group[355-357] examined the polymerisation of anethole by iodine. Again the cocatalytic role of hydrogen iodide generated in situ was shown. While the sequence of reactions leading to the propagating species was essentially the same as in the two previous investigations, the authors concluded that here ionic chain carriers must also be present. The main evidence in favour of such argument was the observation of an appreciable positive effect of applied electric fields on the rate of polymerisation.

The depth and high quality of the work carried out in the Pisa laboratory has contributed decisively to the understanding of the role of iodine in the initiation of cationic polymerisation. Giusti's group has moreover clearly proved that Okamura's method for determining k_p's cannot be applied to systems involving iodine.

A new wave of interest towards the polymerisation of styrene derivatives initiated by iodine has developed in the last few years. Higashimura and coworkers have published several papers on this topic[358-363]. In the first study of this new series, it was found that bimodal distributions were obtained for poly(p-methoxystyrenes) and poly-(p-methylstyrenes) prepared under specific conditions[358]. It was concluded that among the active species present in these systems one might be non-ionic. Copolymerisation studies with vinyl ethers as second monomer seemed to confirm the presence of more than one type of chain carriers[359]. An interesting but incomplete investigation was also carried out on the possible living character of the polymerisation of p-methoxystyrene by iodine in media of low polarity[362,363]. In this situation the active species seem to be long lived and non-ionic, and the molecular weight of the polymers increased with conversion and with second monomer additions after the end of the first polymerisation. It is unfortunate that this study was limited to DP measurements, for concurrent kinetic experiments would have been extremely interesting in testing the validity of the near-living character of this system.

We have already discussed the work of Sawamoto and Higashimura with p-methoxystyrene and methanesulphonic acid (Sect. III-E-10). In those same three papers iodine was also tried as catalyst[89,119,204]. A value of the propagation rate constant was calculated in the first one, using the same considerations as for CH_3SO_3H: at 30 °C in 1,2-dichloroethane k_p was about $5 \times 10^3\,M^{-1}\,sec^{-1}$. As we have already pointed out, this value could be too high if covalent non-absorbing chain carriers are present together with the ionic ones. In the third paper initation rate constants were purportedly determined, but the assumption of bimolecular generation of carbenium ions from iodine and monomer is cleary untenable on mechanistic ground. In fact, the rate constants reported vary and are directly proportional to the iodine concentration (ref. [204], Table 1), which suggests that the ionogenic reaction is second order in iodine. The same criticisms expounded in Section III-E-10 on the criteria employed for determining the mean lifetime of the active species apply in this context.

Stoicescu and Dimonie[364] have studied the polymerisation of 2-vinylfuran by iodine in methylene chloride. This work is an oversimplification of a complex situation where side reactions were not detected despite their important role on the overall kinetic behaviour and on the structure of the polymer. Gandini et al. have shown that the conclusions of the Roumanian authors were erroneous[365].

N-vinylindole and N-vinylcarbazole are readily polymerised by iodine[229,366,367], but no detailed study has been reported on these systems. Gandini and collaborators[366] found that iodine was incorporated in poly(N-vinylcarbazole) probably as C—I bonds in terminal groups.

In the light of recent evidence pointing to the general mechanism of interaction between olefins and iodine, which indicates that diiodides are formed as first product of these complex processes, it would be interesting to restudy the system 1,1-diphenylethylene-iodine originally investigated by Evans et al.[368]. This reversible dimerisation reaction was studied kinetically, but its mechanism was not elucidated, particularly in terms of the intermediates intervening in the process. Thus, no search was made for hydrogen iodide among the products and the nature of the species responsible for the dimerisation remained unclear.

b) Vinyl Ethers

The classical work of Eley and collaborators[348,369,370] described the vinyl ethers-iodine systems as being characterised by the parallel complexation of the monomer by the catalyst and initiation due to the attack of I^+ onto the double bond. It was on this mechanistic basis that Okamura et al.[371] applied their scheme to calculate propagation rate constants in 1,2-dichloroethane and n-hexane. A criticism of that method has already been given in the previous section and the low values of k_p obtained in this study prove that the gross approximations implicit in the method, mainly the neglect of the iodine bound to the polymer, introduce large errors in the estimate of the rate constants.

Parnell and Johnson[372] were the first to propose that a mechanism similar to that developed by Giusti's group for aromatic olefins could be applied to vinyl ethers. Ledwith and Sherrington[198] proved that iodine added rapidly and completely to vinyl ethers to give 1,2-diiodoethanes if the ether was present in excess with resepct to iodine. These reactions were studied at low temperature in methylene chloride and provided the first direct evidence of the real nature of the initial step in the complex mechanism leading to polymerisation. Janjua and Johnson[373] carried these investigations one decisive step forward and proved not only that diiodides are formed but also that they can eliminate hydrogen iodide. From then on the mechanism resembles closely that proposed by Giusti's group for aromatic olefins. The authors also underlined the pseudocationic nature of these polymerisations. Johnson and Young[199] extended these studies in a recent investigation of the phenomena related to the initation of the polymerisation of vinyl ethers by iodine and demonstrated beyond doubt that the addition of iodine to the vinylic double bond is the key process preceding the formation of active species. They also showed that these systems produce polymers containing terminally-bound iodine and that the DP's of these products increase upon further additions of monomer, i.e. that these polymerisations possess some degree of living character. No real termination occurs therefore and the chains retain their potential for propagation. We look forward to future developments in this area, as announced by Johnson and Young in their paper.

The interest of these recent studies lies in the fact that they bring vinyl ethers into the same mechanistic perspective as the aromatic olefins previously investigated. A unifying principle exists today concerning the overall mechanism and the details of the initiation process when iodine is used as catalyst in a cationic polymerisation.

Very recently Higashimura and collaborators[892)] published a paper on the nature of the active species in these systems, but they ignored the evidence gathered by other authors. Their most interesting observation was that the DP's of the poly(isobutyl vinyl ethers) increased with conversion indicating that transfer and termination reactions were not predominant.

F. Conclusions

As we have abundantly seen in the previous sections, the interaction between olefins and Brønsted acids can follow different mechanistic pathways which are often critically determined by the specific chemical composition of a given system. It is therefore objectively impossible to draw a general theory capable of encompassing the very wide range of situations studied, except for elementary and unilluminating considerations about the electrophilic nature of the addition process. This state of affairs reflects similar complications in a much wider area of physical chemistry. Hence, it is not surprising that the initiation mechanisms by protonic acids discussed throughout this chapter are seldom fully understood, and that the structure and properties of the various possible chain carriers remain open to discussion in most investigations.

When a Brønsted acid and an alkenyl monomer are mixed in a given solvent at a given temperature, polymerisation will occur if the lifetime of the resulting intermediate capable of propagating is long enough to allow the growth of a chain. Conversely, if the collapse of this active intermediate (k_t) competes effectively with its propagating potential (k_p), little or no polymer will be formed. If one considers the classical interpretation of electrophilic addition processes of this type based on the rate-determining formation of an intermediate (carbenium ion or $Ad_E 3$ complex) followed by its rapid collapse to give the corresponding ester, a situation very unfavourable to polymerisation results, since any such system where the olefin is a monomer would be characterised by slow initiation and fast termination. Only extremely high values of k_p could make any polymerisation possible in that general situation, given the very short lifetime of the intermediates. In fact, the extrapolation of the mechanistic conclusions referring to esterification and hydration processes to the realm of cationic polymerisation is only partly legitimate for the two following reasons: (i) in most polymerisation systems, or at least in the more successful ones, the marked acidity of the initiator or the high nucleophilicity of the monomer, or both, tend to enhance the lifetime (stability) of the ionic or polar intermediate species, particularly in solvents of medium or high dielectric constant; (ii) often the addition products (esters) are themselves chain carriers or can be activated by specific solvation from the free acid present in the medium. We believe that these two situations are fundamental to the occurence of polymerisation and deserve therefore a detailed discussion in the light of all available experimental information.

The kinetic pattern of these polymerisations can vary appreciably from system to system, but one important feature has been recorded without exception: no induction or acceleration period, but maximum *initial* rate of monomer consumption (note that the behaviour of iodine is different, simply because hydrogen iodide and the monomer iodide must form before polymerisation can start). Indeed, whether the process is very slow, as with styrene and dichloroacetic acid[244], or extremely fast, as with *p*-methoxy-styrene and triflic acid[205], the concentration of active species attains a maximum value very soon after mixing. While this general phenomenon does not tell us anything about the nature of the chain carriers, it indicates that only three alternative kinetic situations can account for its occurrence:

— initiation is fast and not accompanied by any termination; in these stationary-state conditions the activity of the system is maintained throughout the polymerisation and can be revived at the end of it by simply adding more monomer;

— initation is fast and followed by a termination reaction; in these non-stationary conditions the activity of the system is progressively lost and can lead to incomplete polymer yields;

— initiation and termination proceed at approximately the same rate; this second type of steady state, characteristic of free-radical processes, implies the maintainance of activity as long as some catalyst is available, but the concentration of active species is usually much lower than in the previous two altenatives.

Examples of each of these three typical behaviours can be found in the literature and some will be discussed below. But before, it is essential to define the chemical nature of the reactions which give rise and kill the active species and the different types of chain carriers which can be present in these systems.

The polymerisation of styrene by perchloric acid in chlorinated hydrocarbons at different temperatures illustrates most adequately the fundamental aspects of the initiation and the nature of the propagating entities in this type of processes. The behaviour of the system depends markedly upon the temperature and the monomer-to-acid ratio. Three distinct phenomenologies are observed[17,45,118,127,202,335].

(i) At the higher temperatures and with high values of $[M]/[HClO_4]$, a rapid esterification occurs upon mixing and the predominant chain carriers are 1-phenylethyl and oligostyryl perchlorate molecules formed quantitatively in that fast initiation. The polymerisation is pseudocationic, although a minor contribution from ionic species can be observed in the later stages of the process, i.e. when the monomer concentration has decreased considerably[45,342]. The kinetics are simple and together with some specific mechanistic features, they show that the active ester molecules are relatively insensitive to moisture and that they maintain their stability provided they are surrounded by sufficient monomer molecules. The propagation rate constant for these species varies from 0.1 to 20 M^{-1} s^{-1} depending on the temperature and the polarity of the medium. No termination is detected in these reactions. The coordinated propagation mechanism has been discussed by Gandini and Plesch[127].

(ii) At lower temperatures or with low values of $[M]/[HClO_4]$ a more complex situation is encountered. Typically, the conversion-time curves display an initial burst of polymerisation followed by a slower-rate period and finally another acceleration which completes the process. This behaviour is due to the coexistence of ionic and covalent chain carriers, the former being much more active and giving rise to the fast stages. Bimodal DP distributions indicate that the two types of chain carriers are not in fast equilibrium between each other. The kinetics are obviously complex, but no termination exists.

(iii) At very low temperatures only the ionic propagation is detected over short time scales because the ester molecules are too "weak" to induce any appreciable monomer consumption. In these conditions polymerisations occur in a flash but stop at limited yields, unless a large amount of acid is used. The polymers have a unimodal DP distribution. The propagation rate constant for these truly cationic processes is a composite parameter since both ion pairs and free ions contribute to the growth of the chains: its overall value lies between 10^4 and 10^5 M^{-1} s^{-1} between −60 and −80 °C, several orders of magnitude higher than that obtained for covalent species. The dormant ester molecules which are responsible for the "termination" reaction can be reactivated by warming up the mixtures.

On the basis of extensive and accurate information acquired on this system during nearly fifteen years of work at Dublin and Keele, we can conclude that in the polymerisation of an alkenyl monomer by a Brønsted acid one must expect two basic types of active species arising from the initation process: covalent (but probably strongly polarised) entities resulting from the addition of the acid to the monomer double bond, viz. ester molecules, and ionic ones (paired, solvent-separated and free) produced by the protonation of the monomer. The major difference between these two carriers is of course their respective propagating capacity, always very high for the latters and moderate for the formers which are also much more affected by the temperature. We can also conclude that, unless the acid is unstable, the only "termination" reaction one can envisage in these systems is that giving the ester from an ionic chain carriers, provided of course that the covalent species cannot propagate at a measurable rate in the specific conditions in which it is formed. In other words, while ionic chain carriers are always active, the covalent ones may be virtually powerless under certain circumstances, as discussed below.

The above conclusions are amply sustained by numerous studies on other systems as we have tried to indicate all along the previous sections. But these criteria are very general and it is now necessary to analyse some of the other better-known systems to establish more useful and specific principles. The polymerisation of styrene by triflic acid bears an extraordinary resemblance to the above classical system. This similarity of behaviour however, has only become evident after the very instructive work of two Japanese groups[87,313], since earlier studies[305,311] had been marred by ill-conceived experiments and inadequate interpretations. We now know that a purely cationic propagation is observed at room temperature when acid concentrations of the order of millimolar are used. Indeed the propagating carbocations are detected spectroscopically with fast flow techniques and the extremely rapid monomer consumption is due to a propagation rate constant of about 10^5 M^{-1} s^{-1}. Note that here, as with perchloric acid, despite the very large excess of monomer, only a small fraction of the acid is involved in the protonation reaction, i.e. the number of ionic carriers formed is much less than the amount of catalyst used. We think that this is due to two phenomena: a minor one which arises from the necessity of at least two acid molecules *per* ion formed because the anion homoconjugates with its acid (see also the work on 1,1-diphenylethylene[135,306]); and a major one consisting of the parallel consumption of most of the acid in the esterification of the monomer. However, despite the presence of important quantities of ester, the propagation by these species is negligible at the time scale of tens of millisecond during which the ions operate. Thus in these conditions, as in the case of perchloric acid at very low temperature, but for a different reason, the pseudocationic polymerisation contribution is negligible and indeed the polymers obtained have a unimodal DP distribution. With triflic acid concentrations around 10^{-4} M and fairly high monomer concentrations, the polymerisation is essentially pseudocationic, as indicated by the simple kinetics, the relative insensitivity to small amounts of added water and k_p values between 1 and $10 \, M^{-1} \, s^{-1}$ at 0 °C in mixtures of benzene and methylene chloride. Again, the striking similarity with perchloric acid at the higher temperatures indicates that we are witnessing a rapid and complete esterification, the polystyryl triflate being the predominant chain carrier. At the higher dielectric constants (pure methylene chloride) a bimodal distribution is noticed, suggesting that some ions must also be responsible for propagation. Between these two extreme sets of features we find again a situation compatible with propagation by both species with rapid bursts of polymerisation when the ionic ones prevail: Chmelir's experiments with fairly low monomer concentrations give in fact conversion-time curves which simulate almost exactly the behaviour of perchloric acid at intermediate temperatures or with low monomer concentrations at 0 °C, as shown in Fig. 1. Thus, here too some transient ionic carriers are formed upon mixing, they are then substituted by ester molecules and take over again towards the end of the reaction. Given the complexity of such a situation, caused in part by the mechanics of initial mixing (the acid is not dispersed with sufficient rapidity and its local temporary high concentration coupled with the paucity of monomer, favours the transient formation of protonated species, which are subsequently transformed into ester molecules when proper mixing is achieved), and in part by the loss of stabilisation of the ester for lack of sufficient monomer, the quantitative kinetic treatment attempted by Chmelir and collaborators[305,311] appears unrewarding.

If the above two very strong acids exhibit with styrene a close similarity of behaviour, quite different features are instead observed when sulphuric acid is used[288,289,293]. An important termination reaction characterises these polymerisations even at room temperature and incomplete yields are obtained, unless the acid concentration is high. Thus, after an initially rapid monomer consumption, the process is drastically slowed down. Obviously this indicates that the polystyryl bisulphate is inactive for propagation, at least on the time scales for which the apparently limiting yields are obtained: if one leaves these solutions for much longer periods, a slow continuation of the polymerisation is observed suggesting that the ester is not completely passive. As for the nature of the chain carriers responsible for the major initial conversion, we are inclined to believe that they are not carbenium ions, but ester molecules solvated by free acid (see below). As long as ester and acid molecules are present in the system polymerisation will occur, but the esterification reaction is fairly fast and soon it uses up all the free acid leaving an ester which does not possess sufficient polarisation to propagate adequately. As Peniche and Gandini[293] have shown, the addition at this point of a solvating agent for the ester, e.g. trichloro- or trifluoroacetic acid, can revive the polymerisation. An example of such reactivation is shown in Fig. 2. These mechanistic tests and the low values of the apparent propagation rate constant obtained by Pepper and Hayes[289] are, we feel, valid reasons to doubt that ionic species are involved in these polymerisations (see also Sect. E-9, this chapter). Many other systems are characterised

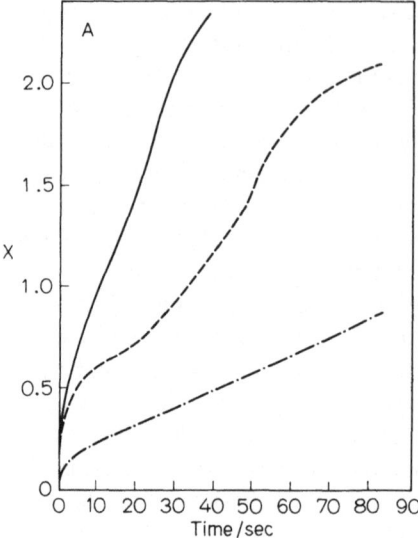

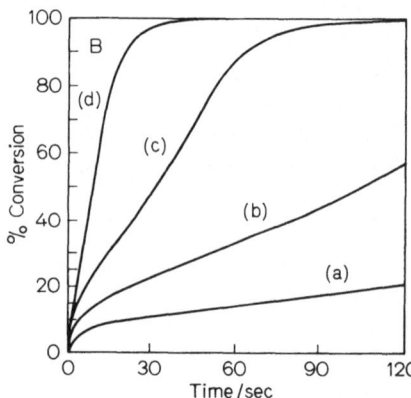

Fig. 1. The similarity of behaviour in the polymerisation of styrene by two strong acids. **A** Time-conversion curves for the polymerisation of styrene by perchloric acid in CH_2Cl_2 at 0 °C with low $[M]_0/[C]_0$: $[C]_0 = 6 \times 10^{-3}$ M; $[M]_0 = 0.05$ (——), 0.1 (————) and 0.2 M (—·—·—·). $X = \ln [M]_0/[M]_t$. Taken from Ref.[17]. **B** Time-conversion curves for the polymerisation of styrene by triflic acid in CH_2Cl_2 at −15 °C: $[M]_0 = 0.1$ M; $10^4[C]_0 = 0.95$ (a), 1.54 (b), 2.1 (c) and 3.1 M (d). (Taken from Ref.[311])

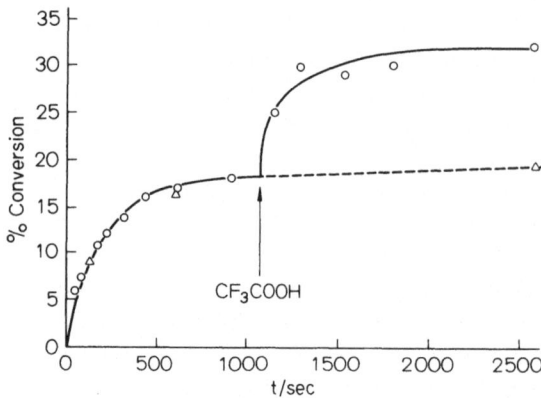

Fig. 2. Time-conversion diagram for a double polymerisation of styrene in ethylene chloride at room temperature. △, reference experiment with $[M]_0 = 0.35$ M and $[H_2SO_4]_0 = 8 \times 10^{-4}$ M. ○, same as above, but CF_3COOH (to give a concentration of 5.5×10^{-2} M) added at the time indicated by the arrow. (Taken from Ref.[293])

by this fast-initiation important-termination pattern, as pointed out in the previous sections.

Continuing our analysis of styrene polymerisation, we move now to weaker acids. With methanesulphonic[297], trifluoroacetic[271−275] and di- and trichloroacetic acids[244,246], one observes a pattern which strongly suggests the absence of carbenium ion carriers (except in the experiments where trifluoroacetic acid is dripped into pure styrene, in which the very fast and non-stationary polymerisation is typical of short-lived ionic intermediates collapsing to inactive ester). We do not known whether polystyryl methanesulphonate is a chain carrier, but we think that this is unlikely, since the ester of the much stronger sulphuric acid is inactive. We known on the other hand that the esters derived from the various haloacetic acids are definitely *not* chain carriers in common solvents[274,271,244]. We believe that in all these systems the real propagating species are ester-acid complexes in which the acid plays a solvating and polarising role for the inactive ester. The presence of ionic species would in fact give rise to different kinetic and mechanistic features to those actually observed.

All the above considerations and the bulk of the studies reviewed previously can be summarised in the following conclusions. The conditions for the formation of ionic chain carriers are seldom achieved and generally require fairly drastic measures such as appreciable concentrations of a strong acid, polar media, preferably low temperatures and/or very nucleophilic monomers or less nucleophilic ones, but present in fairly low concentrations. Even when protonation is attained to give transient carbenium ions, the concentration of the latter is usually a small fraction of the amount of acid used, i.e. the process is far from quantitative. The only recorded exception is the protonation of *p*-methoxystyrene by triflic acid in ethylene chloride at room temperature, in which a high proportion of the acid gave the corresponding carbenium ions when the concentration of the monomer was very low (only 5−10 times that the of acid)[205]. Note however that even in that extremely favourable system, the amount of ester produced became important as soon as the dielectric constant was reduced. It can therefore be said that the protonation reaction is accompanied by rapid esterification. In the systems where carbenium ions have been directly identified, the polymerisation is extremely rapid, even when these ionic chain carriers are only present in concentrations around 10^{-6} M. This implies propagation rate constants at room temperature of about 10^5 M^{-1} s^{-1}, a composite value the resulting from joint action of free ions and ion pairs.

Even with the strongest acids available, it is now well established that propagation by ester molecules is not only possible, but indeed a much more general phenomenon than that due to carbenium ions. The recent reports of "invisible" active species from Higashimura's laboratory[119,205] confirms this statement, and if these species play an important role with strong acids and extremely basic monomers, it seems logical that in less favourable circumstances to the formation of carbenium ions, they should be even more predominant. We must however distinguish between to types of covalent chain carriers. In some instances, the ester as such is capable of propagating: typical examples are the polystyryl perchlorate and triflate, and the esters of *p*-methoxystyrene with triflic, perchloric and methanesulphonic acids. In other systems, the ester alone is virtually inactive i.e. its k_p is extremely low. Polystyryl bisulphate, trifluoroacetate etc. have been shown not to propagate at measurable rates. The polymerisation in the case of acids and monomers which give inactive esters take place under normal conditions because the ester can be activated by the excess free acid present:

$$HM_nB + HB \rightleftharpoons HM_nB \cdot HB \; (Ad_E 3 \text{ intermediate?}).$$

In the above equilibrium HM_nB represents an inactive ester molecule, and $HM_nB \cdot HB$ its acid-activated counterpart. We can envisage that following the mixing of the monomer with the catalyst in such systems, the following events can take place. The equilibrium between inactive and activated ester is rapidly established as soon as some ester is produced, and the polymerisation begins. As time passes, more ester is progressively formed and a corresponding quantity of acid disappears from the medium. The above equilibrium is therefore essentially maintained and the polymerisation proceeds under near steady-state conditions. If however the rate of esterification is high, one might reach the point of total consumption of acid before all the monomer has been polymerised. In that case (sulphuric acid-styrene, for example) the concentration of active species vanishes because of the disappearence of acid in the activation equilibrium and of course the polymerisation ceases at limited yields.

The three kinetic situations described above are therefore all encountered in practice. The actual features of a given system will depend upon the four fundamental parameters defining the reaction: strength of the acid, nucleophilicity of the monomer, polarity of the solvent and temperature. Thus, rapid esterification by all the strong acid present of the fairly basic monomer in a medium and at a temperature which does not favour ionic dissociation will lead to steady state-pseudocationic polymerisation with an active species concentration equal to that of the initial acid. The system will show no termination. But the above situation can also result in no polymerisation at all if the ester is not a chain carrier: it is the case of styrene with hydrogen chloride. A fairly rapid esterification giving a non-propagating entity will produce the non steady-state conditions best characterised by styrene-suphuric acid in chlorinated hydrocarbons. A slow esterification accompanied by activation of the ester by free acid will give the type of stationary state described in the third example, where the concentration of active species is obviously much lower than that of the acid and will depend upon the value of the equilibrium constant of the ester-activation reaction.

A very important paper from Higashimura's school[374] clearly confirms many of the points raised in this discussion. It shows first of all that pseudocationic polymerisation is predominant in media of fairly low dielectric constant for the polymerisation of styrene by three acids of different strength: triflic, methanesulphonic and trifluoroacetic. Up to $\epsilon = 5$, the esters are virtually the only species present and the polymers have a unimodal DP distribution. The second order rate constants derived from the expression

$$- d[M]/dt = k_2 [C]_0 [M]$$

decreased drastically with decreasing acid strength. As shown in Fig. 3, an increase of polarity beyond $\epsilon = 5$ produced in the three systems a sudden increase in k and the corresponding appearence of bimodal polymers. It seems obvious, as already encountered by Dunn et al.[342] for perchloric acid, that this increase reflects the onset of some ionisation, i.e. the concurrent propagation by ester molecules and carbenium ions, the latter present in small proportions, but increasing with the polarity of the medium. Interest-

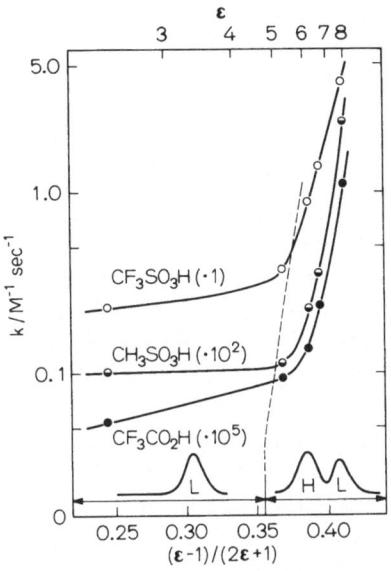

Fig. 3. Effect of the dielectric constant on the second order rate constant ($R_p = k [C]_0[M]$) for the polymerisation of styrene by three protonic acids at 0 °C in mixtures of benzene and CH_2Cl_2 of different composition. $[M]_0 = 1.0$ M; $10^3 [C]_0 = 0.2$ (CF_3SO_3H), 30 (CH_3SO_3H) and 200 M (CF_3COOH). Note also change in polymer distribution (L = low DP, H = high DP) with dielectric constant. (Taken from Ref. [374])

ingly, no such phenomenon was observed with Lewis acid initiators, since in that type of polymerisation ester carriers cannot be formed[374]. The above interpretation of the results is ours, since the authors, as in the case[119,205] of other recent papers, have not committed themselves to giving the name of pseudocationic polymerisation to non-ionic processes. We have already noted Higashimura's school refusal to acknowledge that their invisible species are propagating ester molecules, as first pointed out by Gandini and Plesch fifteen years ago[127,336].

Among the various points raised in the above discussion, one is particularly important because of its partly speculative nature. Indeed, if there is no doubt that ionic species are formed under specific conditions and display a remarkable propagating capability, and that ester molecules are the main products of these interactions and in an isolated form can be active or inactive depending of the particular context, the existence of acid-solvated ester carriers remains to be proved. The only evidence pointing to their presence is, to our knowledge, the experiments of Peniche and Gandini[293] which suggest that polystyryl bisulphate can be reactivated by solvation with haloacetic acids. Obviously, that information alone does not necessarily imply that other systems are subject to a similar activation, nor does it prove anything about the chemistry or the stoicheiometry of this solvation process. Much more work needs to be done in this area before any definite conclusion can be drawn. Meanwhile, we feel that our proposal is a useful tentative explanation of how certain systems display activity which seems incompatible with both a carbocationic mechanism and a simple ester one.

The recent noticeable progress made in the field of Brønsted-acid initiated polymerisations is mainly due to the advent of fast spectroscopic flow techniques. A great deal of important quantitative and qualitative aspects have been and are being clarified. We think that this type of research is extremely useful, but also feel that certain mechanistic aspects, particularly in the realm of pseudocationic polymerisation, should also be tackled from different angles. We refer in particular to the study of the structural and reactivity properties of esters in various media. Too little has in fact been achieved in this field since the discovery of pseudocationic polymerisation, probably because it has not yet been realised that this type of propagation is in fact much more widespread than commonly assumed. We strongly believe that true cationic polymerisation with Brønsted acids is much more the exception than the rule, and is in fact only observed under rather special conditions, i.e. when a series of experimental parameters are properly combined to provide the necessary environment for carbenium ions to survive long enough to propagate. Our belief is based on a very detailed inspection of all the literature dealing with this topic, and we have tried to convey it to the reader through the systematic analysis of the most reliable studies published in the last three decades. Too little is known about chain carriers other than carbenium ions (free or paired) and this situation ought to be rectified in the near future. The recent achievements in the field of the cationic polymerisation of heterocycles by Brønsted acids and esters[18,24−27,895,896] indicate that the role of covalent species can be important and can be assessed quantitatively by the judicious choice and use of modern probing techniques[473].

Dunn[19] has speculated that the origin of bimodal distribution of DP's often encountered in polymers obtained with Brønsted acids (but never with Lewis acids)[375] might be the presence of ester molecules in equilibrium with protonated ester (oxonium ions). Both species were supposed to be active, the neutral ones less than the ions. Moreover, Dunn postulated that the rate of interconversion between the two entities should be low to allow for independent propagation and thus bimodality. This suggestion is interesting and deserves to be taken up by experimentalists. Indeed, the problem of bimodal polymers is difficult and the arguments proposed by Higashimura[375] in favour of two kinds of ion pairs apart from free ions are not convincing, as indicated also by the discussion following his paper. It is in fact hard to conceive that ionic species in different state of association and/or solvation should be related by such slow equilibration processes. Nothing of this type has ever been observed in anionic polymerisation. It seems therefore much more plausible that covalent and ionic species are responsible for independent growth, or that two kinds of active esters can coexist in certain conditions, the simple covalent molecule and a more active entity such as the acid-solvated ester proposed by us above, or the protonated ester suggested by Dunn. It goes without saying that the importance of the "counterion", i.e. the nature of the acid used in terms of the nucleophilicity and leaving tendency of its anion, is paramount throughout in determining the specific behaviour of a system.

Before turning to initiation by Lewis acids, we stress that the capital difference between the two types of acids resides precisely in the fact that while with Brønsted acids the formation of a covalent bond is always possible (collapse of the ionic species to give the ester), no such direct union can occur between a carbenium ion and an anion derived from a Lewis acid. This gives rise to markedly different features, some of which have been underlined in two recent papers of the Kyoto school[374,376]. These qualitative dissimilarities provide additional support for the existence of non-carbenium ion carriers in systems involving protonic acids.

IV. Initiation by Lewis Acids

A. Cocatalysis and Direct Initiation

Ever since it was established that Friedel-Crafts halides can induce the cationic poly-merisation of olefins, researchers in the field have been concerned about finding mechanisms rationalising the appearance of carbenium ions derived from the interaction of these catalysts with various monomers. The direct attack of a Lewis acid molecule onto the olefinic double bond implies the heterolytic cleavage of the latter with the consequent appearance of both a negative and a positive charge on the reaction product (zwitterion). This initiation pathway was proposed in early studies but later criticised on thermodynamic grounds. No fundamental reasons or direct proof were however offered which justified its rejection a priori. In fact, direct initiation was discarded, at least temporarily, when the now classical experiments of Polanyi's school offered an alternative and more plausible way of explaining how the monomer could be protonated in the presence of Lewis acids. The theory of cocatalysis which arose from that work and which was to be confirmed in many later studies, showed that a third component was needed for initiation to take place, namely a proton source, present in small concentrations (water, hydrogen halides, etc.) and activated by the Lewis acid. The widespread success of this mechanism was to make it the only one generally accepted, or even considered, for many years. Recent experimental evidence carefully gathered in several laboratories has however fully revived the theory of direct initiation, without denying the existence of cocatalytic initiation. In the present chapter we intend to show that these two types of initiation are important and to discuss their respective mechanisms.

The following reaction scheme exemplifies the main characteristics of cocatalysis:

(initiation) $TiCl_4 + H_2O + (CH_3)_2C=CH_2 \rightarrow (CH_3)_3C^+ TiCl_4OH^-$

(termination) $\sim\!\!\sim\!\!CH_2-C(CH_3)_2^+ TiCl_4OH^- \rightarrow \sim\!\!\sim\!\!CH_2-C(CH_3)_2OH + TiCl_4$.

Here it is implied that initiation does not take place in the absence of added water and that the consumption of the cocatalyst in the termination reaction can produce incomplete polymer yields if the initial water concentration is sufficiently low. The discovery of these phenomena raised many fundamental questions which were to provide work for a whole generation of polymer chemists. Some of these questions have not yet been fully answered, namely:
- is the scheme given above a general one, applicable to any olefinic monomer and to any Lewis acid?

— does it depend on the polarity of the medium (solvent effect)?

— is the chemistry of a polymerisation influenced by the nature and structure of the cocatalyst?

It is extremely difficult to give a clear answer to these questions essentially because in order to prove that a cocatalyst is involved in a given system, it is necessary to show that no initiation takes place in its absence. As shown above, the yield of a polymerisation depends not only on the concentration of chain carriers, but also on the rate of termination. If the latter is very low, a concentration of active species of 10^{-5} M or even less will be sufficient to give 100% polymer yield. It is therefore clear why reliable studies dealing with this point required the use of sophisticated equipment and high vacuum techniques[146,147]. The use of model molecules to simulate the monomer, i.e., olefins incapable of propagating because of steric hindrance, has been frequent in attempts to understand better the mechanisms of initiation. Thus, the interaction of 2-butene, the isobutene dimers, 1,1-diphenylethylene, 1,1-diphenylpropene, and 3-substituted indenes with Lewis acids have been followed spectroscopically (UV-visible and NMR) and by other probing techniques. Despite the success of these studies, some problems of interpretation remain due to termination and/or transfer reactions, electrophilic substitution side reactions, etc.

Titanium tetrachloride is probably one of the most extensively studied Lewis acid and some attempts at answering the questions posed above can be made. The theory of cocatalysis was firmly established by Polanyi's school and later in Plesch's laboratory with the system isobutene-titanium tetrachloride[377−380]. If one wants to draw any conclusion concerning the possible generalisation of this theory, the details of the reactions among catalyst, cocatalyst and monomer must be understood. This is by no means simple. Given the overall reactions:

$$TiCl_4 + BH + \underset{/}{\overset{\backslash}{C}}{=}\underset{\backslash}{\overset{/}{C}} \rightarrow H-\overset{|}{\underset{|}{C}}-\overset{|}{\underset{|}{C}}{}^+ \ TiCl_4B^-,$$

BH being the cocatalyst, we can consider the interaction

$$TiCl_4 + BH \rightarrow TiCl_4B^-H^+$$

as the primary initiation step which will be followed by the attack of the complex acid formed onto the olefin. This scheme was in fact the one assumed for most classical systems in which cocatalysis was proved. However, since it is well known that hydrochloric acid can function as a cocatalyst, one could postulate that in the case of water cocatalysis the primary reaction is:

$$TiCl_4 + H_2O \rightarrow TiCl_3OH + HCl,$$

followed by the intervention of the hydrochloric acid formed as the true cocatalyst. Another important point can be raised concerning the order of reaction of the two catalytic components with the olefins, if one rejects on probabilistic grounds a termolecular initiation. We can in fact consider the reaction of the cocatalyst with the olefin followed by the activation of this combination product by the catalyst:

$$BH + \overset{\displaystyle \diagdown}{\underset{\displaystyle \diagup}{C}}=\overset{\displaystyle \diagup}{\underset{\displaystyle \diagdown}{C}} \rightarrow H-\overset{|}{\underset{|}{C}}-\overset{|}{\underset{|}{C}}-B$$

$$H-\overset{|}{\underset{|}{C}}-\overset{|}{\underset{|}{C}}-B + TiCl_4 \rightarrow H-\overset{|}{\underset{|}{C}}-\overset{|}{\underset{|}{C^+}} \; TiCl_4 B^-,$$

or alternatively the complexation of the Lewis acid with the olefin followed by the reaction of the complex with the cocatalyst. Few studies have been devoted to the elucidation of this point after the classical work of Williams and Bardsley[381].

A further complication of course crops up in systems where the solvent can also act as a cocatalyst.

The relative cocatalyst efficiency has been examined in a few studies and it has been stated that roughly speaking the more acidic ones are also the more efficient. Thus, in the isobutene-diethylaluminium monochloride system the following order of decreasing activity was reported by Kennedy[382]:

$$HCl, \quad HBr \gg HF, \quad H_2O > CH_3COOH \gg CH_3OH > CH_3COCH_3.$$

We shall in fact see in the discussion below that the situation with respect to this point is probably rather different.

When looking into the behaviour of different Lewis acids one observes that the most acidic ones act in rather peculiar manners especially when they are mixed together. This is particularly the case of Wichterle's catalyst which is prepared by the reaction of boron trifluoride with aluminium tributoxide and by further activation of the solid obtained with titanium tetrachloride[383,384].

The intricacies of the cocatalytic effect are notably evident in the following examples. In the system n-butylvinyl ether-boron trifluoride-hexane added water displaces the catalyst from its complex with the ether[385] to form the active precursor. However, since water is more basic than ether, one would have expected the complex BF_3-H_2O to be less reactive than BF_3-ether. In another system, styrene-stannic chloride-water-toluene, very elaborate schemes have been put forward to explain the behaviour of polymerisations exhibiting induction periods and often a series of accelerations and decelerations. The original results of Colclough and Dainton[386-389] were confirmed in the more recent work of Kucera[344], a fact which does not occur too often in cationic polymerisation. No polymerisation was observed in the absence of water, yet water was reported to deactivate the chain carriers, assumed to be ion pairs. Both chlorine and tin atoms were found in the polymer, the former as a much higher percentage of the amount of catalyst used than the latter. While the chemical incorporation of initiator fragments in the product is a very important observation, the quantitative data are not easy to explain. Kucera has proposed an extremely complex mechanism in an attempt to rationalise all the experimental observations, with not less than twelve reactions and seven equilibria. This mechanism is however highly speculative and open to criticisms. Thus, for example, the author does not even mention the possibility of $SnCl_4$ hydrolysis (considerable amounts of water added at a temperature as high as 35 °C), but invokes such complicated interactions as:

$$Sn_2Cl_8(S)_n + Sn_2Cl_8(H_2O)_m \rightleftharpoons 2\,SnCl_3^+//SnCl_5^-$$

which is after all a disguised version of the well-known selfionisation equilibrium often used as the basis of direct initiation. This controversial reaction is curiously used by Kucera in support of his interpretation of the phenomenology of cocatalysis.

Wichterle and collaborators'[390] reexamination of the role of cocatalysis in the polymerisation of isobutene with various Lewis acids is another example of the delicacy of this subject. While these authors firmly established the need of a cocatalyst with boron fluoride, thus confirming Evans and Meadows' previous findings[391], the work with other catalysts was less straightforward. For instance, with $TiCl_4$ in heptane the chosen temperature of about $-15\,°C$ to test the residual moisture content of the system was inadequate. It would have been much better to work around $-60\,°C$, where the polymerisation behaviour is known to depend much more drastically on the water content. The effect of *added* water in this system was found to vary critically with the order of addition. When water was added to the monomer solution before the addition of the catalyst, a long induction period was observed followed by S-shaped time-conversion curves; when the water was mixed with the catalyst before adding the monomer, the polymerisation displayed maximum initial rates. Does this imply that a (slow) reaction between $TiCl_4$ and water is the rate-determining step in the initiation process? One would be tempted to answer affirmatively but for the puzzling observation that addition of water to a quiescent mixture of monomer and $TiCl_4$ gave immediate polymerisation. With aluminium bromide, direct initiation was claimed but this was in fact the only catalyst for which a chemical purification against residual water was not achieved. Thus, it is difficult to assess the validity of that claim. Also questionable in this otherwise important paper is the use of hypodermic syringes for the addition of reactants.

An interesting paper has been published by Ghanem and Marek[392] showing the difficulties of understanding the mechanism of cocatalysis by water. Isobutene was polymerised by $AlBr_3$ in the presence of tritiated water. It was noticed that whatever the sequence of addition of the reactants the amount of tritium found in the polymer did not exceed the reproducibility errors of these very difficult experiments. Moreover, the same amount of tritium was found in polyisobutenes which has been mixed with tritiated water in the presence of $AlBr_3$. Clearly therefore, neither initiation nor termination involved incorporation of the cocatalyst into the polymer. The same conclusions were reached in the case of boron fluoride. Strikingly, when $TiCl_4$ was used at $-78\,°C$ a very high tritium incorporation was observed both in polymerisation experiments and in blanks carried out on the polymer. The tritium content corresponded to the fixation of one proton per polymer molecule. This phenomenon could be explained by considering that in the presence of $TiCl_4$ the end groups of polyisobutene exchange rapidly with water thus obscuring the phenomenology of the initiation step. Despite the observation that the presence of water was essential for polymerisation to take place, it was therefore impossible to establish the chemical role of this cocatalyst even with the apparently simple system involving titanium tetrachloride.

There seems to be no doubt that a cocatalyst is required to initiate the polymerisation of isobutene with titanium tetrachloride in nonpolar solvents. Plesch[377] found that in hexane the following compounds did not act as cocatalysts: $HCl, SO_2, CO_2, C_2H_5OH$ and diethylether. While it is understandable that ethanol and ether failed to activate initiation, since the former reacts with the catalyst and the latter gives a rather strong complex with it, the inactivity of HCl is less obvious. It is in fact usually assumed that any protogenic impurity can act as a cocatalyst in these systems. Thus, Plesch[393] has shown that trichloroacetic acid and sulfuric acid are efficient cocatalysts and later work by Imanishi et al.[394] extensively confirmed that role for the first compound. The general consensus as to the need of a cocatalyst in systems involving isobutene, $TiCl_4$ and a nonpolar solvent applied to the use of conventional techniques of mixing the reactants, i.e., those involving the mixing of the monomer with the solvent in the liquid phase followed by the introduction of a solution of $TiCl_4$. Cheradame and Sigwalt[395] showed that pure liquid isobutene could be polymerised in high yields with $TiCl_4$ if the unconventional *condensation* technique was used. The peculiarities of these fundamental experiments will be discussed below.

In media of higher polarity the behaviour of the isobutene-$TiCl_4$ system is characterised by the fact that at temperatures between about $-50\,°C$ and $0\,°C$ cocatalysis seems to occur, e.g. in methylene chloride, ethyl chloride, ethylene chloride and isopropyl chloride. The possibility that these solvents could act themselves as cocatalysts must be ruled out because complete polymerisation is not observed. In the specific case of methylene chloride, Cheradame noted that even after the most stringent drying procedure applied to this solvent, he was unable to prevent some polymerisation giving limited yields[396]. If we assume that the above halides are cocatalysts, the incomplete polymerisation could be rationalised by postulating a consumption or complexation of the catalyst leading to its inactivation. However, no evidence is available to show that any interaction between the monomer and the catalyst could be strong enough to prevent the latter from reacting with the solvent in the typical cocatalytic fashion:

$$TiCl_4 + RCl \rightarrow R^+ TiCl_5^-.$$

Although $TiCl_4$ probably forms some kind of complex with isobutene as indicated by the slight yellow colour arising when these two substances are mixed in vacuo, this interaction cannot possibly be a strong one. In fact, attempts to characterise this complex in the gas phase by infrared spectroscopy, using the most stringent techniques, failed[397] since the recorded spectrum was the sum of those of each component without any new band. The ease with which isobutene is polymerised in the liquid phase when $TiCl_4$ is condensed into it makes it extremely difficult to determine the phase diagram of these two substances. Careful cryoscopic studies[398] have demonstrated that $TiCl_4$ is essentially monomeric both in methylene chloride at $-95\,°C$ and in cyclohexane at $6\,°C$, in contradiction with previous claims based on various spectroscopic observations[399,400]. This discrepency is due to the wrong interpretation of slight modifications in the spectra which were not caused by molecular association but rather by much weaker interactions which do not have a measurable concentration effect. Thus, dimerisation of $TiCl_4$ in solution cannot be the cause of its inertness compared with its activity in the gas phase.

Accordingly we can conclude that while in most systems a cocatalyst is probably required for the polymerisation of olefinic monomers by Lewis acids, some exceptions are found particularly in relatively polar solvents. The system isobutene-$TiCl_4$-methylene chloride is one of these exceptions and deserves further comment. Supposing that a

cocatalyst exists in this system and it is consumed in a termination reaction, then incomplete polymer yields would be a proof of a cocatalytic effect. Plesch and coworkers[378−380] demonstrated that the limited yields observed at −60 °C could be
attributed to traces of residual water. Cheradame indeed showed later[396] that these
yields could be reduced by improving the drying techniques. At first glance, these observations seemed to settle the question in favour of the need for a cocatalyst in this polymerisation. However, the reduction of the primary yields did not imply the inactivity
of the system, since for example at −72 °C a low but reproducible percentage of
polymer was always obtained[396]. This unavoidable polymerisation could be attributed
to direct initiation followed by a very efficient termination reaction. Although at first
sight it might seem contradictory to assume direct initiation giving incomplete yields
in the absence of a real consumption of the catalyst, we show in Sect. B-4 that for this
and many other systems direct initiation is a much more plausible interpretation of the
experimental evidence than any mechanistic scheme involving a cocatalyst.

As mentioned above, the possibility of direct attack of a Lewis acid on an olefin
to generate a chain carrier was first proposed more than forty years ago by Hunter and
Yohé[401]. Their zwitterion hypothesis consisted essentially in the following reaction:

$$MtX_n + \underset{/}{\overset{\backslash}{C}}{=}\underset{\backslash}{\overset{/}{C}} \rightarrow Mt\bar{X}_n{-}\underset{|}{\overset{|}{C}}{-}\underset{|}{\overset{|}{C}}{}^{+}.$$

After the discovery of cocatalysis, Gantmakher and Medvedev[402] revived this mechanism claiming that it could be operative in polar solvents. This was however disputed
notably by Colclough and Dainton[389] and by Plesch's school[403]. Later studies have
led to the conclusion that the unmodified Hunter-Yohé mechanism can be reasonably
well applied to direct initiation schemes involving titanium tetrachloride[47].

An alternative way of looking at direct initiation was postulated some years ago
by Kennedy[404] who considered the possibility of hydride ion abstraction as follows:

$$\underset{|}{\overset{|}{C}}{=}\underset{|}{\overset{|}{C}}{-}\underset{|}{\overset{|}{C}}H + MtX_n \rightleftharpoons \underset{|}{\overset{|}{C}}{=}\underset{|}{\overset{|}{C}}{-}\underset{|}{\overset{|}{C}}{}^{+} MtX_nH^{-}.$$

This hypothesis cannot be generalised and its weak points are discussed in the next
section.

The third mechanism which can account for direct initiation involves the self-ionisation of the Lewis acid prior to the attack on the monomer[405]:

$$2\,MtX_n \rightleftharpoons MtX_{n-1}^{+}\,MtX_{n+1}^{-}$$

$$MtX_{n-1}^{+}\,MtX_{n+1}^{-} + \underset{/}{\overset{\backslash}{C}}{=}\underset{\backslash}{\overset{/}{C}} \rightarrow MtX_{n-1}{-}\underset{|}{\overset{|}{C}}{-}\underset{|}{\overset{|}{C}}{}^{+}\,MtX_{n+1}^{-}.$$

A very clear picture of all these alternative initiation mechanisms was drawn by
Sigwalt[47] in his introductory lecture at Rouen.

We conclude this introduction by pointing out that cocatalysis is certainly a much more widespread phenomenon than direct initiation, the latter being relevant only with the stronger Lewis acids and mostly in fairly polar media. Both mechanisms certainly operate, but the details of each one are still open to discussion and will now be closely scrutinised.

B. Direct Initiation

1. Introduction

The number of systems which have clearly been shown not to require the aid of a protogenic cocatalyst in the field of cationic polymerisation of olefins promoted by Lewis acids is so large that direct initiation has become an important alternative mechanism to cocatalysis. Kennedy reviewed some of these systems in 1972[404]. Obviously, before invoking direct initiation in a given system, the work in question must be carefully analysed to insure that maximum care has been taken against the possible presence of residual protogenic impurities and in particular water. Also the experimental observation that water acts as a retarder on a given polymerisation cannot be taken as proof for direct initiation. In fact, as we shall see later, water cocatalysis often gives polymerisation rates which go through a maximum at a specified and often fairly low water concentration. Since the position of this maximum varies appreciably from system to system an apparent retarding effect of water can always be interpreted as a concentration lying beyond the maximum rate, particularly if all experimental precautions have not been taken, i.e., rigorous purification and drying of all reactants and high vacuum manipulations. The systems involving classical monomers for which direct initiation has been proved are listed in Table 8, while the systems involving model compounds display-

Table 8. Olefin-Lewis acid systems giving polymerisation by direct initiation

Monomer	Lewis acid	Solvent	Temperature/°C	Ref.
Isobutene	$AlBr_3 + TiCl_4$	n-Heptane	−13	407)
Isobutene	$TiCl_4$	CH_2Cl_2	−70	395)
Isobutene	$EtAlCl_2$	Heptane	−10	408)
Isobutene	$EtAlCl_2$	n-Heptane	+21, −55	409)
Styrene	$SnCl_4$	$(CH_2Cl)_2$	+25	387)
Styrene	$TiCl_4$	CH_2Cl_2	−90	410)
Styrene	$AlCl_3$	CCl_4	+25	411)
α-Methylstyrene	$AlCl_3$	CCl_4	+25	411)
α-Methylstyrene	$TiCl_4$	CH_2Cl_2	−72	412)
α-Methylstyrene	$TiCl_3OBu$	CH_2Cl_2	−70	413)
Indene	$TiCl_4$	CH_2Cl_2	−72	414)
Indene	$SnCl_4$	CH_2Cl_2	−30	415)

Table 9. Olefin-Lewis acid model systems giving direct initiation

Olefin	Lewis acid	Solvent	Temperature/°C	Ref.
1,1-Diphenylethylene	$TiCl_4$	CH_2Cl_2	$-30, -70$	416)
1,1-Diphenylethylene	$AlCl_3$	CH_2Cl_2	-30	417)
1,1-Diphenylpropene	$TiCl_4$	CH_2Cl_2	-30	418)
3-substituted indenes	$TiCl_4$	CH_2Cl_2	-70	93)

ing steric hindrance to propagation for which the same conclusions have been reached are shown in Table 9.

Many hypotheses have been put forward to explain the occurrence of polymerisation in the absence of an obvious cocatalyst. We will presently examine those which, while probably acceptable in special instances, do not hold general validity.

a) Cocatalysis by Solvent

It is reasonable to expect that under specific experimental conditions certain Lewis acids might promote the ionisation of alkyl halides. Chlorinated hydrocarbons are frequently used as solvents in cationic polymerisation, and one can conceive that in the presence of a Lewis acid they might produce a small but sufficient concentration of carbenium ions capable of attacking the monomer in a solvent-cocatalysed initiation. Indeed, the reaction

$$MtX_n + RX + \diagup C = C \diagdown \rightarrow R - \overset{|}{\underset{|}{C}} - \overset{|}{\underset{|}{C}}^+ MtX_{n+1}^-$$

has been shown to be thermodynamically possible[10]. This type of interactions will be discussed in the section devoted to cocatalysis, but we emphasise here that the presence of solvent fragments in the polymer is not necessarily a proof of its role as cocatalyst, because certain transfer reactions can produce the same result[406]. We feel that solvent cocatalysis has never been indisputably demonstrated, though sometimes assumed reasonably. On the other hand, in many of the studies listed in Tables 8 and 9 the absence of such phenomenon was established beyond doubt. Thus, all systems involving alkanes as solvents are obviously immune from solvent cocatalysis, as is the polymerisation by condensation of undiluted isobutene. A careful search for solvent fragments in systems involving 1,1-diphenylethylene dimerisation in methylene chloride[416] gave negative results. Since that search was made relatively easy by the absence of polymer among the products, it can be safely concluded that methylene chloride, which is the most frequently used solvent in this context, cannot act as cocatalyst in the conditions chosen in the above systems.

b) Initiation Through Electron Transfer

One could envisage as a general mechanism for direct initiation the transfer of an electron from the monomer double bond to the Lewis acid:

$$MtX_n + \underset{/}{\overset{\backslash}{C}}=\underset{\backslash}{\overset{/}{C}} \rightarrow \cdot \overset{|}{\underset{|}{C}}-\overset{|}{\underset{|}{C}}{}^+ + MtX_n^{\bar{}} ,$$

followed by coupling of the radical cations to give dicationic dimers capable of propagating at both ends. An interesting experiment performed in Szwarc laboratory[91] showed in fact that 1,1-diphenylethylene can be activated by an excess of antimony pentachloride in methylene chloride at $-78\,°C$ to give a tail-to-tail dicationic dimer. Although an electron-transfer process can be accepted to be the origin of that transformation given the very high electron affinity of $SbCl_5$[419], it has already been clearly shown that the same mechanism cannot be applied to the dimerisation of that olefin by other Lewis acids under ordinary polymerisation conditions[416].

A similar investigation from Szwarc's laboratory[420] showed that 1,1-di-p-methoxyphenylethylene can be transformed into its carbenium ion by reaction with an excess of antimony pentachloride. However, in this study it was postulated that the ionisation process involved the reduction of antimony through one of the following alternative mechanisms:

$$SbCl_5 + CH_2{=}CR_2 \rightarrow Cl{-}CH_2{-}\overset{+}{C}R_2 \; SbCl_4^{\bar{}} ,$$

or

$$SbCl_5 + CH_2{=}CR_2 \rightarrow SbCl_3 + Cl{-}CH_2{-}CR_2Cl,$$

$$Cl{-}CH_2{-}CR_2Cl + SbCl_5 \rightarrow Cl{-}CH_2{-}\overset{+}{C}R_2 \; SbCl_6^{\bar{}} .$$

This direct "initiation" process is certainly favoured by the very pronounced basicity of the olefin used and moreover must be considered peculiar to antimony-based Lewis acids which combine a marked strength and the capacity of undergoing facile reduction. It is therefore a specific event which obviously does not explain the more general phenomenology of direct initiation in cationic polymerisation.

c) Initiation Through Allylic-Hydride Abstraction

As we have already mentioned, it was Kennedy[404] who proposed that monomers possessing an allylic hydrogen can be converted into the corresponding allylic carbenium ions through a hydride-ion abstraction promoted by Lewis acids. This reaction would not require a cocatalyst and could therefore explain in principle the occurrence of direct initiation. A similar mechanism had previously been claimed by Holmes and Pettit[421] to rationalise the formation of carbenium ions from certain hydrocarbons and antimony pentachloride, e.g.:

$$Ph_3CH + 2\,SbCl_5 \rightarrow Ph_3C^+SbCl_6^- + SbCl_3 + HCl.$$

Although the evidence presented in support of such reaction is not entirely convincing, in its favour is the existence of a lower oxidation state of antimony which can be readily reached by smooth reduction. In the case of other Lewis acids a lower oxidation state is either unavailable or less easily obtained and the likelyhood of hydride abstraction seems therefore more remote, at least from saturated hydrocarbons. Kennedy's scheme implies however the removal of the allylic hydride ion from an olefin. Although plausible in certain cases (but never proved), this mechanism is obviously impossible with styrene[410] and 1,1-diphenylethylene[416]. With 3-phenylindene it would yield an aryl-substituted allylic carbenium ion which would not be expected to be in equilibrium with its precursor; yet, this equilibrium was observed[93]. With 2,3-dimethylindene in the same conditions[93,323] initiation did not take place; yet Kennedy's mechanism should have operated without impediments. Finally, with 1,1-diphenylpropene hydride abstraction would have produced an allylic ion incapable of giving back the precursor by reacting with methanol; yet Bywater and Worsfold[418] showed that this reversible reaction takes place.

On the basis of the above experimental findings, we can therefore conclude that Kennedy's mechanisms of hydride abstraction does not apply to the systems reported in Tables 8 and 9.

Having ruled out other reasonable possibilities, we are left with two mechanisms which can account for direct initiation, viz., the Hunter-Yohé zwitterion reaction and the selfionisation of the Lewis acid followed by the addition of the cation onto the monomer double bond. Before discussing the relevance of each one of them, some remarks must be made on the nature of the interactions between Lewis acids and olefins and on the possible conjugation of Lewis acids with anions.

2. Interactions Between Lewis Acids and Unsaturated and Aromatic Hydrocarbons

Basically, two types of complexes can arise from the interaction of the vacant orbital of a Lewis acid and the π electrons of an olefin: π-complexes and charge-transfer complexes. The first, sometimes called Dewar complexes, are described as involving the donation of the olefin π-bond electron pair to an empty orbital of the acid concurrent with the converse donation of a lone pair of the acid d orbitals to the antibonding π^* orbital of the olefin[422]. The two moieties of the complex are therefore virtually uncharged. Charge-transfer complexes are invoked when the back-donation of electrons from the Lewis acid is impossible for lack of free d electrons, as in titanium tetrachloride and boron fluoride. In these complexes the olefin is strongly polarised by the acid and acquires a net positive electrostatic charge leaving the acid negatively charged. A survey of the literature most relevant to this topic is given below.

$SbCl_3$ is a relatively weak Lewis acid, since in the presence of a mixture of mesitylene and benzoyl chloride it gives a complex with mesitylene[423]. Antimony pentachloride is stronger and gives a 1:1 complex with perylene whose electronic spectrum is similar to that of the perylene cation[424]. Complexation has been claimed between antimony pentafluoride and aromatic olefins such as 1,1-diphenylethylene and stilbene[425].

The presence of aromatic rings was held responsible for complexation, but we feel that ultraviolet and visible spectroscopy alone is not sufficient to give a clear indication about the nature of such complexes. The following order of electron affinity was found by Gerbier in a study of complexes with aromatics: $AlCl_3 > SbCl_3 > I_2$ [426]. The same author showed that $SbCl_3$ is dimeric in certain solutions, and the structure of its complex with the benzene ring has been abundantly discussed[427,428]. The composition and structure of the complexes between aluminium bromide and aromatic hydrocarbons were studied in Brown's laboratory[429,430]. Benzene, toluene, xylenes and mesitylene all gave 1 : 2 complexes, i.e. π-complexes with Al_2Br_6; 1 : 1 complexes were also detected with m-xylene and mesitylene. As expected, the more basic hydrocarbons gave more stable complexes. It was also concluded that monomeric $AlBr_3$ is a weaker acceptor than its dimer in that specific context, since in order to form 1 : 1 complexes the aromatic hydrocarbon must break both bridge bonds of the aluminium bromide dimer. Perkampus and Weiss studied the infrared spectra of binary systems alkene-Al_2Br_6 and cycloalkene-Al_2Br_6 [431], and observed that the C=C stretching frequency was 50 to 70 cm^{-1} lower than in the pure olefins. Addition of a Brønsted acid produced an even larger shift. Reversible interactions between olefins and aluminium bromide had already been detected by Fairbrother and Nixon[432]. Elliot et al.[433] formulated the existence of a 1 : 1 complex between benzene and titanium tetrachloride on the basis of spectroscopic observations. Goates et al.[434] later reinvestigated this and similar systems and from their phase diagrams concluded that no complexation had taken place with benzene, toluene and other aromatic hydrocarbons. We feel that the last two papers are not necessarily in conflict, since it is plausible to assume that a wide range of interactions can exist, from a strong, well-defined adduct to very weak van der Waals associations and that the possibility of detecting them varies from one experimental technique to another. Thus, electronic spectroscopy is certainly a much more sensitive tool than many other physical techniques, and capable of revealing the existence of weak interactions. The problem of course is to establish the chemical relevance of these feeble molecular associations, i.e., whether or not they can be considered as more reactive entities than their independent components. The following situation should illustrate this point. The existence of a complex between 1,1-diphenylethylene and $TiCl_4$ was first proposed by Evans's group to explain the kinetics of dimerisation of that olefin[433]. Very recently, Sauvet et al.[416] vacuum distilled 1,1-diphenylethylene in a $TiCl_4$ solution kept at -76 °C: they observed the formation of a negligible amount of carbocations (chromophore $-CH_2-\overset{+}{C}Ph_2$) and a new absorption band at 340 nm, which they attributed to the olefin-Lewis acid complex. It would be extremely interesting to probe the system further to ascertain the importance of that complex in the mechanism of initiation, i.e., if it is an intermediate leading to the carbenium ion or the product of a side reaction. No such studies have been carried out yet. Plesch and Brackman[435] reported that both cis and trans-stilbene form a red complex with $TiCl_4$ without undergoing isomerisation. A very interesting piece of work was done by Plesch et al.[436,437] concerning the phase diagram of the system isobutene-$TiCl_4$: evidence was obtained for the formation of a 1 : 1 complex. This study must have met with considerable difficulties owing to the occurrence of condensation polymerisation of isobutene[395] and we are somewhat puzzled by the clean behaviour observed. Dijkgraaf[438] examined the formation of charge-transfer complexes between aromatic hydrocarbons and $TiCl_4$, $TiBr_4$ and $VOCl_3$, and the existence of these complexes was later confirmed by Hammond[439]. The spectroscopic characteristics of complexes between olefins and $TiCl_4$ and $VOCl_4$ have also been reported[440]. Stannic chloride gives a 4:1 complex with 1,1-diphenylethylene[441]; substitution of chlorine atoms with alkyl groups reduces progressively the strength of the complex. This is readily explained by elementary considerations about the inductive effect of alkyl substituents. Nakane et al.[442] reported that boron fluoride forms a complex with propene, but claimed that this interaction has no bearing on the mechanism of polymerisation. No evidence was however given in support of this surprising statement. Indeed Elegant et al.[443] suggested that the polymerisation of olefins by boron fluoride in the gas phase is preceded by a complexation reaction. This conclusion seems more reasonable on the basis of (mainly) kinetic evidence gathered in many related systems (see below). Kennedy and Milliman[444] obtained clear evidence by NMR spectroscopy for the existence of a complex between isobutene and trimethylaluminium in cyclopentane. Since trialkylaluminium compounds are the weakest acids in the series $AlCl_3 > AlRCl_2 > AlR_2Cl > AlR_3$ [59], as in the case of stannic chloroalkyls, we can conclude that isobutene gives complexes with all the members of the series. An old experiment reported by Kennedy and Thomas[445] supports the above conclusion: these authors noticed that when a solution of aluminium chloride in methyl chloride was poured into pure pentane, $AlCl_3$ precipitated; however, when the same solution was poured into a pentane-isobutene mixture, no precipitation occurred. The existence of an isobutene-$AlCl_3$ complex seems the logical explanation of that phenomenon. The reaction of ethylene with monomeric trimethylaluminium in the gas phase above 180 °C was studied by Egger and Cocks[446]. The addition of the olefin at the C—Al bond could not be rationalised by a simple four-centre intermediate, and it was instead suggested that complexation preceded the addition reaction.

The above survey does not include those systems in which polymerisation studies led the authors to conclude that complexes between the monomer and the Lewis acid must have formed prior to the appearance of active species. We deliberately restricted our attention to investigations in which some direct evidence for complexation was offered. It can be concluded that while complexes between Lewis acids normally used in cationic polymerisation and aromatic hydrocarbons are common and readily observed, less information is available for similar interactions with the olefinic double bond because of the intrinsic difficulty arising from concurrent polymerisation. However, complexation (probably in the form of charge-transfer associations) seems to be the general behaviour rather than the exception, although its extent (relative strength of the complex) varies appreciably from system to system and is probably low for many olefins.

Two recent studies cast more light on this problem and they are particularly interesting because they relate directly to the context of cationic polymerisation. In the first Bogomolova et al.[447] examined the electronic spectra of mixtures of stannic chloride and styrene or α-methylstyrene in ethyl chloride and cyclohexane. Complexation with the first monomer gave a band at 295 nm with an extinction coefficient of about 1,000 M^{-1} cm^{-1} in ethyl chloride and about 2,000 in cyclohexane. The instability constant for these 1:1 complexes was around 5 M^{-1}. Similar results were obtained with the second monomer. Stannic chloride was also found to associate with non-olefinic aromatic compounds like benzene and polystyrene, but the absorption of these complexes occured at about 270 nm.

Marek[448,449)] has reported preliminary data on a study of the complexation of isobutene with vanadium, titanium and tin tetrachlorides in heptane. Electronic spectra were recorded with variable proportions of reactants and it was ascertained that the complexes involved the participation of one molecule of isobutene and one of Lewis acid. The instability constants were determined at $-80°$: about $7 \, M^{-1}$ for VCl_4 and $TiCl_4$, and about $18 \, M^{-1}$ for $SnCl_4$. While the complex with VCl_4 absorbed in the visible and had an extinction coefficient of about $100 \, M^{-1} \, cm^{-1}$ [448)], those with $TiCl_4$ and $SnCl_4$ gave bands at 350 and 270 nm respectively (no extinction coefficients reported). Unfortunately the full paper describing this investigation has not been published yet, and any detailed analysis of these data must be postponed.

The above reports confirm the essential points: that the interaction of olefins with Lewis acids tend to be weak, but nevertheless complexes do form and can be detected spectroscopically. There is undoubtedly a great need for more studies in this area.

3. Conjugation Involving Lewis Acids

We have already emphasised that homo- and heteroconjugation are commonly observed with Brønsted acids and that the degree of association between the acid and the anion can be very high. We have also stressed in the preceding chapter that these phenomena can play an important role in determining the features of cationic polymerisations. Similar interactions can be postulated with Lewis acids, namely homoconjugation and various types of heteroconjugation:

$$MtX_n + MtX_{n+1}^- \rightleftharpoons Mt_2X_{2n+1}^-$$

$$MtX_n + MtX_nR^- \rightleftharpoons Mt_2X_{2n}R^-$$

$$MtX_n + MtX_m'^- \rightleftharpoons Mt_2X_nX_m'^-$$

$$MtX_n + Mt'X_m^- \rightleftharpoons MtMt'X_{n+m}^-$$

$$MtX_n + Mt'X_m'^- \rightleftharpoons MtMt'X_nX_m'^-.$$

In Plesch's first book several very interesting observations were made concerning conjugation with Lewis acids[450)], without reaching definite conclusions. The conjugated anions derived from the halides of metals belonging to the third column of the periodic table have been well characterised. Brown and collaborators[429,430,451,452)] investigated the formation of π-complexes from aromatic hydrocarbons and $AlCl_3$ and $AlBr_3$ in the presence of hydrogen halides. The anions in these complexes were identified as both AlX_4^- and $Al_2X_7^-$. Later infrared and conductimetric studies[453–456)] have abundantly confirmed the existence of those anions and shown that gallium halides behave likewise. Moreover, the simple and the conjugated anions are in equilibrium[456)] and the relative abundance of each species depends upon the experimental conditions.

The homoconjugation of SbF_5 with its anion leads to several possible complex species. Thus, dilute solutions of this Lewis acid in hydrogen fluoride contain mostly $Sb_2F_{11}^-$, but as the concentration is raised $Sb_3F_{16}^-$ and $Sb_4F_{21}^-$ are also formed[457]. SbF_5 itself exhibits a strong tendency to association: in perfluorocyclobutane its most likely form is a cyclic tetramer[458]. When t-butyl chloride or fluoride are added to an SbF_5 solution in liquid SO_2, dimeric anions appear, but the reactions leading to their formation are not simple[459]. Commeyras and Olah reported that the reaction of SbF_5 with alkyl and acyl halides produces SbF_6^- and $Sb_2F_{11}^-$ in equilibrium with each other[460].

Many other metal fluorides can homo- and heteroconjugate with related anions. Thus, boron fluoride gives readily the complex anions $B_2F_7^-$ and $B_3F_{10}^-$ [461−464], and thallium and niobium fluorides homoconjugate with their corresponding anions[465]. A variety of mixed metal fluride anions (heteroconjugation) have also been characterised in Brownstein's laboratory[466,467], e.g. $BTa_2F_{14}^-$, $TaWF_{10}^-$, and complex fluoroanions derived from vanadium pentafluoride.

Although a lot of ground remains to be covered in this field, Lewis acids do seem to display a marked tendency towards homo- and hetero- conjugation with anions derived from them. Conductivity studies of Lewis acid solutions are difficult to perform and the interpretation of data is often a frustrating exercise, not particularly suited for searching for possible homoconjugation phenomena. For instance, Velichova and Panayotov[468] investigated the conductivity of solutions of boron fluoride etherate in 1,2-dichloroethane. They noticed that above a certain concentration the conductivity levelled off, and attributed this to the formation of triple ions. Homoconjugation of the anion was not considered, but such an equilibrium could explain equally well or better the observed behaviour. Plesch and Grattan[469] studied the conductivity of aluminium bromide solutions in methylene chloride and concluded that a 2 to 2 equilibrium governed the selfionisation of this Lewis acid. However, they could not distinguish between the selfionisation of the monomeric species to give simple ions or that of the dimer to give the homoconjugated anion.

Kramer and coworkers[893,894] recently showed that $AlBr_3$ forms $2:1$ salts with t-BuCl and t-BuBr in halogenated hydrocarbons at low temperatures. These ionogenic interactions were followed by ^{13}C-NMR and clearly proved that homoconjugated anions of the type $Al_2Br_6X^-$ accompanied the t-butyl cation.

The occurrence of a conjugation reaction between a Lewis acid and an anion (derived from it or from another acid) has far-reaching consequences on the course of a cationic polymerisation involving direct initiation. The formation of anions in the early stages of the reaction determines in fact the sequestration of some of the Lewis acid. The homoconjugated anions thus formed are less acidic than the original halide and might not be strong enough to attack the olefin. The net result of this process will therefore be to slow down initiation. However, if the monomer is sufficiently basic to displace the Lewis acid from its conjugated anion, initiation will not slow down. It follows that ideally homoconjugation of Lewis acids should be studied in the absence and in the presence of typical monomers to assure that if homoconjugation occurs, it also persists under polymerisation conditions. The other important corollary is that if homoconjugation is possible between a Lewis acid and the corresponding anion, additions of a common-anion salt to the polymerisation system will result in a decrease of available initiator and any suppression of the free ions in the system will be superseded by the predominant sequestration reaction. This has been expounded at length in the preceding chapter and only needs a brief reminder here.

4. Direct Initiation Mechanisms

a) Experimental Features

We have gathered in this section some of the most important empirical characteristics of cationic polymerisations where the Lewis acids operate through direct initiation. By far the most relevant of these is the fact that in the majority of the systems studied incomplete yields were obtained, at least with typical catalyst concentrations. This of course applies to investigations in which all the necessary precautions were taken to minimise the presence of cocatalytic impurities.

The initiation reaction between $TiCl_4$ and 1,1-diphenylethylene[416] or 3-substituted indenes[93,323] proceeds through a straightforward 1 : 1 stoicheiometry. Yet, the amount of active species obtained as soon as the catalyst concentration is raised above minimal levels is far less than the quantity of Lewis acid used. In the direct initiation of isobutene polymerisation by the same catalyst (prepolymerisation) limited yields have been reported[395-397], although no clear relationship has ever been obtained between the amount of polymer formed and the catalyst concentration.

The system α-methylstyrene-monobutoxytitanium trichloride-methylene chloride[413] gives at low temperature irreproducible prepolymerisation yields. Both cocatalysis by residual moisture (consumed in the process) and direct initiation were assumed responsible for the polymerisation. If this proposal is correct, direct initiation does not lead to complete conversion.

Recent work on the dimerisation of 1,1-diphenylethylene by aluminium chloride[417] produced conclusive evidence that direct initiation does not lead to the total consumption of the catalyst. This excellent piece of research showed that about 2.5 aluminium atoms are needed to give rise to one carbenium ion. Similar indications were reported by Kennedy and Squires[470] for the low temperature polymerisation of isobutene by aluminium chloride. They underlined the peculiar feature of limited yields obtained in flash polymerisations with small amounts of catalyst. The low conversions could be increased by further or continuous additions of the Lewis acid. Equal catalyst increments produced equal yield increments[445]. It was also shown that introductions of small amounts of moisture or hydrogen chloride in the quiescent system did not reactivate the polymerisation. This work was carried out in pentane and different purification procedures for this solvent resulted in the same proportionality between polymer yield and catalyst concentration. Experiments were also performed in which other monomers (styrene, α-methylstyrene, cyclopentadiene) were added to the quiescent isobutene mixture. The polymerisation of these olefins was initiated but limited yields were again obtained. Although the full implications of these observations must await more precise data, we agree with the authors' interpretation that allylic cations formed in the isobutene polymerisation, while incapable of activating that monomer, are initiators for the polymerisation of the more basic monomers added to the quiescent mixture. The low temperature polymerisation of isobutene by aluminium chloride was also studied in Plesch's laboratory, but in methylene chloride[471]. Water was found not to be needed for the polymerisation and other cocatalytic impurities did not seem likely in view of the complete yields obtained. The authors suggested that the solvent acted as a cocatalyst, but no proof was given to support this claim. In view of the recent findings of Priola et al.[472] showing that deuterated methylene chloride does not suffer any deuterid ion abstraction at low temperature in the presence of AlCl$_3$, AlEtCl$_2$, AlEt$_2$Cl and AlEt$_3$, the possibility of solvent cocatalysis seems now untenable. We prefer to think that in methylene chloride, direct initiation by aluminium chloride is very efficient and leads to complete conversion.

The system ethylaluminium dichloride-isobutene-n-heptane in the temperature range −55 to 21 °C was shown to initiate in the absense of any cocatalyst[408]. No mention was given in this paper about the polymerisation yields, but from a figure giving the time-conversion curve for a typical run it can be concluded that reactions did not reach 100% conversion, at least within the first few minutes.

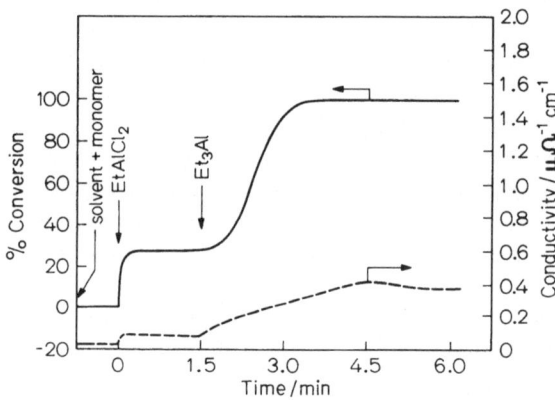

Fig. 4. Time-conductivity and time-conversion curves for a polymerisation of isobutene initiated by EtAlCl$_2$ with subsequent addition of Et$_3$Al. $[M]_0 = 0.1$ M; $[EtAlCl_2] = 1.0 \times 10^{-4}$ M; $[Et_3Al] = 1.1 \times 10^{-4}$ M; solvent: CH$_3$Cl; T = −45 °C. (Taken from Ref. [409])

The mechanism of initiation in the polymerisation of isobutene in methyl chloride at −54 °C seems to follow similar routes whether AlCl$_3$ or AlEtCl$_2$ are used[409]. Reactions tend to stop before reaching complete conversion and further catalyst introduction restores the polymerisation rate nearly to the original value. These experiments clearly show that direct initiation with these two Lewis acids is accompanied by an important deactivation process. In the same paper the behaviour of AlEtCl$_2$ was also investigated in the presence of some preadded Et$_3$Al. This situation readily gives 100% conversion. Inverting the order of addition of the two catalyst components results in a first incomplete reaction followed by total consumption of the monomer when AlEt$_3$ is introduced. This experiment, shown in Fig. 4, clearly indicated that the role of triethylaluminium is not to remove adventitious water, but to form a very effective mixed catalyst with the dichloride. It seems that the dissymmetry of this syncatalytic mixture enhances its degree of selfionisation and thus its initiating potential.

b) Initiation by Direct Acid-Base Reaction

Evidence in favour of direct initiation involving the attack of an undissociated Lewis acid onto an olefin is essentially confined to systems containing titanium tetrachloride. This section is devoted to an analysis and discussion of that evidence and to the proposal of an overall mechanism compatible with it.

The discovery of polymerisation by condensation[395] produced striking indications that a direct initiation process occurred with $TiCl_4$. The most relevant features of those experiments can be summarised as follows. The system isobutene–$TiCl_4$–CH_2Cl_2 always gives some prepolymerisation, even under the most stringent experimental conditions. In part, this prepolymerisation is probably due to the cocatalytic effect of traces of residual moisture. The quiescent mixture obtained after this prepolymerisation can be reactivated by distilling its volatile components onto a side vessel. In fact, upon condensation a new polymerisation is observed, even if the walls of the side vessel are covered with a sodium film. Clearly, water cocatalysis cannot be invoked to explain this second polymerisation. Two other possible sources of reactivation can be excluded, namely light and the solvent, since condensation polymerisation occurs in the *dark* and *without solvent*. If a mixture of isobutene and $TiCl_4$ is made up in the vapour phase and then condensed at temperatures as high a −20 °C, polymerisation takes place almost immediately and the liquid mixture turns yellow. Another important feature is that the catalyst is more efficient in the condensation experiments than mixed in the liquid phase. Again, the solvent is not responsible for this peculiarity, since the same behaviour is observed in pure isobutene. We can readily rule out the possibility of anomalous local overconcentrations of $TiCl_4$ in the case of condensation experiments, because of the relatively low vapour pressure of this compound compared with that of isobutene. If at all, momentarily high local concentration of catalyst would rather be expected when a $TiCl_4$ bulb is crushed in the liquid monomer (solution). The higher efficiency is thus an intrinsic property of the condensation procedure. Interestingly, the DP's of the polyisobutenes are also higher in the condensation polymerisation than in prepolymerisation. A second addition of $TiCl_4$ to the quiescent mixture brings about a totally different reaction depending on how this addition is performed: if a bulb is crushed in the liquid medium, nothing happens; but if more $TiCl_4$ is condensed in the liquid medium from the vapour phase, polymerisation is reinitiated. Here again light is not responsible for the activation in the condensation experiments, because the phenomenon is also observed in the dark.

The occurrence of condensation polymerisation is not unique to isobutene. Sauvet[259] observed it with the system titanium tetrachloride-1,1,-diphenylethylene and Vairon and Sigwalt[474] with cyclopentadiene-butoxytitanium trichloride. A general rationalisation of this phenomenology must therefore be found. This is given at the end of this section after a survey of other systems involving titanium tetrachloride.

A few years after the discovery of condensation polymerisation, the behaviour of prepolymerisation was reinvestigated in the same laboratory[475] using highly purified isobutene, $TiCl_4$ and dichloromethane in the temperature range −60 to −30 °C. It was concluded that direct initiation was responsible for most of the polymerisation observed, and the limited yields were attributed to the consumption of the catalyst in the formation of an inactive complex with the monomer. No indications exist however to support this hypothesis. On the contrary, all available evidence suggests that no *strong* complex is formed between these two compounds[449]. Moreover, that interpretation does not explain the absence of a second polymerisation when more $TiCl_4$ is added in the liquid phase.

No other aliphatic olefin has been investigated in detail with titanium tetrachloride. 2-methyl-2-butene is known to oligomerise in methylene chloride at $-78\ °C$[476], but the mechanism of that process was probably cocatalytic, since no serious attempts were made to exclude the presence of moisture in the system.

With aromatic monomers the situation is different. A large number of studies have been carried out with $TiCl_4$ in an attempt to elucidate the mechanism of initiation. The polymerisation of styrene in ethylene chloride was first investigated by Plesch[477]. At low temperature the yields could not be reduced sufficiently to demonstrate the presence of an adventitious cocatalyst. In alkyl bromide solvents the polymers contained alkyl fragments[477-406] later traced to a transfer with the solvent and not to solvent cocatalysis[478]. Further work in Plesch's laboratory led him to conclude that water cocatalysis was responsible for these polymerisations: the rate of monomer consumption in a series of runs in which successive portions of styrene were added to a solution of $TiCl_4$ in ethylene chloride decreased progressively suggesting that water consumption in each run steadily reduced the level of residual moisture available for cocatalytic initiation[406]. This interpretation was reinforced by the results of some ingenious experiments in which styrene was introduced in a quiescent mixture resulting from the polymerisation of isobutene: no polymerisation was observed (lack of cocatalyst), but upon introduction of some water the system was reactivated and a vigorous polymerisation took place[403]. However, the issue was not settled[479]. Cheradame's work, carried out a few years later[396,552,553], brought about some conflicting evidence concerning the role of water in these systems. He added α-methylstyrene and indene to a quiescent isobutene–$TiCl_4$–CH_2Cl_2 and, as in Plesch's experiment, observed no significant polymerisation but he could not reactivate the systems by adding water, i.e. water did *not* act as a cocatalyst. Cheradame also showed that small quantities of added water had no enhancing effect on the rate of α-methylstyrene polymerisation by $TiCl_4$. Later, this author felt that some of his results[480,481] could not be properly interpreted because the solvent used in those experiments might not have been sufficiently pure. Special attention was therefore devoted to the purification of methylene chloride[72] and it was finally concluded[412,414] that indene and α-methylstyrene are polymerised by $TiCl_4$ in pure methylene chloride at low temperature without the aid of a cocatalyst. Problems of reproducibility are unfortunately quite common in cationic polymerisation, particularly when delicate issues such as the presence of minute traces of active impurities are at stake. It follows that it is sometimes extremely difficult to assess the real impact which certain basic experiments bear on fundamental questions. The following features are noteworthy from the latter studies of the polymerisation of aromatic olefins by titanium tetrachloride[72,412,414,480,481]. At low temperature $(-70$ to $-60\ °C)$ the initial rate of polymerisation is approximately proportional to the monomer concentration, but depends upon the catalyst concentration to a power lower than unity. The kinetics can be interpreted without appeal to a cocatalyst. At higher temperature the yields can be incomplete and reinitiation of these partial polymerisations can be brought about by adding some water or HCl. Thus, direct initiation operates and, as in the case of isobutene, is accompanied by termination reactions although these are not very effective with these systems. The rate of polymerisation increases somewhat as the temperature decreases, indicating that the initiation process is a combination of various elementary interactions which have different temperature dependences. One can visualise for example two or more concurrent reactions, with only one of them producing the active species, viz. that possessing the lower activation energy.

The interaction between titanium tetrachloride and model aromatic olefins has been extensively studied and has afforded a great deal of fundamental information. 1,1-diphenylethylene was the first one to be investigated. Sauvet et al.[482] showed that initiation took place both with and without a cocatalyst. Later work by the same group, focussed on the behaviour of direct initiation, indicated that the concentration of active species decreased when the monomer concentration was increased[259,416]. This effect was interpreted in terms of an inhibiting role of the monomer upon initiation. Selfionisation of the catalyst was ruled out because it led to an expected initiation efficiency which disagreed with the experimental relationship $[R^+] = K[C]_0/[M]_0$, where $[R^+]$ is the concentration of active species (carbenium ions), and $[C]_0$ and $[M]_0$ are the initial catalyst and monomer concentrations. Particular care was taken to purify the monomer and the above relationship is not an artefact. Solvent cocatalysis (CH_2Cl_2) was proved not to be operative in this system and the same applies to initiation through electron transfer. The very interesting mechanism proposed to account for all experimental observations assumed that a complex $TiCl_4 \cdot M_2$ must form. In this complex the $TiCl_4$ is inactive, but can be liberated by in vacuo evaporation. The search for that complex was successfully carried out using ultraviolet spectroscopy[416].

3-Substituted indenes are another group of aromatic olefins which have been used as model compounds in order to gain a better understanding of the mechanisms of initiation in cationic polymerisation. Their behaviour with $TiCl_4$ in methylene chloride is summarised below. 3-Isopropylindene reacts with $TiCl_4$ to give the corresponding carbenium ion, as verified spectroscopically[93,94]. This interaction was studied quantitatively by ultraviolet spectroscopy and it was ascertained that for each molecule of $TiCl_4$ consumed there is one carbenium ion formed, as shown in Fig. 5. Thus, the stoicheiometry of the initiation reaction is unity in both monomer and catalyst. In

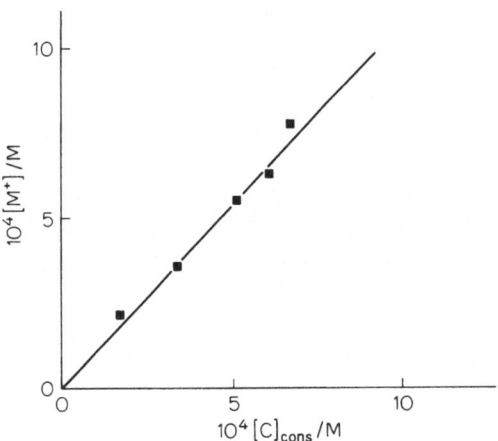

Fig. 5. Reaction of $TiCl_4$ with 3-isopropylindene in CH_2Cl_2 at $-70\,°C$. The concentrations of carbocations were calculated from the optical density at 322 nm, taking $\epsilon = 30,800\,1\,mol^{-1}cm^{-1}$. The concentration of catalyst consumed was calculated from the decrease in optical density at 290 nm. (Taken from Ref.[93])

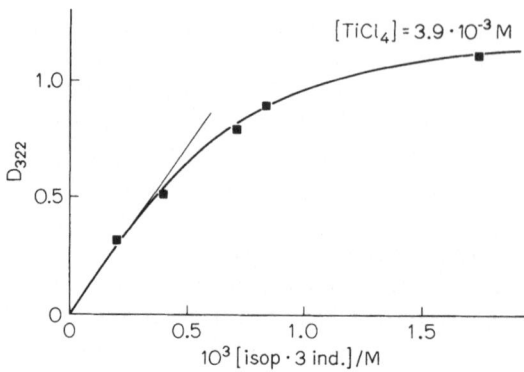

Fig. 6. Reaction of TiCl$_4$ with 3-isopropylindene in CH$_2$Cl$_2$ at $-70\,^\circ$C. Variation of the optical density at 322 nm (carbocation) with the olefin concentration at fixed TiCl$_4$ concentration. (Taken from Ref. [93])

fact, in its initial stages, the reaction rate depends on the first power of both monomer and catalyst concentration. However, limited final yields of carbenium ions are obtained even in the presence of an excess of one of the reactants, suggesting that the initiation step is an equilibrium[93] (Fig. 6). This equilibrium was difficult to characterise in the case of 3-methyl- and 3-isoproypl indene because of the dimerisation reaction, but was clearly verified with 3-phenylindene, since side reactions with this olefin proceeded very slowly[93,323].

All the evidence gathered in this section must now be rationalised into a reasonable mechanism. Titanium tetrachloride is monomeric in methylene chloride[398], even at low temperature. Slight interactions exist between TiCl$_4$ molecules[483], but these do not amount to any strong association. Given the spectroscopic observations concerning the initial stoicheiometry of initiation and the structure of the products of this reaction, we are led to formulate a revival of the old Hunter-Yohé mechanism[401]. Thus, the first step in the direct initiation processes involving TiCl$_4$ and olefins takes the form:

$$TiCl_4 + ^{\backslash}C=C^{/} \rightleftharpoons Ti\overline{Cl}_4-\overset{|}{\underset{|}{C}}-\overset{|}{\underset{|}{C}}\,^+.$$

The above equilibrium generates the primary active species which can propagate to give the polymer (isobutene, styrene, etc.) or the dimer zwitter ion (model olefins):

$$Ti\overline{Cl}_4-\overset{|}{\underset{|}{C}}-\overset{|}{\underset{|}{C}}\,^+ + ^{\backslash}C=C^{/} \rightarrow Ti\overline{Cl}_4-\overset{|}{\underset{|}{C}}-\overset{|}{\underset{|}{C}}-\overset{|}{\underset{|}{C}}-\overset{|}{\underset{|}{C}}\,^+, \quad \text{etc.,}$$

this reaction being equilibrated in the case of 1,1-diphenylethylene[416], but not with 3-substituted indenes[93]. Two types of rearrangements can be postulated which deactivate the chain carriers. One is the migration of a chloride ion to give the corresponding saturated neutral chloroorganotitanium derivative,

$$Ti\overline{Cl}_4-\overset{|}{\underset{|}{C}}-\overset{|}{\underset{|}{C}}\,^+ \rightarrow TiCl_3-\overset{|}{\underset{|}{C}}-\overset{|}{\underset{|}{C}}-Cl$$

$$\mathrm{TiCl_4^-}\!-\!\overset{|}{\underset{|}{C}}\!-\!\overset{|}{\underset{|}{C}}\!-\!\overset{|}{\underset{|}{C}}\!-\!\overset{|}{\underset{|}{C}}{}^+ \;\to\; \mathrm{TiCl_3}\!-\!\overset{|}{\underset{|}{C}}\!-\!\overset{|}{\underset{|}{C}}\!-\!\overset{|}{\underset{|}{C}}\!-\!\overset{|}{\underset{|}{C}}\!-\!\mathrm{Cl};$$

and the other is the expulsion of hydrogen chloride which leaves an unsaturated neutral chloroorganotitanium compound, e.g.,

$$\mathrm{TiCl_4^-}\!-\!\overset{|}{\underset{|}{C}}\!-\!\overset{|}{\underset{|}{C}}\!-\!\overset{|}{\underset{|}{C}}\!-\!\overset{|}{\underset{|}{C}}{}^+ \;\to\; \mathrm{TiCl_3}\!-\!\overset{|}{\underset{|}{C}}\!-\!\overset{|}{\underset{|}{C}}\!-\!\overset{|}{C}\!=\!\overset{|}{C} + \mathrm{HCl}.$$

The only evidence for the occurrence of these two reactions comes from the work with 3-substituted indenes. Both the saturated[93] and the unsaturated[323] products were detected at the end of "large scale" experiments. It is of course possible to assume that the unsaturated compound is formed via dehydrochlorination of the saturated one, in which case only the first deactivation reaction applies to the zwitterion. The lack of pertinent experimental information makes the choice between these two alternative impossible, but in either case HCl can be generated by these systems during the early stages of the initiation process. It is now essential to find a reasonable and general explanation for the often observed limited yields in the presence of unconsumed (vacuum distillable) titanium tetrachloride. As already pointed out, the formation of a strong complex between the olefin and this Lewis acid, a complex which can be decomposed into its original components by heating, does not seem a satisfactory interpretation of the above phenomenology since such complex has often failed to show up in a distinctive manner when its spectroscopic search has been undertaken[397, 449]. We prefer to think that the source of deactivation of these processes is the anion $\mathrm{TiCl_5^-}$. The formation of appreciable concentrations of this species is a natural consequence of the presence of hydrogen chloride generated by the reactions discussed above. In fact, the simultaneous presence of HCl, $\mathrm{TiCl_4}$ and the olefin gives rise to cocatalytic interactions leading eventually to $\mathrm{TiCl_5^-}$. The details of these interactions are fully discussed in the section of this chapter devoted to cocatalysis. The detrimental effects of $\mathrm{TiCl_5^-}$ on these systems, ultimately producing the cessation of their activity (termination) must now be discussed. We believe that the homoconjugation reaction

$$\mathrm{TiCl_4} + \mathrm{TiCl_5^-} \;\rightleftharpoons\; \mathrm{Ti_2Cl_9^-}$$

or similar ones involving more than one molecule of Lewis acid, effectively scavenge free $\mathrm{TiCl_4}$ giving the conjugate anion which is too weak an acid to be able to interact with olefins in an initiation process. Thus, the production of $\mathrm{TiCl_5^-}$ in these systems leads to a progressive "mopping up" of the free $\mathrm{TiCl_4}$ and the consequent decrease in the rate of polymerisation or dimerisation. Evidence in favour of homoconjugation with Lewis acids has already been presented in Sect. III-B-3. Regrettably, the specific case of $\mathrm{TiCl_4}$ has not to our knowledge been studied. There is however an important experiment which corroborates our arguments. It showed[484] that the introduction of small amounts of trityl pentachlorotitanate into a solution of indene in methylene chloride before the addition of $\mathrm{TiCl_4}$ results in much lower rates of polymerisation than those obtained in experiments carried out without the trityl salt. Homoconjugation of free $\mathrm{TiCl_4}$ by

$TiCl_5^-$ coming from the added salt seems to us the most logical explanation for the observed rate depression.

An additional source of $TiCl_4$ sequestration could be the zwitterions acting as chain carriers in these systems. It is conceivable that the following heteroconjugation reaction might take place and contribute to the stabilisation of the charge separation, e.g.:

$$TiCl_4^- -\overset{|}{\underset{|}{C}}-\overset{|}{\underset{|}{C}}{}^+ + TiCl_4 \rightleftharpoons Ti_2Cl_8^- -\overset{|}{\underset{|}{C}}-\overset{|}{\underset{|}{C}}{}^+.$$

Moreover, this new zwitterion could give rise to $TiCl_5^-$ by decomposing according to the reaction:

$$Ti_2Cl_8^- -\overset{|}{\underset{|}{C}}-\overset{|}{\underset{|}{C}}{}^+ \rightarrow TiCl_3 -\overset{|}{\underset{|}{C}}-\overset{|}{\underset{|}{C}}{}^+ TiCl_5^-.$$

This last proposal is purely speculative and must await experimental confirmation. It is not unreasonable and could constitute an additional source of $TiCl_5^-$.

The overall mechanism of direct initiation and termination by conjugation satisfactorily explains all the major peculiarities of systems characterised by the $TiCl_4$-olefin pair in the absence of added or spurious cocatalysts. In particular, it accounts for direct initiation by a one-to-one interaction, the limited yields proportional to the catalyst concentration, the reactivation of polymerisation upon distillation of all volatile components onto another vessel (polymerisation by condensation), the different response of these systems to further additions of $TiCl_4$ depending on whether these are carried out in the liquid or from the gas phase, and the lack of effect of water introduced after cessation of polymerisation. A few comments are in order to substantiate these claims. When the volatile components are evaporated and vacuum condensed in a new vessel, the monomer remaining and the solvent are obviously transferred. As for the catalyst, we can conceive that the $TiCl_4$ will be readily displaced from the inactive conjugated anions under the effect of warming and by the continuous shifting of conjugation equilibria resulting from the removal of free $TiCl_4$ to the vapour phase. The arrival of the three components into the new container will induce a fresh polymerisation because the absence of non-volatile products of the first reaction ($TiCl_5^-$-type species) will allow $TiCl_4$ to operate freely until of course a sufficient amount of sequestrating anions will have again accumulated in the new system. This situation applies equally well to the polymerisation of undiluted isobutene by $TiCl_4$. As to the effect of further catalyst additions, when this is done in the liquid under vigorous stirring one can explain the absence of reinitiation by considering that the terminating non-volatile products of the first reaction are present in sufficient amount to scavenge the new $TiCl_4$ added. However, when the catalyst is condensed into the quiescent mixture from the vapour phase, by cooling the system, an abnormally high concentration of $TiCl_4$ will arise in the upper layer and the contact with the terminating products from the first generation will be less important. There will therefore be a sufficient amount of free $TiCl_4$ to give some new initiation, until again the sequestration of this catalyst takes place and the second polymerisation ceases. Finally, the failure of added water to reactivate the system can be rationalised by the virtual absence of free Lewis acid in the quiescent mixture containing isobotene and an aromatic monomer. Indeed, cocatalytic initiation requires the presence of the Lewis acid in the form of complex with the olefin or free. Obviously, the $TiCl_4$ not bound to related anions by conjugation is complexed by the aromatic rings rather than the olefinic double bond.

Bywater and Worsfold[418] have reported on the direct ionisation of 1,1-diphenyl-propene by $TiCl_4$ in methylene chloride. This very interesting system in which propagation is not observed, allowed a clean study of the initiation mechanism: Two molecules of $TiCl_4$ are necessary to generate one diphenylpropyl carbenium ion. Although no specific data were presented, this conclusion seems to go against our general proposal of a one-to-one Hunter-Yohé mechanism. We believe however that the above result does not reflect the stoicheiometry of the initiation step alone, but the *overall* consumption of $TiCl_4$ (including conjugation) in the process. We have already seen that the real stoicheiometry of the ionogenic reaction was only properly detected in the case of 3-substituted indenes when very low catalyst concentrations were used ([93] and Fig. 6). In that situation the large excess of olefin favoured a rapid and virtually complete consumption of $TiCl_4$ in the initiation step, giving an equal amount of carbenium ions. As the relative concentration of the catalyst is increased, the formation of the zwitterion is accompanied by the reactions leading to sequestration of $TiCl_4$ by conjugation and therefore the *total* amount of catalyst consumed can be as high as three times that of the active species produced. We believe that the value of two, calculated for that ratio by Worsfold and Bywater, does not reflect the stoicheiometry of the initiation reaction (a value which would suggest selfionisation of the catalyst), but an intermediate stage in the range of situations just discussed.

If complexation between $TiCl_4$ and the olefin (and with aromatic dimers and polymers) does take place to such a degree that the catalyst complexed is not available for initiation, as suggested by Sauvet et al.[259, 416, 148] for 1,1-diphenylethylene, then obviously the inverse catalyst efficiency, expressed as the number of its molecules consumed to prepare one active species, will be higher than three, although again the stoicheiometry of the initiation step remains simply one to one.

c) Initiation Following Selfionisation

The selfionisation of Lewis acids in solution proceeds according to two major types of equilibria, depending on whether the halide is monomeric or dimeric in the medium considered, namely:

$$2\,MtX_n \;\rightleftharpoons\; MtX_{n+1}^- MtX_{n-1}^+ \;\rightleftharpoons\; MtX_{n+1}^- + MtX_{n-1}^+$$
$$\Updownarrow$$
$$(MtX_n)_2 \;\rightleftharpoons\; Mt_{n+1}^- MtX_{n-1}^+ \;\rightleftharpoons\; MtX_{n+1}^- + MtX_{n-1}^+.$$

A great deal of speculation concerning both the aggregation state of Lewis acids in solution and their mode and extent of ionic dissociation is to be found in the literature related to the cationic polymerisation of olefins by these catalysts. The number of specific and reliable studies on this topic is however modest in comparison, as illustrated by the brief survey given below.

It seems well established that aluminium bromide is monomeric in methyl bromide, according to some fairly old work on the ebullioscopy of this system[485]. More recent Raman studies[486] have confirmed that the amount of dimerisation of $AlBr_3$ is indeed minimal, but that between -90 and $0\ °C$ this Lewis acid is present in these solutions as a 1:1 n-donor complex with the solvent. Interestingly, $AlBr_3$ was also found to be monomeric in methylene chloride[469], but the authors emphasised that this situation did not exclude the homoconjugation of the anion following selfionisation:

$$AlBr_4^- + AlBr_3 \rightleftharpoons Al_2Br_7^-.$$

Plesch and Grattan[469] also showed that aluminium chloride is monomeric in methylene chloride, but their electrochemical work on the $AlCl_3$-EtCl system giving evidence of a 2:2 ionogenic equilibrium does not provide conclusive evidence for monomeric $AlCl_3$, since one could envisage its dimer being involved in the following solvent-assisted ionisation:

$$2\ Al_2Cl_6 + EtCl \rightleftharpoons Al_2Cl_5^+EtCl + Al_2Cl_7^-,$$

which would also account for the observed conductivity behaviour. Alkylaluminium halides and trialkylaluminium derivatives are the best known examples of Lewis acids which associate into dimeric aggregates in certain solutions and even in the vapour phase[487,488].

Longworth and Plesch[489] examined the conductivity of solutions of titanium tetrachloride in methylene chloride and observed that this was directly proportional to the Lewis acid concentration. This correlation was taken as evidence of a 2:2 ionogenic equilibrium involving the monomeric halide. The presence of a $TiCl_3$ deposit at the cathode seemed to support the presence of $TiCl_3^+$. This type of work is extremely delicate and demands all possible experimental care (purity, dryness, etc.) to provide meaningful results. The interpretation of such results reflecting ionogenic equilibria of various possible stoicheiometry was recently discussed in a very useful paper by Grattan and Plesch[490].

Plesch[491] summarised in his lecture at the Rouen meeting all the data concerning the selfionisation of aluminium-based Lewis acids. Most of the figures he gave have not yet been published and that table is therefore particularly important, but at the same time difficult to analyse in view of the different, and perhaps not equally stringent, experimental conditions. The most relevant deductions are qualitative and concern the fact that (i) the equilibrium concentration of ions derived from these selfionisation reactions is always very small, and (ii) the rate of ionisation (forward reaction) *can* be quite small compared with that of other important steps in the process, such as the subsequent attack of the positive moiety on the monomer and the propagation reaction. These points are particularly important to the discussion of the initiation mechanism developed below.

Eley and collaborators[492] studied the conductimetric behaviour of solutions of antimony pentachloride in methylene chloride kept in a nitrogen atmosphere. The following selfionisation equilibria were postulated:

$$2\ SbCl_5 \overset{K'}{\rightleftharpoons} SbCl_4^+SbCl_6^- \overset{K''}{\rightleftharpoons} SbCl_4^+ + SbCl_6^-$$

and the approximate values $K' = 3 \times 10^3$ M^{-1} and $K'' = 1 \times 10^{-5}$ M obtained at 25 °C. No mention was made about the rate at which the selfionisation proceeded from the moment in which the Lewis acid was dissolved, and unfortunately therefore we cannot establish any comparison of the behaviour of antimony pentachloride with that of aluminium compounds on this aspect.

Aluminium chloride is selfdissociated in nitromethane[505], but this solution and others containing SnCl$_4$, TiCl$_4$, SbCl$_5$ and TaCl$_5$ slowly generate NO$^+$ [506].

It has been shown that solutions containing two Lewis acids display a higher degree of ionic dissociation than expected from simple additivity considerations. This phenomenon has been interpreted by postulating that halide ion transfer is favoured by the presence of two acids of different strength and by the possibility of forming mixed halide anions. Thus, Marek and Chmelir[407] formulated in the case of titanium tetrachloride and aluminium bromide, the following ionogenic interaction:

$$TiCl_4 + Al_2Br_6 \rightleftharpoons TiCl_3^+ AlBr_3 Cl^- + AlBr_3.$$

If there is no doubt that many Friedel-Crafts halides can undergo selfdissociation in certain solvents (particularly halogenated hydrocarbons), and that this ionogenic reaction is not due to basic impurities in the system, the attainment of the equilibrium relating ions to undissociated molecules often requires a long time[469, 491, 493]. This observation may bear important implications on certain peculiarities of polymerisations promoted by direct initiation, as we shall discuss later.

The obvious explanation of the mechanism of initiation arising from the selfionisation of Lewis acids, i.e., in the absence of a cocatalyst, is that consisting of the attack of the positive moiety onto the olefinic double bond, as discussed in detail by Plesch [405, 491].

$$(MtX_n)_2 \text{ or } 2\,MtX_n \rightleftharpoons MtX_{n+1}^- + MtX_{n-1}^+$$

$$MtX_{n-1}^+ + \;\;\overset{\textstyle \diagdown}{\underset{\textstyle \diagup}{C}} = \overset{\textstyle \diagup}{\underset{\textstyle \diagdown}{C}}\;\; \rightarrow MtX_{n-1} - \overset{|}{\underset{|}{C}} - \overset{|}{\underset{|}{C}}{}^+ .$$

The extreme reactivity of such positively charged species towards olefins is well known[494], and even appreciable solvation of the ion should not diminish it to the point of making the attack unlikely. In fact the efficiency of this interaction is clearly shown by the occurrence of rapid initial polymerisation in systems where the concentration of metal halide cations is extremely low.

An important consequence of the above initiation scheme is the presence of a metal-to-carbon bond in some of the polymer molecules formed. Many attempts at searching for this chemical function have been carried out with the intent of proving the occurrence of direct initiation, but most of the results have thus far been rather unsatisfactory and a definite and an unequivocal conclusion as to the general presence of such bonds in the products is still lacking. This is not surprising: the extreme reactivity of these organometallic endgroups makes the problem of their identification a very arduous one, and only the application of sophisticated analytical techniques coupled with an exceedingly careful handling of the polymers can give satisfactory results. The work of Grattan and Plesch [495,496] and of Cheradame et al.[93,484] clearly illustrates the above comments.

Since a direct and crucial proof for the occurrence of an initiation reaction consisting of the addition of a Lewis acid cation to the monomer double bond is not yet available, we will expound some relevant observations which seem to us highly suggestive indications of such an interaction.

Aluminium bromide can polymerise isobutene *via* direct initiation[311,390]. The kinetic behaviour of this system was rationalised in terms of selfionisation of the catalyst and formation of an inactive complex between catalyst and monomer. The latter interaction was invoked to explain why the initial rate of polymerisation was directly proportional to the monomer concentration at low values of [M], but independent of it for higher values of [M]. It seems to us however that such complexation is not necessary, because the simple sequence:

$$2\ AlBr_3 \rightleftharpoons AlBr_4^- AlBr_2^+$$

$$AlBr_4^- AlBr_2^+ + \overset{\backslash}{\underset{/}{C}}=\overset{/}{\underset{\backslash}{C}} \rightarrow AlBr_2 - \overset{|}{\underset{|}{C}} - \overset{|}{\underset{|}{C}}{}^+ AlBr_4^-$$

can account for that change in dependence if it is assumed that at low monomer concentration the second reaction is rate determining, while at high [M], the selfionisation step becomes the slow one. This assumption is in agreement with observations concerning the slowness of some Lewis acid dissociation processes, as discussed above.

When aluminium bromide is mixed with titanium tetrachloride in n-heptane, mixed titanium halogenides are formed as suggested by the appearance of new bands in the infrared spectrum[497]. This catalytic combination is more active than $AlBr_3$ alone in the polymerisation of isobutene[407,498] and this could be interpreted as due to the observed halogen exchange. However, titanium tetrabromide is also effective in enhancing the catalytic activity of $AlBr_3$[407] and this phenomenon can only be explained by the higher effectiveness of mixed ionisation giving $TiBr_3^+ AlBr_4^-$, since only one halogen is present. With the system $TiCl_4$-$AlBr_3$ the authors[407] claimed that some titanium was found bound to the polymer, but this conclusion has been challenged[405]. Chmelir and Marek [498] also observed that $TiCl_4/AlBr_3$ gave limited yields of polyisobutene when the concentration of $AlBr_3$ was small. Non-stationary state conditions were thought to prevail in these processes and the initiation was assumed to proceed according to the reaction:

$$TiCl_3^+ AlBr_3 Cl^- + M \rightarrow TiCl_3 - M^+ AlBr_3 Cl^-.$$

The cessation of polymerisation was not clearly explained since no reason was given to justify why the continuous displacement of the selfionisation equilibrium should not occur thus providing a means of permanent initiation. Our views on this general problem will be given below.

Within the context of mixed Lewis-acid catalysts, Marek and Lopour[499] also showed that if aluminium iodide was used, its best partner in terms of initiation efficiency was titanium tetrachloride.

The work of Marek's group also encompasses an interesting study of the polymerisation of isobutene by ethylaluminium dichloride in heptane between −55 and 21 °C[408]. This system showed all the typical traits of direct initiation through selfionisation. Indeed, addition of small amounts of water produced a decrease in the rate of polymerisation with respect to the dry medium (5 x 10^{-6} M residual water claimed). The initial rate of monomer consumption was directly proportional to the monomer concentration and the square of the catalyst concentration. The course of the first part of the polymerisations (up to about 40% yield) was internally first order. As the reaction proceeded further, a sensible deceleration was noticed and indeed incomplete yields were often obtained, confirming a general feature of direct initiation. The authors discussed the mode of initiation in terms of selfionisation of the catalyst followed by the attack of the positive ion onto the isobutene molecule, but did not comment about the origin of the anomalous rate decreases and of the limited conversions.

Many of the peculiarities reported in the studies of the Czech group were found again in similar investigations involving isobutene and $AlCl_3$, $AlEtCl_2$ and $AlEt_2Cl$[409,493]. Thus, the formation of asymmetric syncatalytic aggregates resulting from the addition of $AlEt_3$ increases the degree of selfionisation and therefore the initiation efficiency. Also, the slowness of the ionogenic process, particularly marked in the case of $AlEt_2Cl$[493] which is in fact inactive because its selfionisation is too small, seems to be a common feature in this family of catalysts[491]. The latter phenomenon may explain why direct initiation in these polymerisation often produces incomplete yields despite the presence of free catalyst. Indeed, the consumption of the available initiating cations following the addition of the monomer could result in a temporary absence of the former with cessation of polymerisation due to the slowness with which selfionisation of the catalyst proceeds. Although other, perhaps more important reactions are also responsible for the limited conversions, as discussed below, this lack of fast reequilibration of the dissociation of the Lewis acid must also play a detrimental role.

Grattan and Plesch[495,496] recently restudied the polymerisation of isobutene by $AlBr_3$ and $AlCl_3$ and carried out some interesting original experiments. Very slow distillation of the monomer into the catalyst solution held at −78 °C resulted in no polymerisation and a progressive decrease in conductivity as the isobutene condensed in the mixture. This slow monomer addition obviously produced a deactivation of the catalyst, since further slow or rapid introductions of variable amounts of isobutene did not give any polymerisation. In complete contrast, when the monomer was rapidly added to the catalyst solution, polymerisation took place and the conductivity increased appreciably. The authors interpreted this paradox by postulating that isobutene forms an inactive complex with the aluminium halides and that this interaction predominates over direct initiation if the concentration of olefin is very low (slow addition). One crucial point however suggests an alternative explanation. The authors state that a negligible amount of isobutene dimers was produced in the slow distillation experiments. A close inspection of the results reveals that the yield of diisobutene was in fact far from negligible[496], and ought therefore to be taken into account as an important element of the general phenomenology. Initiation must have taken place even in the presence of very low concentrations of isobutene. We think that following that initiation, the propagation reaction is strongly disfavoured by the paucity of monomer arround the active species, and termination reactions can actively compete with it. More specifically, if unsaturated dimers are the major product of the limited propagation, allylic ions could form in a Kennedy type termination, i.e. carbenium ions too stable to attack the monomer. This deactivation process might also be reinforced by the sequestration of free aluminium halide through homoconjugation with the anions accompanying the allylic cations. The above speculations are offered not as a substitute interpretation of the mechanism, but as a complement to the complexation theory. We believe that the formation of an inactive complex rationalises adequately most of the experimental observations related to the slow addition runs, but some other concurrent interaction must be advocated to explain the presence of a non-negligible amount of dimers at the end of these runs.

Sigwalt's group carried out a very detailed study of the dimerisation of 1,1-diphenylethylene initiated by aluminium chloride in methylene chloride[417]. This process was characterised by three distinct stages. The first one was a very fast "burst" of dimerisation and was attributed to minute traces of residual moisture acting as a co-catalyst and being consumed thereby. The second one was much slower and the rate of dimerisation exhibited a first order dependence on both monomer and catalyst concentrations. However, it was found that 2.5 aluminium atoms were consumed *per* carbocation produced. This figure has recently been reassessed to a value close to 2[500]. These results could be reconciled with the type of direct acid-base initiation mechanism proposed in the previous section, i.e. a fast addition of $AlCl_3$ onto the double bond giving the zwitterion and the consumption of a second molecule of catalyst in a conjugation reaction with the negative end of the active species or in another of the interactions envisaged for $TiCl_4$. But in view of the large body of evidence related to the selfionisation of aluminium trichloride[469], initiation by $AlCl_2^+$ seems a more plausible alternative. This scheme is also compatible with the experimental observations (kinetics and stoicheiometry). The reactions following the formation of the active species in this system are essentially dimerisation and monomer transfer. The latter generates an aralkylaluminium dichloride, a potential catalyst for further, if slower, initiation. In fact, these molecules can selfionise in the same way as the well-known ethylaluminium dichloride, and the resulting cation attack the monomer. The lower acidity of these species with respect to $AlCl_3$ would make this second type of direct initiation slower. We suggest that the third very slow stage observed in this investigation arises in this way. Thus, for stage II:

selfionisation $\qquad 2\,AlCl_3 \;\rightleftharpoons\; AlCl_2^+ + AlCl_4^-,$

initiation $\qquad AlCl_2^+ + \;\overset{\backslash}{\underset{/}{C}}{=}\overset{/}{\underset{\backslash}{C}}\; \rightarrow AlCl_2-\overset{|}{\underset{|}{C}}-\overset{|}{\underset{|}{C}}{}^+,$

homoconjugation $\qquad AlCl_4^- + AlCl_3 \;\rightleftharpoons\; Al_2Cl_7^-$ (probably minimal),

dimerisation $\qquad AlCl_2-\overset{|}{\underset{|}{C}}-\overset{|}{\underset{|}{C}}{}^+ + \;\overset{\backslash}{\underset{/}{C}}{=}\overset{/}{\underset{\backslash}{C}}\; \rightarrow AlCl_2-D^+,$

transfer $\qquad AlCl_2-D^+ + \;\overset{\backslash}{\underset{/}{C}}{=}\overset{/}{\underset{\backslash}{C}}\; \rightarrow AlCl_2R + H-\overset{|}{\underset{|}{C}}-\overset{|}{\underset{|}{C}}{}^+,$

where D indicates two monomer units and R an unsaturated dimer radical. For stage III:

selfionisation $\qquad 2\,AlCl_2R \rightleftharpoons AlClR^+ + AlCl_3R^-,$

initiation $\qquad AlClR^+ + \ \diagdown\!\!\!\!C\!\!=\!\!C\!\!\!\diagup \ \rightarrow AlClR\!-\!\overset{|}{\underset{|}{C}}\!-\!\overset{|}{\underset{|}{C}}{}^+,$ etc.

The original authors reported that transfer reactions proceeded at a low rate. Their chemical nature was not studied and we have written one plausible mechanism involving the dimer cation and the monomer. The nature of R is therefore not certain. We feel however that the mechanism proposed for stage III explains well the observation that if the system is left for several days (very slow dimerisation stage), the carbenium ion yield increases from 50% to close to 75% of the $AlCl_3$ used. Indeed while half of the $AlCl_3$ is used in stage II if homoconjugation is negligible, stage III as we have envisaged it would eventually use up half of the $AlCl_2^+$ formed previously, i.e. 25% more of the initial $AlCl_3$.

Interestingly, it was reported in the same publication that a very slow dimerisation of 1,1-diphenylethylene[417] was also observed with titanium tetrachloride following the major process which had been previously studied[416] and which we discussed in the preceding section. It seems therefore likely that the $TiCl_3R$ species formed in that first stage can reinitiate the dimerisation, but with a much reduced efficiency. Recent unpublished work[484] indeed showed that CH_3TiCl_3 is able to initiate the polymerisation of indene. These systems would certainly deserve more attention since they could provide valuable information about the mechanism of direct initiation.

On the basis of existing evidence it can be concluded that numerous valid arguments are available supporting the theory of direct initiation by ionic species formed from the selfdissociation of Lewis acids. Direct proof of the occurrence of the crucial addition reaction is however extremely difficult to obtain (e.g., metal-carbon bonds as end groups in the products), and more work is obviously needed to establish this point irrefutably for many systems. Plesch has discussed this topic in depth and given a critical appraisal of arguments and facts[405,491]. In the specific context of initation by aluminium halides, Plesch[405] suggested two alternative routes to that involving the addition of AlX_2^+ to the double bond, viz.:

$$AlX_2^+ + \ \diagdown\!\!\!\!C\!\!=\!\!C\!\!\!\overset{\displaystyle CH_3}{\diagup} \ \rightarrow AlX_2H + \ -\overset{|}{\underset{|}{C}}\!\cdots\!\overset{|}{\underset{|}{C}}\!\cdots\!CH_2$$
$$\underset{+}{}$$

$$AlX_2^+ + \ \diagdown\!\!\!\!C\!\!=\!\!C\!\!\!\diagup \ \rightarrow AlX_2^\cdot + {}^\cdot\overset{|}{\underset{|}{C}}\!-\!\overset{|}{\underset{|}{C}}{}^+.$$

The first reaction would produce allylic cations whose initiating capability must be very limited, as shown by Kennedy[404]. The second one is an electron transfer which, as we have already pointed out, is possible with antimony (V) halides under special conditions [91,420], but unlikely with elements which do not possess lower oxidation states.

5. The Possible Role of Halonium Ions

Dialkylhalonium ions are known to be effective alkylating agents for aromatic compounds and their reactions with olefins have also been studied[62,63]. Dimethylbromonium hexafluoroantimonate has been shown to initiate the cationic polymerisation of butene-1 and isobutene in liquid SO_2:

$$CH_2=C(CH_3)_2 + (CH_3)_2Br^+ \rightarrow C_2H_5(CH_3)_2C^+ + CH_3Br.$$

It is therefore important to examine the possible involvement of these species in systems containing Lewis acids and halogenated solvents. The evidence available has been discussed by Olah[63,501] and we will only touch upon the most salient points. The very strong acid SbF_5 is capable of interacting with alkyl halides to give the corresponding dialkylhalonium ions, which were identified spectroscopically. The formation of dimethylbromonium ions has been inferred from kinetic evidence in the system $CH_3Br–GaBr_3$[63] and followed by NMR spectroscopy in the case of $CH_3Br–AlBr_3$[63,501]. A similar ionisation was postulated when CH_3I was mixed with AlI_3[63,501]. It is important that the chemical shifts for these halonium ions differed appreciably from those corresponding to simple donor-acceptor complexes. Thus, under certain conditions, the addition of a metal halide to an alkyl halide can result in the appearence of detectable amounts of halonium ions according to the following equilibrium:

$$2\,RX + MtX_n \rightleftharpoons R_2X^+ + MtX_{n+1}^-.$$

Among the mechanisms discussed to explain direct initiation, we have not taken into account the above equilibrium and thus the possibility of a primary interaction consisting of the attack of the halonium ion on the monomer double bond. Such an alternative mode of initiation seems most unlikely, simply because the concentration of halonium ions, in the specific conditions under which pertinent cationic polymerisations were studied, must have been negligible. Grattan and Plesch[469] have discussed this point in the context of their work on aluminium halides in alkyl halides; they showed that their conductimetric correlations were incompatible with the formation of halonium ions. Most other studies on direct initiation involved the use of methylene chloride as solvent. From a chemical standpoint the formation of the ion $(CH_2Cl)_2X^+$ seems extremely improbable. Moreover, initiation by such a cation would produce chloromethylated polymers or dimers, but no report to this effect has ever been made in investigations even where the structure of the products was carefully analysed. This type of end group would have been easily detected, particularly when working with non-polymerisable olefins.

While initiation by halonium ions is certainly an interesting and little studied field in cationic polymerisation, it can be safely concluded that the mechanism of direct initiation in the cationic polymerisations discussed in this chapter cannot be ascribed to the formation of such species.

6. Conclusions

The two mechanisms for direct initiation of cationic polymerisation by Lewis acids which we have proposed owe their main difference to the very first interaction, i.e. whether the attack on the double bond is carried out by the neutral metal halide or by the cation resulting from its selfdissociation. Indeed, the reactions following the generation of active species which must be considered as side events detrimental to propagation, are essentially the same in both schemes. Specifically, both mechanisms give rise to products containing carbon-metal bonds and one end of the chain and a halogen atom or a double bond at the other. The latter structure can be associated in both instances with production of hydrogen halide which can produce some further polymerisation by cocatalysis. Conjugation of the acid with its (or another) anion can play a detrimental role upon the yield in both mechanisms.

Consequently, it is extremely hard to establish criteria capable of discriminating between the two mechanisms. Only accurate determinations of the kinetics and the stoicheiometry of the reaction leading to carbenium ions (the active species) coupled with a precise knowledge of the state of the Lewis acid in the polymerisation medium (degree of association and selfionisation) can provide the necessary indications as to which type of primary attack prevails in a given system. Moreover this type of investigation is much more informative if conducted with model olefins, as indicated by the results already obtained with 1,1-diphenylethylene[416,417], 1,1-diphenylpropene[418] and 3-substituted indenes[93]. Finally, the occurrence and extent of homo- and heteroconjugation with various Lewis acids ought to be studied to establish the real importance of these interactions and their effective role in the events following the primary attack.

Given the present lack of precise information about the quantitative aspects, the division we established between the behaviour of $TiCl_4$ (acid-base initiation through a Hunter-Yohé mechanism) and that of aluminium-based Lewis acids (initiation following selfionisation), is not entirely corroborated by indisputable experimental evidence. Certain facts led us to differentiate between the two situations, although the main reason for this separation was to present the two alternatives in a clear, if schematic, fashion. The complex phenomenology of these systems and the wide spectrum of primary and secondary interactions possible make that division rather premature, but we hope that our speculations will stimulate further work in the field. It is conceivable that future studies might indicate that both mechanisms can in fact operate concurrently with a relative specific contribution depending on the extent of selfionisation of the catalyst.

Concerning the possible causes of the premature cessation of polymerisation in most systems characterised by direct initiation, we believe that the accumulation of anions derived from the Lewis acid is the most important single factor responsible. We have already discussed the possible reactions giving rise to an increasing concentration of these anions in the case of direct initiation by the neutral Lewis acid. To these must be added the formation of carbenium ions which cannot propagate either because they are sterically hindered, as in the case of isobutene unsaturated dimers and trimers[502], or because they are too stable to attack the monomer, as in the case of termination by formation of allylic cations. The accumulation of these species is inevitably accompanied by an equivalent rise in anion concentration. Whatever the origin(s) of this rise, a higher anion concentration can determine various detrimental effects on the survival of active species:

sequestration of free Lewis acid by conjugation, shifting of the selfionisation equilibrium towards undissociated Lewis acid, and/or reduction of the amount of free propagating ions due to the common ion effect on the free ions-ion pairs equilibrium. The last effect would be inhibiting on the polymerisation if it is assumed that ion pairs are incapable of propagating or that, as proposed by Kucera[343,344], ion pairs can be solvated by the monomer and thus made inactive. Since no information is available at present on the relative activity of free ions and ion pairs in these systems, the common anion effect cannot be assessed. As to Kucera's mechanism, it must be regarded as highly speculative, although non-inhibiting monomer solvation of propagating species has been considered fairly common in ionic polymerisations[503], and more specifically in the case of the dimerisation of 1,1-diphenylethylene in methylene chloride[148].

There are no compelling reasons against a third mechanism of direct initiation, which we wish to suggest for the sake of completeness. It is in fact conceivable that a concerted, Ad_E3-type attack could take place between the Lewis acid and the olefin. This mechanism would resemble closely the one we propose in the next section to explain the way cocatalysis operates. Thus:

complexation, $\text{MtX}_n + \overset{\backslash}{\underset{/}{C}}=\overset{/}{\underset{\backslash}{C}} \rightleftharpoons \text{MtX}_n \leftarrow \overset{\backslash/}{\underset{/\backslash}{\overset{C}{\underset{C}{\|}}}}$

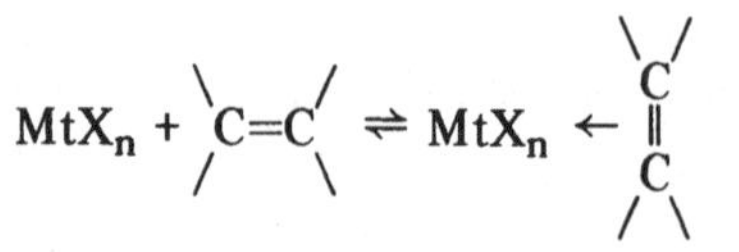

selfionisation, $2\,\text{MtX}_n \rightleftharpoons \text{MtX}^+_{n-1}\text{MtX}^-_{n+1}$

concerted attack, $\text{MtX}_n \leftarrow \overset{\backslash/}{\underset{/\backslash}{\overset{C}{\underset{C}{\|}}}} + \text{MtX}^+_{n-1}\text{MtX}^-_{n+1} \rightarrow \overset{\text{MtX}_n}{\underset{\text{MtX}^-_{n+1}\ \ \text{MtX}^+_{n-1}}{C{=\!=\!=\!=}C}} \rightarrow$

$\rightarrow \text{MtX}_{n-1}{-}\overset{|}{\underset{|}{C}}{-}\overset{+}{\underset{|}{C}}\ \text{Mt}_2\text{X}^-_{2n+1}.$

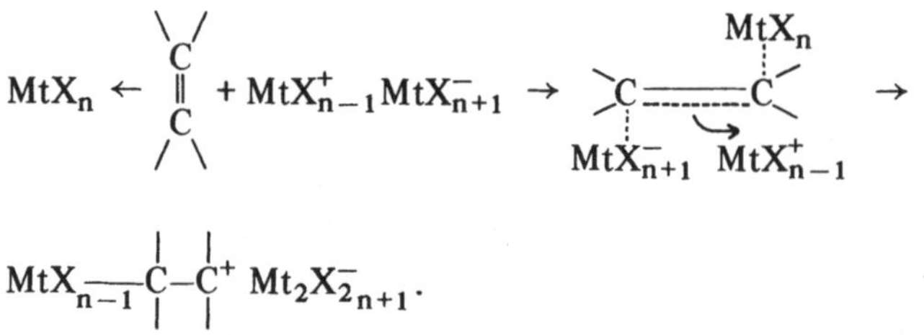

If the rate-determining step of this process is the formation of the carbenium ion, a second or third order dependence on Lewis acid concentration would be expected depending on whether the species $\text{MtX}^+_{n-1}\text{MtX}^-_{n+1}$ is in the form of ion pairs or free ions. This third mechanism is not inconsistent with the general phenomenology of direct initiation, but no specific evidence in its favour is available.

C. Cocatalysis

1. Introduction

This section is devoted to the analysis of two main chemical routes which lead to the formation of carbenium ions when an olefin is mixed with a Lewis acid in the presence of a third reactant capable of providing, through its critical intervention at some stage of the process, a substantial role in the activation of the mechanism. The term *cocatalyst* for this third component remains in our view a most appropriate one. In fact, we have seen throughout Section B of this chapter that Lewis acids can interact with olefins to give the corresponding carbenium ions (propagating species) *without* the assistance of other substances. In these direct initiation processes the catalyst may not display a vigorous or durable activity, but is nevertheless capable of performing its essential task. The addition of a third substance must therefore be viewed as a means to improve both on the rate and the continuity of the initiation reaction, but not as it was believed formerly, to determine its occurrence. Of course, the presence of a cocatalyst alters substantially the mechanism through which the active species are generated, but nonetheless in all the systems which will be discussed below, the Lewis acid plays the dominant function of catalyst, without which initiation is impossible.

The two fundamental reaction pathways available for the preparation of carbenium ions, which have already been examined in different contexts above, apply equally well to cocatalysis with Lewis acids. The first is the addition of an electrophile to the double bond of the monomer: here the electrophile is a proton arising from a cocatalyst whose acidity is enhanced by the Lewis acid. Thus, the first three parts of this section are devoted respectively to cocatalysis by hydrogen halides, water and weak organic acids. We intend to show that the mechanism of these cocatalytic interactions is generally consistent with the Ad_E3 scheme of electrophilic addition to olefins (see Sect. III-B). But before reaching this conclusion it is necessary to inspect the peculiarities of specific interactions involving two of the three components at a time, to assess their relative importance and therefore their possible role prior to the ionogenic reaction. Since we do not believe in the likelyhood of a termolecular initiation step with the simultaneous participation of olefin, Lewis acid and cocatalyst, we shall analyse the three alternative bimolecular schemes:

$$\begin{cases} MtX_n + BH \rightleftharpoons MtX_nB^- H^+ \\[2mm] MtX_nB^- H^+ + \underset{/}{\overset{\backslash}{C}}=\underset{\backslash}{\overset{/}{C}} \rightarrow H{-}\overset{|}{\underset{|}{C}}{-}\overset{|}{\underset{|}{C}}{}^+ MtX_nB^-\ ; \end{cases}$$

$$\begin{cases} MtX_n + \underset{/}{\overset{\backslash}{C}}=\underset{\backslash}{\overset{/}{C}} \rightleftharpoons \underset{\underset{/\backslash}{C}}{\overset{\overset{\backslash/}{C}}{\|}} \rightarrow MtX_n \\[4mm] \underset{\underset{/\backslash}{C}}{\overset{\overset{\backslash/}{C}}{\|}} \rightarrow MtX_n + BH \rightarrow H{-}\overset{|}{\underset{|}{C}}{-}\overset{|}{\underset{|}{C}}{}^+ MtX_nB^-\ ; \end{cases}$$

$$\left\{ \begin{array}{l} BH + \ \overset{\diagdown}{\underset{\diagup}{C}} = \overset{\diagup}{\underset{\diagdown}{C}} \ \rightleftharpoons \ \overset{\diagdown / C \diagdown}{\underset{\diagup \backslash}{\underset{C}{\parallel}}} \ \rightarrow HB \quad (\text{or } H - \overset{|}{\underset{|}{C}} - \overset{|}{\underset{|}{C}} - B) \\[4ex] \overset{\diagdown / C \diagdown}{\underset{\diagup \backslash}{\underset{C}{\parallel}}} \ \rightarrow HB + MtX_n \ \rightarrow \ H - \overset{|}{\underset{|}{C}} - \overset{|}{\underset{|}{C}}{}^+ \ MtX_n B^- ; \end{array} \right.$$

where BH symbolises the protogenic cocatalyst, e.g. HCl, H_2O or PhOH. The possible interactions between Lewis acids and olefins were discussed in the previous section devoted to direct initiation. Our conclusions were that, apart from addition reactions on the double bond by both the undissociated Lewis acid and/or the cation resulting from its selfionisation, complex formation is quite likely although the concentration and strength of these complexes varies appreciably from case to case and can often be very low. The interactions between Brønsted acids and olefins have also been already discussed (see Chap. III) and therefore only some reminders and complementary observations will be given here. Our main attention will be focussed on the types of interaction arising from the encounter of a Lewis acid with a Brønsted one (the latter taken in its broader sense). The proposal of a plausible mechanism describing the overall initiation process with these protogenic cocatalysts will stem from these basic considerations and from an inspection of the extensive experimental evidence concerning polymerisation systems.

The second reaction pathway leading to carbenium ions involves the heterolytic cleavage of a carbon-halogen bond. In the present context a Lewis acid can induce such cleavage on certain alkyl and aryl halides, which are therefore the third component in a polymerisation system. Accordingly, the cocatalytic action of these halides consists in providing the source of primary carbenium ions which add onto the monomer double bond in the second stage of the initiation mechanism. The fourth part of this section deals with cocatalysis by organic halides and discusses the likelihood of such phenomenon and of the frequently invoked initiation mechanism:

$$MtX_n + RX \ (\text{or } ArX) \ \rightleftharpoons \ R^+ \ MtX_{n+1}^- \quad (\text{or } Ar^+ \ MtX_{n+1}^-)$$

$$R^+ \ (Ar^+) \ MtX_{n+1}^- + \ \overset{\diagdown}{\underset{\diagup}{C}} = \overset{\diagup}{\underset{\diagdown}{C}} \ \rightarrow \ R - \overset{|}{\underset{|}{C}} - \overset{|}{\underset{|}{C}}{}^+ \ MtX_{n+1}^- .$$
$$\qquad\qquad\qquad\qquad\qquad\qquad\qquad (Ar)$$

In all the schemes written above the final products (active species) are assumed to be carbenium ions, and although they have been shown as ion pairs, it is implicit that their other forms, free ions, solvent-separated ion pairs, etc., are also present. Details about the precise nature of these species will only be discussed whenever some evidence is available (mostly from kinetic data) to suggest the predominance of a specific form. Due to the high reactivity of carbenium moieties, we think that in general all available states are in fact reactive, but of course the propagation rate constant may vary considerably from one to another (see Sect. II-J and other quantitative aspects in Chaps. III and V).

An enormous amount of literature describes systems composed of Lewis acids and cocatalysts. Most of it however is not concerned with the mechanism of initiation, but simply describes the convenient use of a given active pair. Those papers will not be surreyed here, since as already mentioned only papers dealing with some fundamental aspect of this problem are of interest to us in the context of this review.

2. Cocatalysis by Hydrogen Halides

a) Basic Interactions

Hydrogen halides add onto olefins through electrophilic mechanisms which have been briefly discussed in the preceding chapter. Some specific examples of acid-olefin complexation have been reported, but the intimate nature of these interactions is not always well understood. The addition of a Lewis acid to such binary combinations brings about a marked increase in the reactivity of the system either in terms of a higher rate of addition or by promoting polymerisation. As we pointed out in Sect. III-E-1, only the most basic alkenyl monomers can be polymerised by hydrogen halides alone. The presence of a Lewis acid however, modifies that situation, since the cocatalytic pair metal halide-hydrogen halide can be an excellent initiator of the cationic polymerisation of a wide variety of monomers. It is important therefore to establish which reaction pathway predominates among the three schemes considered above. Since water also plays a very effective cocatalytic role in these systems, we will concentrate on publications in which care was taken to minimise its presence.

Homo- and heteroconjugation of the hydrogen halides with halide ions is well established[210-212] in several solvents commonly used in cationic polymerisation. What is less clear is the association state of the pure acids in those sovents and the extent of their ionisation. NMR studies indicated that the proton chemical shift of HCl is practically the same in the pure acid and in methylene chloride[504], suggesting that hydrogen-bond associations exist in both media. Addition of traces of water induces specific interactions, namely the formation of a complex whose acidic proton is strongly deshielded to lower fields relative to the proton resonance of the dry acid peak in solution[211]. Electrochemical studies of HCl in methylene chloride showed that freshly prepared solutions in the carefully dried solvent are very poor conductors, but that as expected, the introduction of moisture induces a considerable increase in conductivity[507]. Acetonitrile is protonated by hydrogen halides[211] and a similar interaction is observed between HCl and nitromethane[165]. Halide exchange reactions are known to occur between halides and methylene chloride[211]. Indeed, it is well known that refluxing methylene chloride in the presence of potassium iodide gives methylene iodide in high yields. These exchange reactions are however very slow at low temperature and the concentration of the intermediate must therefore be extremely low and possess insufficient charge separation to induce initiation of cationic polymerisation.

The solubility of hydrogen chloride in a variety of liquids has been determined[508]. Interestingly, it was found that the system $HCl-CH_2Cl_2$ followed approximately the ideal behaviour:

$$P_{HCl} = P^0_{HCl} \cdot x_{HCl}$$

where P_{HCl} is the hydrogen chloride pressure above the solution, P^0_{HCl} is the acid pressure over its own pure liquid at the same temperature, and x_{HCl} is the acid molar fraction in the solution. A slight deviation on the negative side of Raoult's law was observed in carbon tetrachloride, and slight positive deviations in titanium tetrachloride and stannic chloride. As expected, very large positive deviations were obtained in water, carboxylic acids, esters, etc. Positive deviations can be attributed to interactions with the solvent stronger than those existing in the pure acid. When Raoult's law is followed, as in the case of methylene chloride, the interactions between solvents and solute are roughly of the same magnitude as those among pure solute molecules. It is known that in the pure liquid, hydrogen chloride molecules are largely associated through hydrogen bonding. According to the solubility results[508], one can conclude that at least at high concentrations, this acid is also associated in methylene chloride, a conclusion corroborated by the NMR studies quoted above[504]. There is however no evidence indicating the extent of this association or whether the acid can be considered to exist in a dimeric form. The solubility parameter in various media relevant to cationic polymerisation are given below for 1 atmosphere at $-30\ °C$[508]: x_{HCl} in $SnCl_4 = 0.115$ (mole/mole), in $TiCl_4 = 0.105$, in $CH_2Cl_2 = 0.100$, in $CCl_4 = 0.044$, and in $BCl_3 = 0.027$. The similarity of values in CH_2Cl_2, $SnCl_4$ and $TiCl_4$ would tend to indicate that the acidity of HCl is not enhanced by its interaction with these two Lewis acids. Unfortunately, no details were given in that study of the purification techniques used for the metal halides and one is left with some doubts about the presence of residual moisture, which could have affected the results, given the very high affinity of water for hydrogen chloride.

Concerning the interactions between olefins and hydrogen halides, we must first refer to the reports dealing with complex formation. Among these, we will recall the classical work of Cook et al.[174] on the freezing-point diagrams of mixtures of HCl and simple olefins such as ethene, propene and cis-butene. Two molecular complexes were detected for each system, containing one and two acid molecules, respectively. 2-Methylbutene-2 gave the addition product instantaneously, even at temperatures lower than $-60\ °C$. An infrared study of solid matrices of ethene doped with hydrogen chloride at $20°K$ showed that strong interactions arise between the two species, suggesting the formation of a complex[509]. Stable complexes between hydrogen halides and ethene, propene, isobutene and trifluoropropene were also identified at low temperature in a recent investigation based on positron formation[510]. All these observations confirm older work and the generally accepted hypothesis that the addition of hydrogen halides to the olefinic double bond is preceded by complex formation[165–172]. Similar studies on the interaction of aromatic hydrocarbons and HCl and HBr[511] proved the existence of fairly weak π-complexes, whose stability increases with the basicity of the hydrocarbon.

The addition of hydrogen halides (particularly HCl and HBr) to olefins has been thoroughly studied in a variety of situations, as discussed in the preceding chapter. The work of Pocker's and Fahey's schools[165–171,224,225] and other investigations[172] showed that the reaction is particularly rapid and effective with such basic olefins as styrene, acenaphthylene and indene, but of course less so in the case of simple alkenes, some of which are unreactive towards HCl at -78 °C in toluene[173], unless small amounts of the Lewis acid GaCl$_3$ are added to the systems. As to the mechanism of the addition, it seems well established that the concerted Ad$_E$3 pathway can best account for the kinetic and structural rseults. The details of this mechanism have already been dealt with (see Sect. III-B). Suffice it here to underline that in this type of process two or three acid molecules are involved in the activated complex and that conjugated anions HX$_2^-$ can play a key role in assisting the reaction. The relevance of these details on the intimate chemistry of cocatalytic initiation will be expounded in each specific section below.

In a very important series of experiments, Pocker investigated both the addition of hydrogen chloride to isobutene and its elimination from t-butyl chloride[512]. He noted in particular that at 75 °C the elimination reaction only occurred in the absence of free acid, and was unimolecular in the presence of pyridine or Et$_4$NCl. While the role of pyridine could easily be rationalised, it was found that the effect of tetraethylammonium chloride could not be ascribed to changes in the ionic strength of the medium, but to homoconjugation to the liberated acid to give HCl$_2^-$ [513]. Pocker also demonstrated that t-butyl chloride undergoes chloride ion exchange with Et$_4$NCl in nitromethane. A similar exchange between t-butyl chloride and HCl took place, and its rate decreased when Et$_4$NCl was added. The following reaction scheme was given to explain all these observations:

$$t\text{-BuCl} + \text{HCl} \underset{2}{\overset{1}{\rightleftharpoons}} t\text{-Bu}^+\text{HCl}_2^- \underset{4}{\overset{3}{\rightleftharpoons}} (\text{CH}_3)_2\text{C=CH}_2 + 2\,\text{HCl},$$

where the rates follow the order $4 > 3 > 2 > 1$. This accounts for the often quoted fact that neither HCl nor t-butyl chloride are cocatalysts for the polymerisation of isobutene by Lewis acids. Indeed, the rate of acid addition is so high that there is insufficient time for the transient ionic species to attack the monomer.

Pocker also reported that hydrogen chloride can induce the ionisation of aralkyl chlorides in media where the hydrogen dichloride anion is a relatively stable entity, i.e., nitromethane[223], nitrobenzene[514] and benzene[515]. Such situations would therefore be expected to give olefin polymerisation in the presence of Lewis acids, owing to the higher lifetime of the carbocations formed as intermediate, but as we shall see this is not always observed experimentally.

b) Cocatalysis with Stannic Chloride

The studies pertinent to this section involve only hydrogen chloride as cocatalyst, i.e., SnCl$_4$–HCl–monomer. Given a system made up of these three reactive components, we wish to propose a reasonable initiation mechanism and compare it with the available experimental evidence. If we take the Ad$_E$3 scheme as a guide, we can imagine that stannic chloride interacts first with the olefin and the resulting complex is then modified by hydrogen chloride:

$$\underset{/}{\overset{\backslash}{C}}{=}\underset{\backslash}{\overset{/}{C}} + SnCl_4 \;\rightleftharpoons\; \underset{/\;\backslash}{\overset{\backslash\;/}{\underset{C}{\overset{C}{\|}}}} \;\rightarrow\; SnCl_4$$

$$\underset{/\;\backslash}{\overset{\backslash\;/}{\underset{C}{\overset{C}{\|}}}} \;\rightarrow\; SnCl_4 + HCl \;\rightleftharpoons\; \overset{\overset{\textstyle SnCl_4}{\cdots}}{>}\underset{\underset{\textstyle Cl-H}{\searrow}}{C{\cdots}C}{<} \qquad \rightarrow \text{ ionisation.}$$

This sequence implies the release of a chloride ion in the ionisation step, and this elimination usually requires the assistance of a molecule of HCl in the solvents used in cationic polymerisation. Also, stannic chloride is not consumed in the overall process. The interactions leading to the active species can therefore be written as:

$$\underset{\underset{\textstyle Cl-H}{\cdots}}{\overset{\overset{\textstyle SnCl_4}{\cdots}}{>\overset{\delta^+}{C}{\cdots}C<}} \;+\; HCl \;\rightleftharpoons\; \underset{\underset{\underset{\textstyle Cl-H}{\cdots}}{\textstyle Cl{-}H}}{\overset{\overset{\textstyle SnCl_4}{\cdots}}{>\overset{\delta^+}{C}{\cdots}C<}} \;\rightarrow\; HCl_2^- \; {}^+\overset{|\;\;|}{\underset{\underset{\textstyle H}{|\;\;|}}{C{-}C{-}}} + SnCl_4.$$

It can be seen that at equilibrium, the concentration of carbenium ions is independent of the stannic chloride concentration in this scheme. However, if chloride ion exchange can occur between the Lewis acid and HCl_2^-, the sequence of equilibria is shifted to the right and a dependence is established between carbenium ion concentration and amount of $SnCl_4$ used:

$$SnCl_4 + HCl_2^- \;\rightleftharpoons\; SnCl_5^- + HCl.$$

The existence of the pentachlorostannate anion in well known, but to the best of our knowledge this species has never been characterised in a polymerising system. If we consider its presence in this context as plausible, the concentration of carbenium ions at equilibrium will be determined by the overall relationship:

$$K = \frac{[C^+]\,[SnCl_5^-]}{\left[\underset{/}{\overset{\backslash}{C}}{=}\underset{\backslash}{\overset{/}{C}}\right][SnCl_4]\,[HCl]} \quad;$$

and if the concentration of HCl_2^- is low, i.e., if the last equilibrium above is strongly shifted to the right, $[C^+] \simeq [SnCl_5^-]$. In the above treatment the equilibrium between free ions and ion pairs has not be taken into account. In dynamic terms, assuming that the rate-determining step is the protonation reaction and that all available carbenium ions can propagate, the rate of monomer consumption will be given by:

$$R_p = k\,[C^+]\,[M] = k'\,[SnCl_4]\,[HCl]^2\,[M]^2$$

if all ions (free or paired) have the same propagating capability, and:

$$R_p = k'' \, [SnCl_4]^{0.5} \, [HCl] \, [M]^{1.5}$$

if propagation takes place predominantly on free ions.

An alternative mechanism can also be proposed where the initial interaction occurs between the monomer double bond and hydrogen chloride, while the role of stannic chloride consists in providing the driving force for the release of the anion, Cl^- or HCl_2^- :

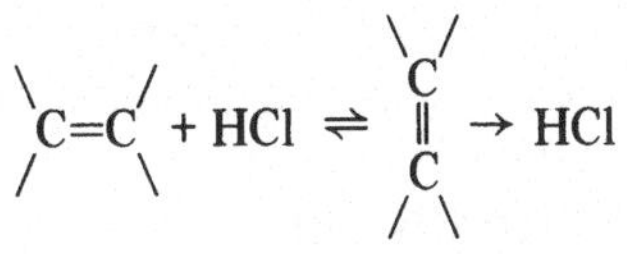

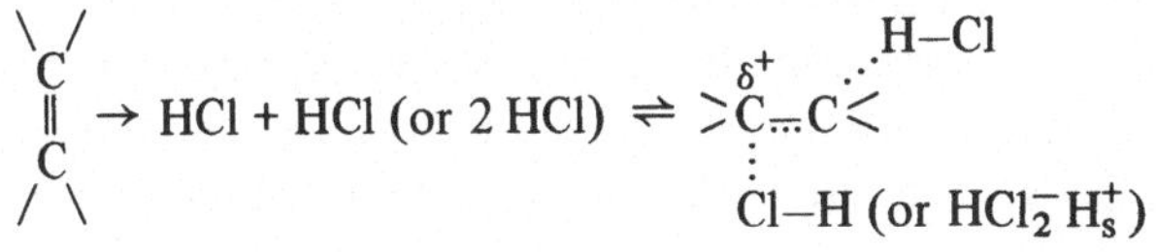

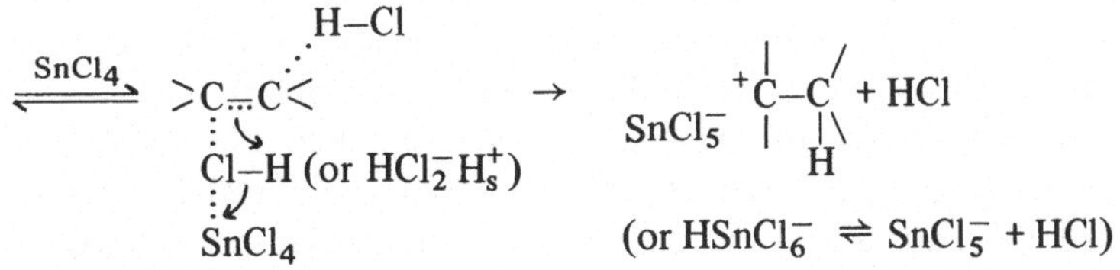

If, as in the preceding scheme, the hydrogen dichloride anion is the relevant species, the equilibrium concentration of (paired) carbenium ions formed will be governed by the same equation, and the rate of polymerisation will also follow the same kinetic laws, assuming that protonation is rate determining.

Finally, we can envisage that HCl_2^- is never formed, i.e. stannic chloride assists in all mechanisms the release of the chloride anion in the step leading to the formation of the active cationic species. In this event, only the first step is different between the two alternatives (complexation with the double bond by either acid) and of course the acid reformed at the end will be the one responsible for the initial interaction. The equilibrium concentration of carbenium ions will be given by the expressions:

$$[C^+] = K' \, [M] \, [SnCl_4] \, [HCl] \quad \text{for ion pairs and}$$

$$[C^+] = K'' \, [M]^{0.5} \, [SnCl_4]^{0.5} \, [HCl]^{0.5} \quad \text{for free ions,}$$

irrespective of the nature of the initial complexation. The rate of polymerisation will be:

$$R_p = k \, [SnCl_4] \, [HCl]^2 \, [M]^2 \quad \text{for ion pairs and}$$

$$R_p = k' \, [SnCl_4]^{0.5} \, [HCl] \, [M]^{1.5} \quad \text{for free ions,}$$

if the initial complexation is with HCl; and:

$$R_p = k \, [SnCl_4]^2 \, [HCl] \, [M]^2 \quad \text{for ion pairs and}$$

$$R_p = k' \, [SnCl_4] \, [HCl]^{0.5} \, [M]^{1.5} \quad \text{for free ions,}$$

if the initial complexation is with $SnCl_4$.

The reasons for our proposal of an Ad_E3-type mechanism of cocatalytic initiation instead of a simpler one consisting of HCl-olefin complexation followed by the intervention of $SnCl_4$ in a transition state determining the release of Cl^- to give $SnCl_5^-$ and the carbenium ion, can be synthesised as follows: (i) the Ad_E3 mechanism operates in many systems involving olefins and hydrogen halides and there is no indication that the combination Lewis acid-Brønsted acid should behave differently; (ii) cocatalysis by water and other oxygen-containing proton sources can also be adequately rationalised by an Ad_E3-type mechanism; (iii) the interaction between olefins and Lewis acids are generally strong, to the point that, as we have seen, direct initiation often results from them; (iv) the experimental results obtained in several studies of reliable quality during the last thirty years can all be adequately reinterpreted by our generalised mechanism, as indicated by the following analysis.

The fact that hydrogen chloride alone is able to induce the polymerisation of styrene in polar solvents[218] demonstrates that carbenium ions are produced in the addition reaction; but the low yields of polymer and the poor reproducibility of these systems also show the very short lifetime of these intermediates and their sensitivity to the specific conditions under which they are generated, in the absence of stabilising species. Traces of stannic chloride considerably accelerate those polymerisations and make them more manageable, a phenomenon which can be interpreted in terms of stabilisation of the active species by the Lewis acid. Alternative explanations can however be offered. Thus, one could think that the interactions between $SnCl_4$ and HCl produce a more active catalyst, through the formation of ionic electrophilic entities. But it is also conceivable that the formation of a complex styrene—$SnCl_4$ capable of reacting further with HCl through the Ad_E3 mechanism is the main reason for the enhanced reactivity of the cocatalytic mixture. The same comments can be made about the behaviour of the system styrene—$SnBr_4$—HBr[516−518].

In a classical series of papers, Williams and collaborators reported many observations about the cocatalytic role of HCl (and HBr) in the polymerisation for styrene catalysed by $SnCl_4$ (and $SnBr_4$)[516−519]. These studies were carried out in a non-polar solvent, CCl_4, and termination reactions giving inactive chloride oligomers dominated over propagation. At 25 °C and with HCl concentrations as high as the monomer ones, the major product of the reaction was 1-phenylethyl chloride, its rate of formation being proportional to the $SnCl_4$ and the HCl concentrations. Hydrogen chloride alone was found not to add to styrene in these conditions. On the other hand, low concentrations of HCl inhibited the polymerisation of styrene "*temporarily, but completely*"[516]. Finally, the system styrene—$SnCl_4$-1-phenylethyl chloride in the absence of HCl gave unsaturated dimers and some oligomers. Some residual moisture must have remained in the above solutions, since extreme care was not taken to exclude it, and a discussion of these old results must necessarily take this into account. Thus, the reported polymerisation in the absence of HCl was most probably due to water cocatalysis, while the inhibiting effect of small quantities of HCl can be interpreted as reflecting the very high reactivity of the acid towards the styrene-SnCl$_4$ π-complex, to give 1-phenylethyl chloride. This addition reaction must be very fast, i.e., the collapse of the carbenium ion with the chloride ion following the attack of HCl on the complex, does not leave enough time for any appreciable propagation on the too-short lived intermediates. Therefore, in the absence of HCl, water is a cocatalyst for polymerisation, but in its presence the dominant event is formation of 1-phenylethyl chloride, an addition process accelerated by the Lewis acid. The kinetic results must be viewed as reflecting the overall addition of HCl, including Cl$^-$ incorporation by the carbenium ion, and not simply an "initiation process" dealing exclusively with the formation of active species. We can conclude that termination (ion pair collapse) is too fast in this system to allow a proper study of the interactions associated with the mechanism of cocatalytic initiation, although there is enough evidence to suggest the presence of a strong complex between styrene and SnCl$_4$. The system styrene–SnBr$_4$–HBr was found to respond more vigorously both to addition and to polymerisation, undoubtedly because of the higher acidity of both catalyst and cocatalyst, but no definite mechanistic conclusions can be drawn from this shorter study. The above work would be worth restudying to-day with more sophisticated experimental techniques.

Another classical study deserving discussion is the investigation of the dimerisation of 1,1-diphenylethylene by SnCl$_4$–HCl in benzene at 40 °C, carried out by Evans et al. [520,521]. The initial rate of dimerisation was found to be proportional to the first power of both catalyst and cocatalyst and to the second power of monomer concentration. The kinetics of cyclisation of the linear unsaturated dimer of 1,1-diphenylethylene were also studied under the same conditions and the initial rate of this process was found to be directly proportional to the concentration of each of the three components, as shown in Fig. 7. We believe that in solvent benzene the active species formed, i.e. the monomer and linear dimer carbenium ions, existed essentially as ion pairs, and that the counterion was most probably SnCl$_5^-$ resulting from the interaction of HCl$_2^-$ with SnCl$_4$ (see above). Since both the dimerisation and the cyclisation reactions were slow, equilibrium was reached in the various interactions leading to the generation of the carbenium ions, the concentration of which can therefore be expressed by the relationship

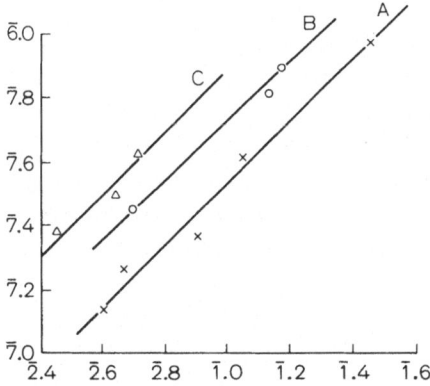

Fig. 7. Dependence of the rate of formation of the cyclic dimer of 1,1-diphenylethylene on the concentrations of linear dimer LD (A, [HCl] = 0.1028 M, [SnCl$_4$] = 5.146 x 10^{-2} M), HCl (B, [LD] = 0.1774 M, [SnCl$_4$] = 3.565 x 10^{-2} M) and SnCl$_4$ (C, [LD] = 0.107 M, [HCl] = 0.1039 M). Solvent: benzene; T = 54.9 °C. *Vertical axis:* log$_{10}$ (rate of cyclisation); *horizontal axis:* log$_{10}$ (concentration). (Taken from Ref. [521])

$$[C^+] = K \left[\begin{array}{cc} | & | \\ C=C \\ | & | \end{array} \right] [SnCl_4] \, [HCl]$$

for a generalised Ad_E3 mechanism. It follows that the rate of dimerisation will be proportional to the above expression multiplied by the monomer concentration, while the rate of cyclisation will just be proportional to the concentration of active species. The experimental results of Evans' group show precisely these kinetic laws. An alternative way of looking at these processes consists in assuming that they proceeded through the addition of HCl to the double bond of both monomer and dimer, followed by ionisation of the tertiary chloride by $SnCl_4$ (chloride ion abstraction). Two specific kinetic situations can arise from such mechanism: if HCl addition is fast, the corresponding chlorides would have been identified by Evans and coworkers, who examined the products carefully; if the Cl^- abstraction by $SnCl_4$ is the fast step, the kinetic study would have given a reaction rate independent of $SnCl_4$. It seems therefore plausible to invoke a process initiated by the complexation of the olefin with $SnCl_4$ and assisted by HCl in a typical Ad_E3 fashion. Of course we still lack some important information about the interaction between HCl_2^- and $SnCl_4$ and our suggestion that $SnCl_5^-$ is the main counterion needs confirmation.

The system indene–$SnCl_4$–HCl in methylene chloride was the object of a reliable study by Polton and Sigwalt[415]. Reactions were followed in high vacuum dilatometers at $-30\,^\circ$C, using carefully purified and dried reagents. The cocatalytic efficiency of HCl was clearly demonstrated in a plot of the maximum polymerisation rate vs. the [HCl]/[SnCl_4] ratio: the rate increased steadily with that ratio and unfortunately became too fast to be measured at [HCl] = [SnCl_4]. The authors interpreted their experimental results in terms of the following initiation scheme:

$$SnCl_4 + HCl \rightleftharpoons SnCl_4, HCl$$

$$SnCl_4, HCl + M \rightarrow HM^+ SnCl_5^-,$$

but did not offer any experimental evidence showing a first power dependence of the polymerisation rate on catalyst and cocatalyst concentration. We have replotted the relevant results (Fig. 8) and found a square root dependence on the hydrogen chloride concentration (note that the last point is off the straight line, but the authors underlined in their paper that that experiment had been difficult to follow because of its high rate). This behaviour could be indicative of an initiation scheme of the type proposed above involving free ions as chain carriers. Regrettably, the influence of stannic chloride concentration was not studied and a more detailed discussion is therefore impossible. However, our scheme would predict a progressive decrease in the active species concentration due to the corresponding increase in the amount of counterions produced (shifting of the equilibria away from the carbenium ion side). This phenomenon was indeed observed by Polton and Sigwalt in their analysis of the internal order of the polymerisations.

Bywater and Worsfold[418] recently studied the initiation mechanism in the system 1,1-diphenylpropene–$SnCl_4$–HCl in methylene chloride at $-30\,^\circ$C. Mixtures of the monomer and the catalyst or the cocatalyst alone gave negligible amounts of carbenium

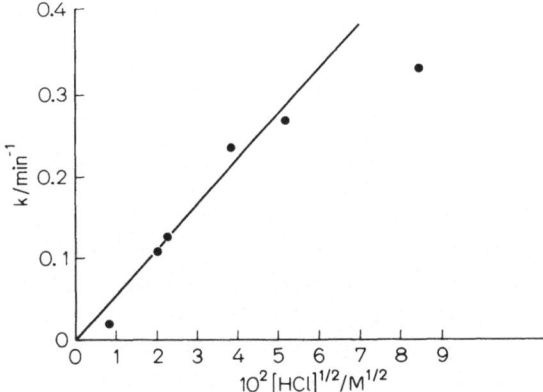

Fig. 8. Dependence of the first order rate constant for the polymerisation of indene by $SnCl_4$–HCl on the square root of the acid concentration, at fixed $[SnCl_4]$ (0.01 ± 0.0005 M). Solvent: CH_2Cl_2; T = -30 °C. (From data published in Ref. [415])

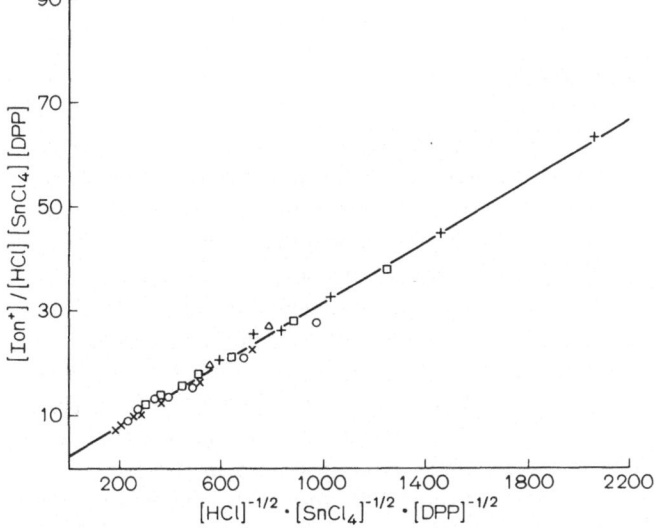

Fig. 9. The concentration of diphenylethylcarbenium ion as a function of the concentration of HCl, $SnCl_4$ and 1,1-diphenylethylene (DPP) at -30 °C in CH_2Cl_2. +, $[SnCl_4] = 3.5 \times 10^{-3}$ M, $[DPP] = 5.2 \times 10^{-2}$ M; □, $[SnCl_4] = 1.0 \times 10^{-2}$ M, $[DPP] = 5.1 \times 10^{-2}$ M; x, $[SnCl_4] = 3.0 \times 10^{-2}$ M, $[DPP] = 5.1 \times 10^{-2}$ M; ○, $[DPP] = 5.8 \times 10^{-2}$ M, $[HCl] = 1.0 \times 10^{-2}$ M; △, $[SnCl_4] = 1.1 \times 10^{-2}$ M, $[HCl] = 1.0 \times 10^{-2}$ M. (Taken from Ref. [418])

ions. Additions of HCl to a monomer–$SnCl_4$ solution resulted in a progressive increase in ionic concentration (detected spectroscopically) and this increase was proportional to the acid concentration raised to a power between 0.5 and 1. Additions of $SnCl_4$ to a monomer-HCl solution raised the carbenium ion concentration with a dependence on $SnCl_4$ slightly larger than 0.5. The authors proposed the following equilibria:

$$SnCl_4 + HCl + \underset{/}{\overset{\backslash}{C}}=\underset{\backslash}{\overset{/}{C}} \underset{}{\overset{K'}{\rightleftharpoons}} H-\overset{|}{\underset{|}{C}}-\overset{|}{\underset{|}{C}}{}^+ SnCl_5^- \overset{K''}{\rightleftharpoons} H-\overset{|}{\underset{|}{C}}-\overset{|}{\underset{|}{C}}{}^+ + SnCl_5^-$$

leading to the expression

$$\frac{[C^+]}{[HCl]_0 \, [SnCl_4]_0 \left[\begin{matrix}\backslash\ \ /\\C=C\\/\ \ \backslash\end{matrix}\right]_0} = K' + \frac{(K'K'')^{0.5}}{[HCl]_0^{0.5} \, [SnCl_4]_0^{0.5} \left[\begin{matrix}\backslash\ \ /\\C=C\\/\ \ \backslash\end{matrix}\right]_0^{0.5}} .$$

The experimental verification of this relationship is shown in Fig. 9, and its quality seems to corroborate the authors' mechanism. It must be pointed out however that the termolecular interaction proposed by Bywater and Worsfold does not give any detail about the specific steps involved in the generation of carbenium ions. Their results are compatible with and Ad_E3 mechanism in which free ions and ion pairs coexist, but the former are present in a larger proportion (reactants' exponent close to 0.5), as discussed above for the equilibrium concentration of active species.

c) Cocatalysis with Titanium Tetrachloride

As in the case of stannic chloride, the only work dealing with cocatalysis by hydrogen halides in cationic polymerisations using titanium tetrachloride as Lewis acid involves the use of hydrogen chloride, and even those studies are few and far between. Careful NMR experiments performed under high vacuum have shown that there is no measurable interaction between HCl and $TiCl_4$ in different solvents and over a wide range of temperatures[504]. This conclusion is borne out by the following observations: (i) the proton chemical shift for HCl in carbon tetrachloride is not affected by increasing amounts of added $TiCl_4$, and it occurs at a higher field than that for HCl in methylene chloride, even at temperatures as low as $-60\ ^\circ C$; (ii) the proton chemical shift for HCl in pure carbon tetrachloride is slightly downfield with respect to that in pure titanium tetrachloride; (iii) successive additions of titanium tetrachloride to a solution of HCl in methylene chloride (-70 to $10\ ^\circ C$) induce proton shifts to higher fields, and the same behaviour is noted if carbon tetrachloride is added instead of the Lewis acid. It follows that the interactions between HCl and $TiCl_4$ are even weaker than those of these acids with the two solvents used. The solubility data concerning HCl in $TiCl_4$, CH_2Cl_2 and CCl_4 respectively[508] have already been discussed and confirm the absence of any appreciable interaction between hydrogen chloride and titanium tetrachloride. Any cocatalytic effect of HCl in polymerisations of olefins catalysed by $TiCl_4$ cannot therefore be ascribed to the formation of an adduct or a complex between the two acids.

Next let us consider if the cocatalysis proceeds through an initial interaction between hydrogen chloride and the monomer, followed by activation by $TiCl_4$:

$$\begin{matrix}\backslash\ \ /\\C=C\\/\ \ \backslash\end{matrix} + HCl \ \rightarrow \ H{-}\overset{|}{\underset{|}{C}}{-}\overset{|}{\underset{|}{C}}{-}Cl$$

$$H{-}\overset{|}{\underset{|}{C}}{-}\overset{|}{\underset{|}{C}}{-}Cl + TiCl_4 \ \rightleftharpoons \ H{-}\overset{|}{\underset{|}{C}}{-}\overset{|}{\underset{|}{C}}{}^+ \ TiCl_5^- .$$

This mechanism implies that initiation should also proceed if, instead of HCl, the mono-mer hydrochloride were used as cocatalyst. However, it was shown that t-butyl chloride is not a cocatalyst for the polymerisation of isobutene by $TiCl_4$ in methylene chloride at -60 °C, while hydrogen chloride is a (weak) cocatalyst in the same situation[475]. The above reaction sequence cannot therefore explain the way HCl cocatalyses the cationic polymerisation of olefins with $TiCl_4$.

It remains to examine the third alternative, i.e., formation of a $TiCl_4$-monomer com-plex, followed by intervention of HCl to give the active species. Olefins such as isobutene and butene-1 interact with $TiCl_4$ as indicated by the appearence of a pale yellow colour in these mixtures prepared in vacuo. However, the complexes must be very weak and in low concentration, since they have never been detected by other techniques, and their apparent extinction coefficients are minute. Given the existence of these rather labile entities, we propose that hydrogen chloride intervenes in the second step of the initia-tion process through a Ad_E3-type mechanism:

$$\begin{array}{c} \diagdown \diagup \\ C \\ \parallel \\ C \\ \diagup \diagdown \end{array} \rightarrow TiCl_4 + HCl \rightarrow \overset{\overset{\displaystyle TiCl_4}{\vdots}}{\underset{\underset{\displaystyle Cl-H}{\vdots}}{>\overset{\delta^+}{C} \cdots C<}} \quad etc.$$

This intervention is successful, but not very powerful in the system involving isobutene in methylene chloride at -60 °C[475], but not when hexane is used as solvent, down to -90 °C[377]. The polarity of the medium obviously plays a discriminating role in deter-mining the occurrence of cocatalysis by HCl. Interestingly, hydrogen chloride[504] is also inactive in promoting this activation in the system butene-1–$TiCl_4$-methylene chloride, probably because that monomer lacks the basicity necessary for promoting the final ionisation, i.e. its double bond is not sufficiently polarised in the complex with $TiCl_4$ to be able to interact with HCl (its chlorine atom) or with its conjugate base HCl_2^-. In the latter study it was also found that water is a cocatalyst and this must be ascribed to the basicity of its oxygen atom which would provide the necessary driving force for ionisation in the Ad_E3 mechanism. These specific aspects will be discussed in the section devoted to cocatalysis by water.

d) Cocatalysis with Aluminium Compounds

Detailed studies of systems involving aluminium-based Lewis acids and hydrogen ha-lides are scarce. Fontana and Kidder[522] investigated the polymerisation of propene initiated by the pair aluminium bromide-hydrogen bromide. The cocatalytic role of the latter acid was clearly proved since no polymerisation could be detected in its absence. The dependence of the rate of polymerisation upon the cocatalyst concen-tration and the induction periods observed make this system similar to those in which stannic chloride induces the polymerisation of olefins in the presence of variable quan-tities of water (see Sect. IV-C-3-b). With relatively large quantities of added hydrogen bromide, addition of this acid to the monomer to give iso-propyl bromide must have constituted an important side reaction.

Kennedy and Squires[523] studied the effect of hydrogen chloride on the course of isobutene polymerisation catalysed by aluminium chloride. They showed that at −78 °C HCl increased the polymer yield if introduced before the catalyst, but had no effect if added to a quiescent mixture obtained by direct initiation giving a limited conversion. These observations are entirely consistent with our interpretation of the phenomenology of direct initiation: HCl is a cocatalyst in the presence of free aluminium chloride, i.e. when added at the beginning of the experiment, but is ineffective if the Lewis acid is tied up in conjugation products which, as we have seen, bring the polymerisation to a halt before all the monomer is consumed (cf. Sect. IV-B).

Kennedy and Sivaram[524] carried out some interesting model experiments on the interaction of cycloolefins with trimethylaluminium in the presence of hydrogen chloride at −50 °C in ethyl chloride. Hydrogen chloride did not add to cyclohexene in the absence of Me_3Al. Chlorocyclohexane *was* obtained in its presence and the yield of this addition product depended markedly on the order of addition of the two catalysts. Thus, when hydrogen chloride was introduced before trimethylaluminium, the reaction proceeded to higher conversions than if the order was inverted. Methylcyclohexane, an alternative reaction product, was only detected in experiments performed at room temperature and after prolonged periods. In contrast, 1-methylcyclohexene readily gave 1,1-dimethylcyclohexane in high yields at −50 °C with the mixing sequence olefin + HCl + Me_3Al. In this experiment, no 1-chloro-1-methylcyclohexane was obtained. It seems likely that this product can easily exchange a chlorine atom for a methyl group with trimethylaluminium. In this paper, it was also stated that the cationic polymerisation of isobutene initiated by the pair Me_3Al-*t*-butyl bromide occurred more rapidly if the alkyl aluminium compound was added as a last component.

All the above investigations do not provide sufficient quantiative information to allow the drawing up of a reasonable initiation mechanism for the combination aluminium compounds-hydrogen halides. The order of mixing of these two species seems to be quite critical. We have already explained why this is so in the case of the polymerisation of isobutene by aluminium chloride since that system also gives direct initiation and limited yields and the cocatalytic role of HCl can only be exploited if the latter compund is present from the beginning of the reaction. When trimethylaluminium was used, it is clear that the order HCl + Me_3Al provided a better opportunity for the production of mixed methylaluminium chlorides or even $AlCl_3$ than the order Me_3Al + HCl, since in the latter case trimethylaluminium probably complexes with the olefin. The higher catalytic activity observed with the former procedure stems from the higher reactivity of the mixed organoaluminium halides compared with that of trialkylaluminium compounds.

e) Cocatalysis with Antimony Trichloride

Evans et al.[525] showed that the initial rate of conversion of the linear dimer of 1,1-diphenylethylene into an equilibrium mixture of monomer and dimer (D), catalised by the pair $SbCl_3$−HCl in benzene follows the kinetic law:

$$R = k [D] [HCl] [SbCl_3]^3.$$

Antimony trichloride was believed to be dimeric in the medium used for these experiments. In the absence of HCl, no reaction was observed. It is difficult to envisage how three molecules of dimeric SbCl$_3$ can be necessary to produce the polarisation of the olefin double bond. However, it is known that this catalyst is a weak Lewis acid as testified by the fact that in a mixture of benzoyl chloride and mesitylene, antimony trichloride forms a one-to-one complex with mesitylene[423]. It is known moreover, that this Lewis acid readily complexes with aromatic hydrocarbons[426,428]. We believe therefore, that antimony trichloride was involved in a specific complexation with the linear dimer of 1,1-diphenylethylene where more than one molecule of the Lewis acid intervenes by interaction with the phenyl rings of that olefin. This peculiar type of complexation has already been postulated by Sauvet[259] with the same olefin and titanium tetrachloride. It is plausible that the above complex can interact further with antimony trichloride through the olefinic double bond thus forming the reactive entity capable of being ionised in the presence of hydrogen chloride. This mechanism is consistent with an Ad_E3 reaction pathway, the role of the base being played by the chlorine atom in hydrogen chloride.

f) Cocatalysis with Other Lewis Acids

Cocatalytic features have been reported for the conjoint action of gallium chloride and hydrogen chloride in the addition of the latter to various olefins[173,169]. No detailed study of the mechanism of these reactions was however carried out and the interpretation of the results is somewhat complicated by the fact that in both studies toluene was used as solvent, and it is known that hydrogen chloride gives a complex with this aromatic hydrocarbon.

Sal'nikov et al.[526] investigated the polymerisation of butyl vinyl ether by ferric chloride in the presence of hydrogen chloride. The products had a high degree of polymerisation, in contrast with the black resinous liquids produced when hydrogen chloride alone was added to the monomer. Regrettably, these experiments were carried out in the presence of butanol, which makes any discussion about the details of the initiation mechanism impossible.

g) Conclusions

Cocatalysis by hydrogen halides most probably arises from a primary attack on the monomer which produces a polarised intermediate, precursor to the carbenium ions, through an Ad_E3 mechanism. The best supporting evidence for this general interpretation comes from the studies of polymerisation and simple ionisation using the stannic chloride-hydrogen chloride pair. The concentration of active species is usually very low because all the interactions leading to their formation are in fact equilibria some of which lie heavily on the reactants' side, e.g., the interaction between the monomer double bond and the Lewis acid. We propose that the generation of carbenium ions in a system involving an alkenyl monomer, a Lewis acid and a hydrogen halide can therefore be rationalised by the following sequence of equilibrated reactions:

$$\overset{\backslash}{\underset{/}{C}}=\overset{/}{\underset{\backslash}{C}} + MtX_n \rightleftharpoons \overset{\backslash \ /}{\underset{/ \ \backslash}{\overset{C}{\underset{C}{\|}}}} \rightarrow MtX_n$$

$$\overset{\backslash \ /}{\underset{/ \ \backslash}{\overset{C}{\underset{C}{\|}}}} \rightarrow MtX_n + HX \rightleftharpoons \overset{MtX_n}{\underset{X\!-\!H}{\overset{\backslash\delta+ \ \ \ :/}{\underset{/ \ \ \ \backslash}{C\cdots C}}}} \overset{MtX_n}{\underset{\rightleftharpoons}{\longrightarrow}}$$

$$\overset{MtX_n}{\underset{MtX_n\cdots X\!-\!H}{\overset{\backslash\delta+ \ \ \ :/}{\underset{/ \ \ \ \backslash}{C\cdots C}}}} \rightarrow MtX_{n+1}^{-} \ \overset{+/}{\underset{H}{\overset{}{C}}}\!\!-\!\!\overset{/}{\underset{\backslash}{C}} + MtX_n$$

Variations to the above scheme can of course be envisaged, particularly in terms of the intervention of HX_2^{-} as the assisting entity for the intermediate concerted complex (as in some Ad_E3 mechanisms proposed for the addition of Brønsted acids to olefins), and of the formation of homoconjugated Lewis acid anions of the type $Mt_2X_{2n+1}^{-}$.

3. Cocatalysis by Water

a) Introduction

Water is the most frequently invoked cocatalyst in Friedel-Crafts reactions and in particular in cationic polymerisations promoted by Lewis acids. This is not only due to the almost inevitable presence of traces of water as an impurity in most reagents and solvents, but also to its high efficiency compared to many other cocatalysts. Thus, when a polymerisation system employing a Lewis acid as catalyst is examined, it is always indispensable to check and assess the possible role of adventitious moisture as the main cocatalyst.

As we have already mentioned in the preceding sections, water is often a more powerful cocatalyst than hydrogen halides. This observation clearly excludes the possibility that the general cause of cocatalysis by water originates from the hydrolysis of the Lewis acid,

$$MtX_n + H_2O \rightarrow MtX_{n-1}OH + HX,$$

followed by the cocatalytic intervention of the hydrogen halide formed thereby. It is also obvious that in polymerisation systems containing appreciable concentrations of Lewis acids and much lower concentrations of water, the latter will be present mostly as a complex with the former, a lone pair of the oxygen atom being responsible for this type of association, and therefore the amount of free water in these situations will be minimal.

We intend to show that the details of the initiation processes characterised by water cocatalysis can be generally rationalised by the intervention of an Ad_E3 mechanism in which the basic properties of water or of its complexes with the catalyst are a determining driving force in providing the nucleophilic moiety necessary to the formation and stabilisation of the concerted intermediate species. Two different interactions can be envisaged in this context: the first involves the intervention of simple catalyst and cocatalyst molecules to build the transition state,

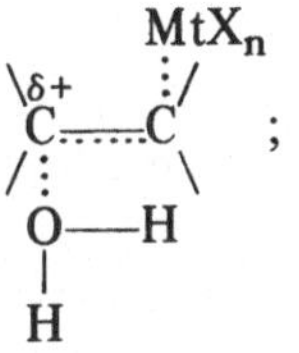

$$
\begin{array}{c}
\text{MtX}_n \\
\overset{\displaystyle \vdots}{\underset{\text{H}}{\overset{\displaystyle \backslash^{\delta+}}{\text{C}}\!\!-\!\!-\!\!-\!\!\overset{\displaystyle \vdots/}{\text{C}}}} \\
\end{array} \; ;
$$

while in the second the catalyst-cocatalyst complex acts as nucleophile in the $\text{Ad}_E 3$ intermediate,

$$
\text{MtX}_n + \text{H}_2\text{O} \rightleftharpoons \text{MtX}_n \cdot \text{OH}^{\delta-}\text{H}^{\delta+},
$$

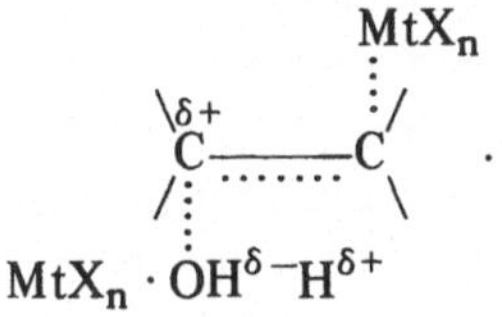

$$
\begin{array}{c}
\text{MtX}_n \\
\overset{\displaystyle \backslash^{\delta+}}{\text{C}}\cdots\cdots\overset{\displaystyle \vdots/}{\text{C}} \\
\text{MtX}_n \cdot \text{OH}^{\delta-}\text{H}^{\delta+}
\end{array} \; .
$$

Generally speaking, one would expect the above complex to be less basic than the water molecule, and thus less effective as a cocatalyst. On the other hand the ability of the complex to coordinate with the monomer double bond is probably higher and this would give a more stable intermediate resulting also in an enhanced cocatalytic efficiency. As we shall see below, the order of addition of catalyst and cocatalyst to the solution results in a marked difference in the rate of initiation in many systems. These peculiarities are related to the considerations made above and it seems that the specific response of a given system depends critically on the Lewis acid used and the actual experimental conditions chosen for the polymerisation. The same conclusions can be drawn concerning the occurrence of a maximum rate of initiation for a given water/catalyst concentration ratio. This phenomenon is in fact quite general, but the actual value of the critical ratio varies considerably from system to system.

The following sections are devoted to a discussion of the most relevant publications dealing with the kinetics and mechanism of initiation processes proceeding through water cocatalysis.

b) Cocatalysis with Stannic Chloride

Stannic chloride is a relatively weak Lewis acid in cationic polymerisation and gives rise to rates of monomer consumption which are sensibly lower than those observed with TiCl_4, BF_3 and AlCl_3 and other "strong" catalysts in the same conditions. This allows the use of dilatometry, a convenient technique which can be readily adapted to high vacuum manipulations. However, in order to ensure the thorough drying of the bulb placed below the capillary, it is advisable to rinse it with some dry solvent after evacuation and before introducing the reaction components. Only by resorting to such expedients can one test the delicate question of whether a cocatalyst is indispensable for initiation. There are good reasons to believe that stannic chloride can induce direct initiation in some systems, particularly in media of moderate-to-high dielectric constant. Perhaps the most convincing evidence to this effect is the work of Polton and Sigwalt[415] on the polymerisation of indene in methylene chloride at −30 °C. These authors showed that it was impossible to obtain a non-reactive mixture even after the most stringent purification and drying procedures, since a low, but measureable rate of polymerisation could be detected with $SnCl_4$. But similar observations had been made in earlier investigations. Thus, Pepper[527] found that styrene was polymerised in dry ethylene chloride and that added moisture had a detrimental effect on the rate of this reaction. He concluded that traces of hydrogen chloride present in the catalyst might have played the role of cocatalyst. Evans and Lewis[528] reported that the dimerisation of 1,1-diphenylethylene by $SnCl_4$ in benzene at 40 °C occurred at exceedingly low rates in a carefully dried system and concluded that water was necessary for the process to take place. Colclough and Dainton found that styrene was slowly polymerised by $SnCl_4$ in ethylene chloride[387] and nitrobenzene[388] at 25 °C, in the absence of added water. They proposed solvent cocatalysis for the former and did not comment on the latter situation. Overberger et al.[529] studied the polymerisation of styrene in carbon tetrachloride containing increasing quantities of nitrobenzene at 25 °C. They noticed that the rate of the reaction in the absence of added water was much lower with 15% nitrobenzene than with 30%. Finally, Metz[530] claimed that ionic species were formed in the reaction of $SnCl_4$ with styrene in a mixture of CCl_4 and ethylene chloride at 30 °C, but his experimental techniques were somewhat inadequate to guarantee the absence of residual moisture.

The ensemble of results described above suggests that $SnCl_4$ can indeed promote the direct initiation of aromatic olefins polymerisation, but that this is very slow and occurs if the polarity of the medium is not too low. We have however refrained from discussing the mechanism of this process in the section specifically devoted to direct inititation because we feel that the slowness of these particular systems always leaves the doubt that minute amounts of residual cocatalysts might in fact be responsible for initiation. More studies like the one on indene[415] are necessary before a firm conclusion on this question can be reached.

In media of low dielectric constant it is well established that stannic chloride needs a cocatalyst to initiate the cationic polymerisation of alkenyl monomers. This point has been clearly proved for carbon tetrachloride[386,529] and benzene[531,532] solutions. In the present section the role of water as cocatalyst in these processes will be examined, but given the substantial amount of literature available on this topic, we will concentrate on important kinetic and mechanistic aspects, relevant to our new interpretation. We remind the reader that much of the older work has been thoroughly reviewed in Plesch's second book[533].

Most of the studies of water cocatalysis with $SnCl_4$ were performed at room or higher temperatures and it is surprising that the various authors concentrated on the formation of the hydrates without considering the possible reaction of the Lewis acid with water to give hydrogen chloride. The latter can in fact be a cocatalyst in solvents of medium and high polarity, while it does not fulfil such a role in low dielectric-constant media[533]. Indeed, the fact that induction periods were frequently observed in the latter situation[386,529,534], strongly suggests that the initiation process is complicated by the formation of hydrogen chloride, which, as Williams showed[534], inhibits the polymerisation of styrene and gives rise to 1-phenylethyl chloride.

Concerning the $SnCl_4-H_2O$ adducts and their activity as possible initiators, various hypothesis have been put forward. Colclough and Dainton[389] proposed the following scheme:

$$SnCl_4 \cdot 2 H_2O \rightleftharpoons SnCl_3OH \cdot H_2O + HCl$$

$$SnCl_3OH \cdot H_2O + CH_2=CHPh \rightleftharpoons SnCl_3OH \cdot H_2O \cdot CH_2=CHPh$$

$$SnCl_3OH \cdot H_2O \cdot CH_2=CHPh + CH_2=CHPh \rightarrow CH_3-CHPh-CH_2-CHPh^+$$

$$SnCl_3(OH)_2^-,$$

which is highly speculative, although the acid $SnCl_3OH$ seems to have been isolated once from its ethereal solution[535], i.e. in the presence of a nucleophile. These authors observed that the maximum rate of polymerisation occurred at $[H_2O]/[SnCl_4] = 1/500$ in CCl_4 and 2 in ethylene chloride. Overberger et al.[529] found a ratio of unity in carbon tetrachloride (70%)-nitrobenzene (30%) and of about 0.3 in a 85%—15% mixture. For the dimerisation of 1,1-diphenylethylene Evans and Lewis[528] obtained a ratio of 2 in benzene, and for the polymerisation of indene Sigwalt and Polton a value of 0.25[415] in methylene chloride at $-30\,°C$. Kucera et al.[536] studied the variation of that critical ratio as a function of the solvent composition (mixtures of toluene and 1,1,2,2-tetra-chloroethane) and monomer concentration (up to 50% in volume); its value increased from about 0.1 in pure 1,1,2,2-tetrachloroethane with 15% styrene to about 2 in pure toluene with higher styrene concentrations. Finally, Gantmakher and collaborators[537—539] have claimed that the complex $SnCl_4 \cdot H_2O$ is more active than the $SnCl_4 \cdot 2 H_2O$ for styrene and α-methyl styrene polymerisation in various solvents, but their arguments are not very convincing.

A close inspection of all the above literature revealed to us a probable general explanation of the superficially different phenomenologies. We think that the complex $SnCl_4 \cdot 2 H_2O$ is the predominantly active promoter of initiation (see below for mechanism), but its solubility is very limited in many of the solvents used for these studies. Thus, the often-observed cloudiness of the reaction medium reflects a saturation with respect to that active complex and obviously further addition of water produces a decrease in rate, since this compound is a cocatalyst *but* also a poison (terminating agent) if added in excess. In the solvents in which the $SnCl_4 \cdot 2 H_2O$ complex is more soluble the expected critical ratio of 2 was indeed observed: these media seem to be preferably of aromatic nature (benzene, toluene or a high proportion of styrene).

We turn now to an analysis of the internal and external orders obtained in these investigations dealing with aromatic monomers. Since many of those processes were

marred by induction or acceleration periods and other kinetic complications, we have
selected the clearest systems which gave understandable results. The best of these is
undoubtedly the study of the dimerisation of 1,1-diphenylethylene conducted by
Evans and Lewis more than twenty years ago[528]. The initial rate of this reaction was
found to depend on the square of the monomer concentration in all experiments, on
the concentration of water and of $SnCl_4$ when the $H_2O/SnCl_4$ ratio (N) wass less than
2, and on the square of $[SnCl_4 \cdot 2 H_2O]$ when that ratio was close to 2. Thus:

$$R_0 = k [SnCl_4 \cdot 2 H_2O] [SnCl_4]_0 [M]^2, \quad \text{for N} < 2,$$

and

$$R_0 = k [SnCl_4 \cdot 2 H_2O]^2 [M]^2, \quad \text{for N} = 2.$$

Overberger and coworkers[529] found that the rate of styrene polymerisation at very
low conversions was directly proportional to the square of the monomer concentration, the total concentration of stannic chloride (for N < 2), and the concentration
of water (approximately, for low values of N). Similar monomer and catalyst dependences had also been previously reported by Pepper[540] in ethylene chloride (claimed
to be the cocatalyst) and by Colclough and Dainton[387] in the same solvent, who
also added t-butyl chloride and found the reaction rate to depend also on its concentration. Polton and Sigwalt[415] noted that the polymerisation of indene was internally first order in monomer if sufficient water was added to eliminate the induction
periods, i.e. with N greater than about 0.02. No external orders were measured by
these authors, but very important experiments were carried out to show that the
addition of a premixed catalyst-cocatalyst solution to the monomer solution always
gave a much slower polymerisation than when the normal procedure of adding the
catalyst to the wet monomer solution was used. This point is particularly relevant
to the discussion below. The external orders obtained by Gantmakher and coworkers[537-539] in the polymerisation of styrene and α-methylstyrene are not too
clear, but an order in monomer between two and three seems to apply to those
systems. Finally, Okamura and Higashimura[541] had quite early found a third-order
dependence on styrene concentration in carbon tetrachloride, just like Dainton and
Colclough in the same solvent (with $N > N_{crit}$)[386]. In benzene, the Japanese authors
reported an order in styrene close to two[542].

In order to shed some light on the nature of the initiation process in these systems,
we will first consider the more straightforward studies. All the evidence gathered by
Evans and Lewis[528] points to the following situation. If N < 2, the solution contains
both free $SnCl_4$ and its dihydrate, but when N = 2 the formation of the dihydrate
seems to reach completion. The olefin can be complexed by either $SnCl_4$ or $SnCl_4 \cdot$
$2 H_2O$ and this complex is then activated by the dihydrate to give an Ad_E3-type
intermediate, precursor to the ionised active species:

$$SnCl_4 \text{ (or } SnCl_4 \cdot 2 H_2O) \; \rightleftharpoons \; \overset{\displaystyle \backslash \;/}{\underset{\displaystyle /\;\backslash}{\overset{C}{\underset{C}{\|}}}} \; \rightarrow \; SnCl_4 \text{ (or } SnCl_4 \cdot 2 H_2O),$$

$$
\begin{array}{c}
\begin{array}{c} \backslash \quad / \\ \text{C} \\ \parallel \\ \text{C} \\ / \quad \backslash \end{array}
\rightarrow \text{SnCl}_4 \text{ (or SnCl}_4 \cdot 2\,\text{H}_2\text{O}) + \text{SnCl}_4 \cdot 2\,\text{H}_2\text{O} \rightleftharpoons
\begin{array}{c}
\qquad\qquad\qquad \text{SnCl}_4 \text{ (or SnCl}_4 \cdot 2\,\text{H}_2\text{O}) \\
\qquad\qquad\qquad \vdots \\
\backslash \overset{\delta+}{\text{C}}\cdots\text{C} \\
/ \vdots \qquad \backslash \\
\text{H}_2\text{OSnCl}_4\text{O} - \text{H} \\
| \\
\text{H}
\end{array}
\end{array}
$$

If the ionisation of the concerted intermediate is the slow step in this process, the kinetic observations of Evans and Lewis[528] are adequately explained by the proposed mechanism.

The results obtained by Overberger's group and by Coclough and Dainton in ethylene chloride are also compatible with the above scheme, although the kinetic evidence is less complete. A supporting argument in favour of a mechanism involving prior complexation of the monomer with the catalyst (or the catalyst-cocatalyst adduct) is given by published evidence showing the existence of such complexes with styrene and α-methylstyrene[447].

In other studies, precipitation of the dihydrate would complicate matters and particularly, would give a maximum rate as a function of N which varies according to the solubility of that species in different media, as observed. We wish to note here that the interpretation of initiation proposed by Kucera and collaborators, based on direct initiation and solvation by water molecules[344] is untenable. In fact, Dainton et al.[543] showed irrevocably that the role of water is cocatalytic (chemical), by proving that when D_2O replaces H_2O deuterium is found in the polymer and the rate of initiation is lower (kinetic isotope effect).

Monomer orders higher than two could be due to the intervention of a second monomer molecule in the initiation process to solvate the intermediate Ad_E3 complex, and favour ionisation (note that these higher orders were found in low dielectric-constant mixtures). Such monomer complexation is not uncommon and has bee proposed in similar contexts[148,503].

The initiation scheme proposed by Colclough and Dainton in the presence of added alkyl chlorides and invoked also for solvent cocatalysis with ethylene chloride[389] seems most unlikely. Indeed, the possibility of RCl cocatalysis in these systems is remote, due to the high instability of the carbenium ions involved which would hardly be expected to form or at least to have a sufficient lifetime to attack the monomer. It seems much more plausible to us to envisage that dehydrochlorination of the alkyl chloride provided the HCl cocatalysis needed for polymerisation to occur, and of course the kinetic dependence on RCl observed in both ethylene chloride and nitrobenzene.

The mechanism proposed by Gantmacher's school[539] is, as we have already stated, not very clear, and certainly not very well sustained by a set of complicated experiments in which solvents are often a mixture of two or three components and the second additions of stannic chloride do not seem to us to prove that the monohydrate is more active than the dihydrate. The reported incomplete yields in ethyl chloride with $N > 1$ are also difficult to explain since termination reactions were not properly studied.

Another piece of evidence which comes to support our mechanistic interpretation is the lower catalytic activity of preformed mixtures of water and stannic chloride reported by Polton and Sigwalt[415]. Obviously, in the absence of monomer these two compounds interact to give ultimately some species which are rather poor initiators.

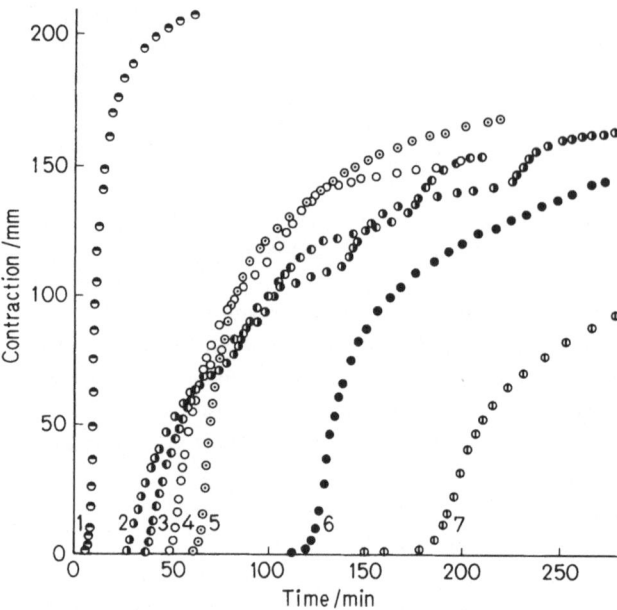

Fig. 10. Dilatometric curves for the polymerisation of styrene (3.0 moles /kg) by SnCl$_4$ (6.6 mmoles/ kg) at 35 °C. *1*, in 1,1,2,2-tetrachloroethane with 2.0 mmoles/kg of added water. all other polymerisations in toluene with increasing water concentrations: *2*, 0.3; *3*, 0.5; *4*, 3.5; *5*, 8.8; *6*, 15.0; *7*, 30.0 mmoles/kg. (Taken from Ref. [344])

Any interpretation of these polymerisations based on this type of interaction as the basic initiation step (i.e., without the participation of the monomer) is therefore incorrect. On the other hand, our mechanism involving the intervention of a monomer-catalyst complex justifies the necessity of the presence of the monomer when catalyst and cocatalyst are mixed, if initiation rates passing through a maximum are to be obtained.

However, if our mechanism explains a certain number of kinetic and chemical features of SnCl$_4$–H$_2$O systems, other remain more difficult to rationalise. Two major phenomena characterise the polymerisations in low dielectric-constant media: induction (or acceleration) periods and "wavy" conversion-time curves. Colclough and Dainton[386] reported these phenomena in their study of the polymerisation of styrene in carbon tetrachloride. Fifteen years later, Kucera et al.[344,536] observed the same unusual behaviour using toluene as solvent. As shown in Fig. 10, the induction period increased with water concentration and on the other hand the waves seemed to become less pronounced as the amount of added water was increased. Kucera's interpretation of these peculiarities encompassed a rather immaginative, but unsubstantiated set of eleven reactions focussed around the selfionisation of stannic chloride and initiation by the resulting SnCl$_3^+$, and termination by chloride ion abstraction from SnCl$_5^-$ to give a C-Cl end group. The polymers were said to contain tin and chlorine, but the analytical determination of these elements showed considerable inconsistency, particularly in that the polystyrene prepared in 1,1,2,2-tetrachloroethane was free of tin, but contained a very high amount of chlorine, probably coming from traces of solvent left in the polymer. In view of this anomaly we strongly suspect the quoted amounts of tin and chlorine in the polymers prepared in toluene. As for the mechanism proposed, we fail to see any valid evidence to support it. We also find it surprising that in the maze of interactions proposed, the author failed to consider the possible hydrolysis of the catalyst, which must have been rather important at 35 °C during reaction times up to several hours.

Unfortunately, no detailed kinetic study is available concerning these peculiar systems, and indeed the proper concentration dependences would be hard to obtain with such oddly-shaped curves. Moreover, a further complication arises from the fact that water intervenes both as a cocatalyst and as a terminating agent when its concentration is higher than the critical value giving maximum rate, and the chemistry of this termination or inhibition processes is not clear. It is probable however that both the induction periods and the acceleration-deceleration phases are linked to the detrimental action of hydrogen chloride[534] and of 1-phenylethyl chloride[389] on the polymerisation of styrene by $SnCl_4$–H_2O in carbon tetrachloride, since those two compounds could be generated in the early stages of the process.

The mechanism of inhibition could therefore be summed up in the following series of interactions:
— Hydrogen chloride is produced in the hydrolysis of $SnCl_4$.
— This acid readily attacks the styrene-stannic chloride (or stannic chloride hydrate) complex and, through an Ad_E3 process, gives 1-phenylethyl chloride.
— This product retards the beginning of polymerisation by scavenging the stannic chloride in a complexation equilibrium presumably strongly shifted to the right hand side:

$$CH_3-CHPhCl + SnCl_4 \rightleftharpoons CH_3-\overset{\delta+}{C}HPh----\overset{\delta-}{S}nCl_5.$$

Note that the polarised complex formed in this interaction is not active for initiation, but (–)-1-phenylethyl chloride is racemised by $SnCl_4$ in CCl_4[519].
— While these processes occur, the removal of $SnCl_4 \cdot 2\,H_2O$ to form the products described allows some of the precipitate to reenter the solution, until there is sufficient catalyst not bound to 1-phenylethyl chloride for polymerisation to start.
That the induction period is linked specifically to 1-phenylethyl chloride is not only proved by the fact that addition of this compound to polymerising mixtures with styrene stops the process[389], but also because no induction period is observed in the dimerisation of 1,1-diphenylethylene by $SnCl_4$–H_2O in benzene[528], while if styrene is polymerised by the same system induction periods are noticed. We believe that initiation with the latter monomer needs one molecule of free $SnCl_4$ and one of dihydrate.

Our tentative explanation of the waves in the conversion-time curves observed in the polymerisation of styrene in CCl_4 or toluene is also based on the presence of a reservoir of undissolved initiator. We assume that the precipitated $SnCl_4 \cdot 2\,H_2O$ reenters the solution at a slow constant rate. The presence in the system of 1-phenylethyl chloride and/or chain ends possessing a similar structure, and of HCl, does not allow the entering initiator to generate a sufficient number of active species and one observes therefore an induction period. Only when all the inhibiting substances have been consumed can the concentration of chain carriers increase following the slow but continuous arrival of $SnCl_4 \cdot 2\,H_2O$. Polymerisation occurs, but this process regenerates some or all of the original inhibiting products. It follows that the rate of monomer consumption will progressively decrease leading to a new pause. After the elimination of the inhibiting substances by the $SnCl_4 \cdot 2\,H_2O$ arriving in solution another polymerisation period will take place. In this mechanism a stationary state is never attained owing to the slow accumulation and removal of poisonous species. This interpretation of the sinuous character of the recation curves implies that longer induction periods would be noted if one added to the system some polystyrene possessing terminal 1-phenylethyl chloride-type groups. Such experiments were however never carried out.

We wish to stress that the above conjectures are not thoroughly backed by experimental evidence, but based on rather circumstantial information. They should therefore be considered as working hypotheses to be verified by future investigations, particularly concerning the role fo 1-phenylethyl chloride.

The low temperature polymerisation of isobutene by $SnCl_4$ in ethyl chloride is one of the classical studies of the golden era of cationic polymerisation. Norrish and Russell[544-546] found that with no added water an extremely slow reaction period was followed by a sudden acceleration. A similar phenomenon was later reported by Polton and Sigwalt[415] for the polymerisation of indene in a dry system. It seems reasonable to suppose that the slow initial process reflects direct initiation in both systems, and that the sudden acceleration arises from the internal production of a cocatalyst, probably hydrogen chloride formed from the dehydrochlorination of active species.

The work of Norrish and Russell with added water has been thoroughly reviewed by Plesch in his second book[547]. The rate laws appeared to follow the forms:

$$R_p = k\,[SnCl_4]\,[H_2O]\,[M], \quad \text{when} \quad [SnCl_4] > [H_2O]$$

and

$$R_p = k'\,[SnCl_4]\,[M]\,[SnCl_4 \cdot H_2O], \quad \text{when} \quad [SnCl_4] < [H_2O].$$

The order in monomer concentration was however difficult to establish with confidence due to the poor solubility of the growing polymer chains. The above kinetics results can be rationalised by an initiation mechanism based on the formation of an Ad_E3 intermediate, as in the case of cocatalysis by phenols (see Sect. IV-C-4-b), if one assumes that in these specific conditions isobutene and stannic chloride form a strong complex. In other words, if the concentration of complex is practically independent of monomer concentration because the reaction

$$SnCl_4 + (CH_3)_2C{=}CH_2 \ \rightarrow \ SnCl_4 \cdot (CH_3)_2C{=}CH_2$$

is virtually complete, then initiation is independent of monomer concentration, and the first order observed experimentally is due to the propagation step. The intermediate species leading to active centers could be formed by the following interaction:

$$SnCl_4 \cdot (CH_3)_2C{=}CH_2 + SnCl_4 \cdot H_2O \;\rightleftharpoons\;$$

$$
\begin{array}{c}
CH_3 \quad\quad SnCl_4 \\
\searrow^{\delta+} \quad \vdots\, \diagup \\
C {-\!\!-\!\!-} C \\
\diagup \quad \cdots \quad \diagdown \\
CH_3 \quad \vdots \\
SnCl_4O{-\!}H \\
| \\
H
\end{array}
$$

Since that major investigation by Norrish and Russell, no more work has been published on this system, except a very interesting piece of information obtained in Russell's laboratory by Laflair[548]. He proved that water is chemically involved in the cocatalytic initiation mechanism, since the rate determining protonation step takes place with a proton derived from water. This proof came from rate comparisons between water and heavy water, which gave the expected kinetic isotope effect.

c) Cocatalysis with Titanium Compounds

The cocatalytic role of water in polymerisations promoted by titanium tetrachloride has been extensively investigated. Indeed, the very theory of cocatalysis was formulated on the basis of the decisive intervention of added water in non-polymerising solutions of isobutene in non-polar solvents containing $TiCl_4$[377]. In that pioneering study it was also shown that hydrogen chloride was not a cocatalyst, a fact which is probably best interpreted by assuming that the the combination $TiCl_4–HCl$ promotes a very fast addition reaction on isobutene to give t-butyl chloride thereby consuming all the HCl present (note that t-butyl chloride is not a cocatalyst in this system). The cocatalytic intervention of moisture was also thoroughly demonstrated in the polymerisation of isobutene by titanium tetrachloride in solvent of medium polarity (CH_2Cl_2, C_2H_5Cl, etc.) where the consumption of the cocatalyst leads to incomplete conversions, unless present in amounts greater than a given critical concentration $[H_2O]_c$[378,403,549]. The substantial contribution of Plesch's laboratory to the understanding of this polymerisation system has been clearly outlined in his second book[550], but the most detailed kinetic study was published a few years later[380]. It is, quite simply, an example of technical mastery. The authors showed that the order of catalyst and cocatalyst introduction was unimportant, since all procedures gave the same polymerisation rate. At $-60\,°C$, and with $[H_2O] > [H_2O]_c$, the rates of monomer consumption followed a first order law, indicating that, as in the case of stannic chloride catalysis, the rate of initiation was independent of monomer concentration. Chlorine atoms and hydroxyl groups were found in the polymers. While the presence of the latter moieties is readily explained by a termination reaction with the counterion $TiCl_4OH^-$, the incorporation of the former is less obvious. Transfer with solvent (CH_2Cl_2) seems unlikely in view of the lack of cocatalytic activity of added t-butyl chloride. The reactions of the growing species with traces of HCl produced in the system, or with the counterion,

$$
\sim\!\!\!\sim\!\!\overset{|}{\underset{|}{C}}{}^+ + TiCl_4OH^- \longrightarrow \sim\!\!\!\sim\!\!\overset{|}{\underset{|}{C}}{-}Cl + TiCl_3OH \,,
$$

as suggested by Plesch, seem more plausible alternatives, but more work would be needed to clear this point. The external orders in catalyst and cocatalyst were also determined. At $-60\,^\circ$C and below that temperature, the first order rate constant was directly proportional to the concentration of $TiCl_4$ when the water concentration was above the critical level and depended on the square of the water concentration when it was below the critical level. The latter correlation is less meaningful than the former because at low cocatalyst concentrations the reproducibility was not very good and the contribution of polymerisation due to direct initiation must have been an important fraction of the overall measured rate. We therefore limit our present discussion to the better defined situation in which the amount of added water was sufficient to insure complete conversion. The experimental observations suggest that a mechanism entirely similar to that we proposed in the preceding section for stannic chloride catalysis applies adequately to the present system. The only major difference stems from the enhanced reactivity of the counteranion $TiCl_4OH^-$ which favours an important termination reaction leading to hydroxyl end groups and probably a less frequent one giving chloride ion abstraction by the growing carbenium ion (see above). The possible production of hydrogen chloride by a mechanism similar to that proposed for $SnCl_4$ is less relevant here, because this acid is known to play a very minor role (if any) in the initiation processes involving $TiCl_4$. As for the actual number of water molecules involved in the intermediate Ad_E3 complex, we can only speculate that since titanium compounds exhibiting both penta and hexacoordination are known, the mono and dihydrates of $TiCl_4$ can probably be formed in the system under study, and therefore two different active species might coexist with different propagating capacity.

Recently, Sigwalt and collaborators[397,475] showed that the cocatalytic efficiency of water with titanium tetrachloride decreases considerably if the two catalytic components are mixed before being added to isobutene solutions. We interpret this important observation as further evidence in favour of an Ad_E3-type concerted mechanism of initiation. In fact when $TiCl_4$ and water are added directly to the monomer solution, the former gives a π-complex with the double bond and the latter (or a polarised addition product $TiCl_4OH$ H) approaches that complex as a nucleophile to induce the formation of the carbenium ion precursor following the Ad_E3 scheme. If however $TiCl_4$ and water are allowed to interact for a long time (relative to the polymerisation time scale) before they are added to the monomer, the lack of a suitable base necessary to assist in the formation of the concerted intermediate makes the initiation process much slower. One of the most likely product of this slow reaction between catalyst and cocatalyst is obviously hydrogen chloride, which as we have seen repeatedly, is a poor coinitiator in systems involving $TiCl_4$.

Plesch's group also studied the polymerisation of styrene by the combination $TiCl_4$–H_2O[410,533,551]. The internal order in monomer was found to be close to two and the external order, based on initial polymerisation rates, also two. The initial rate of polymerisation depended on the first power of water concentration for $[H_2O] < [H_2O]_c$, but became independent of it for $[H_2O] > [H_2O]_c$, $[H_2O]_c$ being an empirical value. Further monomer additions at the end of the first polymerisation resulted in a new reaction proceeding at the same rate as the first one. The effect of temperature on the initial polymerisation rate was unusual in that a decrease was observed between $-30\,^\circ$C and about $-45\,^\circ$C, and an increase below this range. All the above observations must be considered in the specific context of the simultaneous occurrence of the already discussed direct initiation at low temperature (with slow termination with aromatic monomers) and the water cocatalysed initiation (with no water consumption, as in the case of $SnCl_4$), more important at higher temperatures, i.e., above about -30 °C. Indeed Plesch had already concluded that two initiation reactions, "one dependent on water, the other not", coexisted in this system. A detailed interpretation of the various features of the polymerisation of styrene by $TiCl_4$ and water would need a quantitative knowledge of the relative contribution of each mechanism at different temperatures and reagents' concentration. Such knowledge is unfortunately not available as yet.

A few comments are in order concerning the peculiarities of experiments conducted on the copolymerisation of isobutene with aromatic olefins both by Plesch's and Sigwalt's groups[410,552,553]. In the first study[410] addition of water to a quiescent mixture resulting from the incomplete copolymerisation of isobutene and styrene promoted by the $TiCl_4$–H_2O pair, reinitiated the copolymerisation, and this was taken as evidence for the need of water cocatalysis in the polymerisation of styrene. Cheradame and Sigwalt carried out similar experiments, but using indene and α-methylstyrene as comonomers[552,553], two monomers for which they established that no cocatalyst is needed to initiate their polymerisation by $TiCl_4$ in methylene chloride (although direct initiation is accompanied by slow termination; this is also true for styrene[410], but at low temperature termination is so slow that often complete conversion is attained). We visualise the course of these copolymerisations as follow: water is consumed in the cocatalysed polymerisation of isobutene by the well known termination reaction giving hydroxyl end groups; this phenomenon, coupled with the intrinsic termination reactions characteristic of the direct initiation mechanisms for both comonomers (see Sect. IV-B), lead to the interruption of the copolymerisation before complete conversion is achieved; further addition of water to this quiescent mixture reactivates the initiation of isobutene and thus the copolymerisation; the reactivation does not necessarily involve the aromatic comonomer because of the presence of added water. We think in fact that at least for indene and α-methylstyrene no cocatalytic effect of water has ever been proved at low temperature in methylene chloride, and the fact that with styrene below -30 °C the rate of polymerisation is independent of water concentration suggests that direct initiation predominates too, although a minor contribution arising from water-cocatalysed intitiation cannot be ruled out.

Mondal and Young[554] investigated the polymerisation of *cis-cis*-cyclooctadiene using $TiCl_4$–H_2O in methylene chloride at 25 °C. The initial rate of monomer consumption was found to be proportional to monomer, catalyst and cocatalyst concentration, the latter being a few percent of the amount of $TiCl_4$ added. An interesting study was made of the interaction between catalyst and cocatalyst, and of the effect of aging of the resulting mixture on the rate of polymerisation. A precipitate was formed which had no influence on the course of the initiation process, since the same activity was obtained from the total heterogeneous mixture and from the filtrate alone. Also, a few hours' aging did not alter the efficiency of the solution. Obviously, only the soluble species display a catalytic role. We think that at this relatively high temperature the reactions between titanium tetrachloride and water are fast and reach an equilibrium leaving a certain amount of water (free or complexed) in solution. The excess of $TiCl_4$ and this residual water constitute a stable complex mixture of species possessing all that is needed to induce the formation of chain carriers when added to the monomer, probably through an Ad_E3 pathway. These results differ from those reported by Pham Minh Chau and Cheradame[397-475], who, as reported above, observed a detrimental effect of aging. The major experimental difference between these two sets of data is the temperature, which was much lower (about $-70\ ^\circ C$) in the latter work. It seems plausible to assume that the reaction of the two catalytic species at low temperature is much slower so that the concentration of available water, and thus the initiation efficiency, decreases steadily with time.

The polymerisation of α-methylstyrene by $BuOTiCl_3$ in methylene chloride at $-70\ ^\circ C$[413] has already been discussed in the context of direct initiation, of which it is a typical example. In that study the effect of added water was also investigated. The rate of initiation was found to depend on the first power of water and catalyst concentration, but to be independent of monomer concentration; these relationship applied when $[H_2O] < [BuOTiCl_3]$. When the amount of water exceeded that of the catalyst, the initial polymerisation rate increased more than expected. This coupled with the facts that hydrogen chloride was found to be a more efficient cocatalyst than water and that the rate of polymerisation with water reflected a continuous increase in the number (or activity) of the chain carriers, suggests that the formation of HCl by the reaction of the catalyst with water slowly builds up the initiation potentials of the system. With excess water that acid production is accelerated and higher and higher initial rates are observed. Apart from these peculiarities, we believe that the behaviour of this polymerisation with small amounts of water can be rationalised by the following initiation scheme, valid before appreciable quantities of hydrogen chloride are produced:

$$BuOTiCl_3 + H_2O \rightleftharpoons BuOTiCl_3 \cdot H_2O$$

$$BuOTiCl_3 + \overset{\backslash}{\underset{/}{C}}{=}\overset{/}{\underset{\backslash}{C}} \rightleftharpoons \ \substack{\backslash / \\ C \\ \| \\ C \\ / \backslash} \rightarrow BuOTiCl_3$$

$$BuOTiCl_3 \cdot H_2O + \ \substack{\backslash / \\ C \\ \| \\ C \\ / \backslash} \rightarrow BuOTiCl_3 \rightleftharpoons I_{Ad_E3} \rightarrow \text{active species;}$$

I_{Ad_E3} indicating the concerted intermediate. In this sequence of equilibria the rate determining step is probably the proton transfer on the double bond to give the active species.

The authors of this very interesting paper[413] also carried out a detailed kinetic study of the course of the polymerisation and showed that the concentration of active species went through a maximum value and then decreased with time both at -50 and $-30\ ^\circ C$. This behaviour accounts for the fact that at those temperatures the conversion-time curves were S-shaped.

d) Cocatalysis with Boron Halides

The interaction of boron fluoride with water was the object of a very careful study by Clayton and Eastham[555]. They found that in ethylene chloride two hydrates, $BF_3 \cdot H_2O$ and $BF_3 \cdot 2 H_2O$, are formed at 25 °C in equilibrium with free boron fluoride. In the same paper, the isomerisation of butene-2 by this catalytic mixture was examined kinetically. The rate of isomerisation was found to be proportional to the free BF_3 as well as to its water-complex concentration, and to the olefin concentration. No complex between boron fluoride and monomer was identified, but in their conclusions the authors noted that "a mechanism involving attack of the complex $BF_3 \cdot H_2O$ on the complex butene-BF_3 would satisfy the kinetics, but seems to have little else to recommend it". In fact we feel that precisely that mechanism was operative, through an Ad_E3 intermediate, the Lewis acid-olefin complex being formed, but as in many other instances being too weak to be detected. In the light of present-day knowledge, this conclusion seems quite acceptable, but twenty years ago it probably seemed far-fetched. In any case, the authors clearly realised this possibility, but were too cautions about proposing it outright.

Bywater and Worsfold[92] examined the role of water cocatalysis in the dimerisation of 1,1-diphenylethylene catalysed by boron fluoride at 20 °C in methylene chloride. A temperature-dependent equilibrium between the carbenium ions produced and the reactants was observed and the quantitative relationship expressing this correlation was given as:

$$[C^+]^2 = aK [H_2O]_0 ([BF_3]_0 - [H_2O]_0) [M]_0,$$

the ions being highly dissociated into free species. The difference between the initial concentrations of boron fluoride and water obviously reflect the concentration of free boron fluoride remaining after its rapid complexation with water to give the monohydrate. This relationship therefore confirms the previous observations of Clayton and Eastham[555]. The dependence of carbenium ion concentration on the initial monomer concentration was in effect somewhat lower than 0.5, indicating that the catalyst (free or complexed with water) is strongly complexed with the monomer. This is not surprising in view of the pronounced acidity of the Lewis acid used and the high basicity of 1,1-diphenylethylene. All the above evidence can readily be interpreted in terms of an Ad_E3 mechanism of initiation involving the intermediate

the monohydrate acting as nucleophile. The carbenium ion produced by proton transfer to the double bond then reacts rapidly with the monomer to give the corresponding dimeric ion. As usual, transfer reactions bring about subsequent dimerisation. The square

root dependence with water and catalyst concentrations is a consequence of the predominance of free carbenium ions among the products of this process. In this same
paper, Bywater and Worsfold also studied the ionisation of 1,1-diphenylpropene by the
same initiating system. They found that in the same operating conditions, the amount
of carbenium ions produced with this olefin, which cannot dimerise, was smaller than
with 1,1-diphenylethylene. However, the law relating the concentration of these ions
to the concentrations of the various reagents was basically the same as in the preceding
system.

Kennedy and coworkers recently studied the different catalytic behaviour of boron
fluoride, chloride and bromide in the polymerisation of isobutene cocatalysed by
water[556]. With BF_3–H_2O, the monomer is readily polymerised and the product yield
increases with increasing temperature; this process does not require a polar medium
to be effective. With BCl_3, the situation changes drastically. This Lewis acid alone is
incapable of initiating the polymerisation of pure isobutene at $-78\ ^\circ C$, and addition
of water is not sufficient to activate the system. However, when an alkyl halide is
introduced in the dormant mixture (CH_2Cl_2, CH_3Cl or C_2H_5Cl), polymerisation takes
place and the yield is higher the greater the amount of solvent added. Thus, the
BCl_3–H_2O pair requires a relatively polar medium to become an effective initiator,
at $-78\ ^\circ C$. At $-50\ ^\circ C$ the polymerisation of isobutene is not observed even in methylene chloride. Another complication in these systems is that because of the fast hydrolysis of BCl_3, water must be added to the monomer-catalyst solution. This hydrolysis
only occurs in fairly polar media, since a mixture of catalyst and cocatalyst in pure
isobutene is stable. With boron bromide the polymerisation was extremely inefficient
even with water and chlorinated hydrocarbons. These results were interpreted assuming
that the hypothetical acid $H^+BF_3OH^-$ is stronger than $H^+BCl_3OH^-$ and that BBr_3
hydrolyses too rapidly to give the complex acid. An alternative explanation was based
on the occurrence of an addition of HX (from the hydrolysis of the Lewis acid) to
isobutene to give the corresponding t-butyl halide. It was postulated that t-butyl fluoride is a cocatalyst with BF_3, but that t-butyl chloride and bromide are not with
their respective boron halides. Indeed it was later shown[557] that the system isobutene–
BCl_3–H_2O in heptane at $-20\ ^\circ C$ gives t-butyl chloride and no polymerisation because
the formation of the alkyl halide must be considered as a dominating termination reaction. The decrease in polymer yield with increasing temperature is well explained by
the enhanced rate of hydrolysis followed by addition of HCl to isobutene. These results also suggest that HCl is obviously not a cocatalyst for the polymerisation of isobutene by BCl_3, because the predominant reaction would be the formation of t-butyl
chloride (termination). The behaviour of these various systems can be explained qualitatively by the Ad_E3 mechanism when initiation can procede. The positive role of
increased polarity arises from a better stabilisation of the concerted intermediate and
of the active species formed thereby. Also, the monomer-Lewis acid complex is
evidently favoured in a medium of higher dielectric constant. Unfortunately these
interesting studies were not complemented by a more quantitative approach including
the establishment of the relevant kinetic relationships. If we assume however that the
boron halides investigated can all give the corresponding hydrates, the following general mechanism can be tentatively offered, on the basis of the previous findings by
Clayton and Eastham, and by Bywater and Worsfold:

$$
\begin{array}{c}
\overset{\displaystyle BX_3}{\underset{\displaystyle \quad}{\vdots}} \\
\overset{\delta+}{C}\!-\!\overset{\vdots}{C} \\
\end{array}
\quad\longrightarrow\quad
\overset{+}{C}\!-\!C \;+\; BX_3.
$$

The ionic product can either rapidly collapse into the t-butyl halide by reaction with the anion (termination) or propagate with the monomer. When X = F, the anion is stable and the sustained ionisation favours polymerisation. When X = Cl, the anion is bulky and tends strongly to release a chloride ion to the carbenium ion, unless its stability is enhanced by a higher polarity of the medium. When X = Br, the instability of the anion is so marked, that formation of BBr_2OH and Br^- is practically instantaneous and the carbenium ion pairs collapses without any chance to propagate, even in a polar surrounding.

e) Cocatalysis with Aluminium Compounds

The mechanism of water cocatalysis in polymerisations promoted by aluminium chloride has never been elucidated, because of the complicated behaviour of these systems. The intricacies found with isobutene[558] are a good illustration of the difficulty in formulating a reasonable explanation for all the experimental observations. This Lewis acid often gives rise to heterogeneous systems, a fact which makes fundamental studies almost impossible. Moreover, direct inititiaon is readily achieved with $AlCl_3$ and its alkyl derivatives and a distinction between this phenomenon and cocatalyis by moisture or added water would require very complex investigations.

The asymmetric selective polymerisation of olefinic monomers has been studied in the context of coordinated anionic catalysis, a type of process related to Ziegler-Natta polymerisation. Preferential polymerisation occurs with the monomer possessing the same absolute configuration as the catalyst[559]. Cationic polymerisations induced by an asymmetric catalyst are not numerous and only certain specific monomers can be polymerised selectively to give optically active products. It has been shown for instance that only monomers bearing a β-methyl substituent undergo stereoselective propagation[560]. The asymmetric α-substituent may be an alkoxy or an arylalkoxy group, or a group possessing the asymmetric center close to the olefinic double bond. In the latter instance, the effect of asymmetry is more marked than in the former[560]. Stereoselective action was achieved in the polymerisation of racemic 1-methylpropylpropenyl ether by (−)menthoxyaluminium dichloride, and only with the *cis* isomer of the monomer. The stereoselectivity decreased with decreasing temperature and with increasing medium polarity. All these observations suggest that the counterion played a determining role in the process. In this study no particular care was taken to exclude moisture from the system, so that water cocatalysis was most probably responsible for the initiation, giving the optically active counterion through the Ad_E3 mechanism.

Ueshima et al.[561] have attempted to demonstrate that water is a cocatalyst in polymerisations induced by triethylaluminium. They showed that tritiated water reacted with the catalyst to give species capable of activating the polymerisation of styrene and *iso*butylvinyl ether in methylene chloride at -78 °C. They hoped to be able to distinguish between two possible effects of water: (i) transformation of the catalyst into a more acidic entity, and (ii) creation of protons for the initiation of cationic polymerisation. Some tritium was found in the polymers, and the highest yields were obtained with a $[H_2O]/[AlEt_3]$ molar ratio of one. These experiments were however inconclusive, because water may also have played a part as terminating agent. It is nevertheless certain that triethylaluminium is activated by water, although the mechanism of this "cocatalysis" is unclear.

Saegusa et al.[861,862] studied the polymerisation of various monomers promoted by the catalytic combination Et_3Al-H_2O. They concluded that highly acidic entities are formed in the interaction of these two compounds and that their amount is maximum for a 1:1 molar ratio. Such a situation is quite different from ordinary cocatalysis since the initiating species were in fact a novel chemical combination possessing the proposed formula $Et_2AlO-(Al-O-)_{\overline{n}}AlEt_2$.
$$\underset{\text{Et}}{\overset{|}{}}$$

f) Cocatalysis with Miscellaneous Halides

Masuda et al.[562] studied the polymerisation of phenylacetylene catalysed by tungsten hexachloride and molybdenum pentachloride in benzene at 30 °C. Water cocatalysis was qualitatively established, although the authors claimed that the monomer itself could also act as a cocatalyst. This conclusion is however in apparent contradiction with the fact that polymer yields were incomplete. The only way to reconcile these two phenomena would be that terminating species were produced capable of inhibiting further initiation.

The kinetics of the polymerisation of *iso*butylvinylether catalysed by zinc chloride in a mixture of ethylene chloride and heptane were investigated by Bhattacharya and Mukherjee[563]. The initial rate of monomer consumption was found to be proportional to the square of the monomer concentration and to the catalyst concentration. The experimental techniques employed certainly left residual moisture in the systems and water cocatalysis seems most likely. The kinetic results are compatible with an initiation mechanism involving the Ad_E3 intermediate.

The polymerisation of benzene through repeated nucleophilic substitutions on the rings was studied by Kovacic et al.[564] using ferric chloride as catalyst and water as cocatalyst. This system is of course outside the realm of cationic polymerisation through the double bond of an olefin, but illustrates well the role of water in Friedel-Crafts polycondensations. The authors showed that the rate of this reaction went through a maximum at a catalyst/cocatalyst ratio of one and attributed this observation to the high activity of ferric chloride monohydrate:

$$FeCl_3 + H_2O \rightleftharpoons H^+ FeCl_3OH^-$$

$$H^+ FeCl_3OH^- + C_6H_6 \rightarrow C_6H_7^+ FeCl_3OH^-.$$

This initiation mechanism was thought to be followed by propagation involving *para*-polyphenylation with reduction of ferric chloride to ferrous chloride and formation of an equivalent amount of hydrogen chloride.

4. Cocatalysis by Organic Oxygen Compounds

a) Introduction

This section examines the the details of cocatalytic initiation processes involving three classes of compounds: (i) alcohols and phenols, resembling water in their structure and behaviour; (ii) carboxylic acids, probably acting in a similar way to hydrogen halides; and (iii) ethers, a somewhat special type of cocatalyst. As in the preceding sections, we will attempt to establish at what point in the initiation sequence these cocatalysts intervene and whether they react as independent molecules or as complexes with another component of the system. Our aim is to show that much evidence is available in favour of a general mechanism resembling those already proposed for hydrogen halides and water, i.e. the formation of Ad_E3-type intermediates immediately preceding the protonation of the monomer.

b) Cocatalysis by Alcohols and Phenols

In some of the first studies connected with the discovery of the general phenomenon of cocatalysis, *t*-butanol was found to play such a role in the polymerisation of isobutene by boron fluoride[564,565]. There is an apparent contradiction between this observation and the fact that polyisobutene chains bearing hydroxyl end groups[566], i.e. the same structure as *t*-butanol, do not exhibit a cocatalytic function (see all the evidence of limited yields due to water consumption by termination with counterion to give those end groups). The likely explanation of this dichotomy is that the OH groups on the polymer molecules are embedded in the macromolecular coils and therefore much less reactive towards the Lewis acid-monomer complex.

 Methanol is an effective cocatalyst in conjunction with boron fluoride. Okamura et al.[567] ascribed this initiation capability to the formation of a partly dissociated complex between Lewis acid and alcohol. This suggestion was recently verified in a specific study on this interaction involving vibrational spectroscopy[568]. Boron fluoride was found to give well-defined 1:1 and 1:2 adducts with methanol and the Raman spectrum of the latter complex indicated some dissociation according to the reaction

$$BF_3 \cdot 2\,H_2O \; \rightleftharpoons \; BF_3OCH_3^- \; CH_3OH_2^+.$$

 Eastham and collaborators carried out a thorough study of the isomerisation and polymerisation of butenes by the BF_3–CH_3OH pair[569,570]. They found that the initial rate of isomerisation of butene-2 was proportional to the concentration of both the catalyst-cocatalyst complex and the free catalyst. We think that these findings are consistent with the Ad_E3 mechanism, passing through a highly polarised intermediate of the type:

$$
\begin{array}{c}
BF_3 \\
\vdots \quad / \\
\overset{\delta+}{C}\!\!-\!\!-\!\!-\!\!-\!\!C \\
\vdots \qquad \backslash \\
CH_3\!-\!O\!-\!H \\
\vdots \\
BF_3
\end{array} \quad ,
$$

the role of the base being played by the oxygen atom. The fact that the isomerisation rate was independent of the olefin concentration suggests that the actual isomerisation reaction is the slow step in the above process. In other words, the butene concentration was high enough in these experiments to complex virtually all of the boron fluoride not bound to methanol, thus making the concentration of the Ad_E3 intermediate independent of the amount of olefin added. If the rate of formation of the intermediate had been the determining step, the rate of isomerisation would have depended on the monomer concentration to the first power. The rate of polymerisation was also measured as a function of the reagents' concentration and the following relationship obtained (Fig. 11):

$$-d\,[M]/dt = k\,[BF_3]\,[BF_3 \cdot CH_3OH]\,[M],$$

where $[BF_3]$ was assumed to be the concentration of free boron fluoride, but was in fact most probably the concentration of its complex with the monomer. These results are readily interpreted in terms of the mechanism proposed above for the isomerisation reaction, the kinetic dependence on monomer arising from the propagation step.

Zlamal and Kazda[571)] investigated the effect of various alcohols and phenol on the polymerisation of isobutene catalysed by aluminium chloride in ethyl chloride at $-78\,°C$. They obtained evidence for the formation of complexes between this Lewis

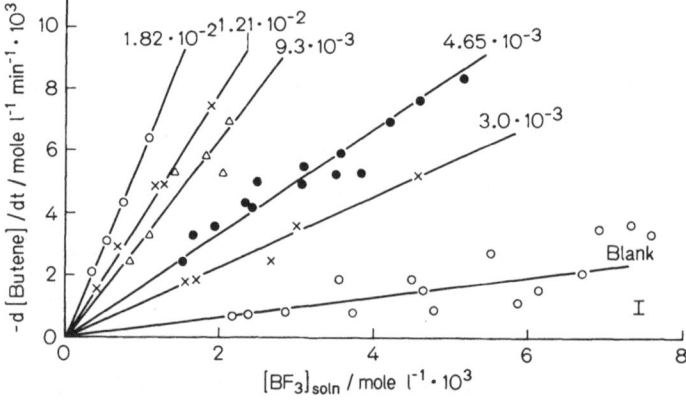

Fig. 11. Dependence of the rate of disappearance of butene-2 (0.40 M) on the concentration of free BF_3 in the presence of $BF_3 \cdot CH_3OH$ (concentrations as indicated on the curves). Solvent: ethylene chloride; $T = 0\,°C$. (Taken from Ref. [570)])

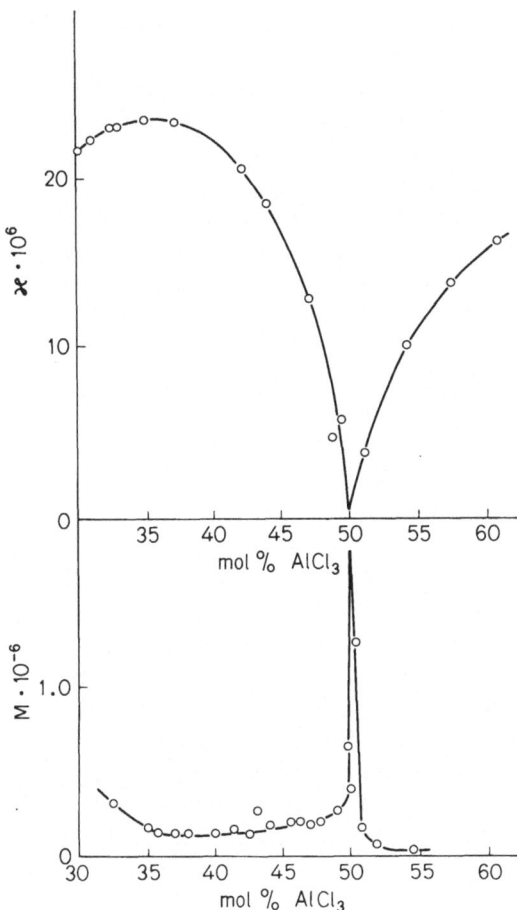

Fig. 12. Variation of electrical conductivity and polymer molecular weight in the polymerisation of isobutene by $AlCl_3-C_2H_5OH$ in ethyl chloride at $-78.5\,^{\circ}C$, as a function of the catalyst/cocatalyst ratio. (Taken from Ref. [571])

acid and the hydroxyl compounds, possessing the general formula $(ROH)_x(AlCl_3)_y$, where the ratio x/y may vary from 0.5 to 4. These complexes are ionised and their ability to initiate the polymerisation of isobutene was directly related to the acidity of the alcohol used, phenol giving by far the highest activity. It was also observed that the rate of polymerisation gradually decreased with increasing excess of alcohol above $[ROH]/[AlCl_3] = 1$. This conclusion was however drawn on roughly estimated rate values and it is therefore difficult to provide a mechanistic interpretation of the initiation process in these systems for lack of quantitative correlations. It seems nevertheless that the role of all hydroxyl compounds used can be ascribed to typical cocatalysis, based on the concerted attack of the ionised complex on the monomer double bond, probably assisted by free $AlCl_3$ or another molecule of complex. Another interesting result obtained in this work was the correlation between the conductivity of the complex solutions and the DP's of the polymers formed, as a function of the $[ROH]/[AlCl_3]$ ratio. A conductivity minimum was found for ethanol at a 1:1 ratio and this corresponded to a maximum in the polyisobutene DP, as shown in Fig. 12. The implications of these findings have already been thoroughly discussed by Plesch[572].

We feel that the correlation between conductivity and number of initiated chains ought to be accepted with caution, unless the details of transfer and termination reactions are well understood and can be taken into account quantitatively, if necessary, to correct the original correlation. In other words, a given number of ions derived from the initiating complex does not necessarily lead to an equal or proportional amount of chain carriers, owing to possible complicating factors intervening in the actual polymerisation.

Russell and coworkers have studied in great details the role of phenols as cocatalysts, transfer and termination agents in the polymerisation of isobutene catalysed by stannic chloride. This very interesting series of papers[573−575] deserves a close inspection. The equilibrium constant of the interaction

$$\text{ArOH} + \text{SnCl}_4 \; \rightleftharpoons \quad \overset{\displaystyle Ar}{\underset{\displaystyle H}{\diagdown\diagup}} O \rightarrow \text{SnCl}_4$$

was determined over the temperature range −45 to 30 °C. The cocatalytic effect of phenol and various substituted phenols was tested and it was found that many gave positive results, but the more bulkily substituted ones were inert because of steric hindrance. The influence of phenol on the initial rate of polymerisation was studied quantitatively. At low phenol concentration a first order dependence was observed, but then the rate went through a maximum and decreased to reach a value approximately independent of phenol concentration. The maximum occurred at a catalyst-cocatalyst ratio of about 10. This behaviour is similar to that reported by Polton and Sigwalt[415] for the system indene−SnCl₄−HCl. Russell's group also studied the kinetics of polymerisation as a function of monomer concentration, keeping the dielectric constant of the medium fixed by using isobutane-isobutene mixtures. It was found that the initial rate of polymerisation was proportional to the first power of the monomer concentration, and this was taken as evidence for a rate of initiation independent of the monomer concentration. The interpretation of this kinetic correlation consisted in formulating an initiation mechanism based on the attack on a π-complexed monomer molecule by the stannic chloride-phenol adduct, the π-complexation between monomer and catalyst being important (i.e. little free SnCl₄ left). Previous work by Norrish and Russel on the same pair but using water as a cocatalyst[544−546] had given, as we have already discussed, a similar behaviour. The proposal of an Ad_E3 mechanism seems to us quite reasonable in the present context too:

$$\text{SnCl}_4 + \overset{\diagdown}{\underset{\diagup}{}}C{=}C\overset{\diagup}{\underset{\diagdown}{}} \; \longrightarrow \quad \overset{\diagdown\diagup}{\underset{\diagup\diagdown}{\underset{C}{\overset{\|}{C}}}} \rightarrow \text{SnCl}_4$$

$$\text{SnCl}_4 + \text{ArOH} \; \longrightarrow \; \text{ArOH} \cdot \text{SnCl}_4$$

with a rate-determining ionisation of the above intermediate to give the initiating t-butyl cation coupled with $SnCl_4OAr^-$. Kinetic isotope effect experiments using OD deuterated phenol showed irrefutably that proton (deuteron) transfer is the basic step of cocatalysis by phenols and that this reaction is the rate-determining step of the initiation process. The value of 2.5 obtained for k_H/k_D is in fact typical of a proton transfer reaction. The assumption that the monomer-Lewis acid complex is attacked by the phenol-Lewis acid adduct was adequately supported by independent studies of the adduct formation, which showed that in the conditions of polymerisation most of the phenol is bound to stannic chloride. Finally, Russel and collaborators obtained detailed evidence concerning the participation of phenols in transfer and termination reactions, as in the case of water cocatalysis. Of particular interest is the occurrence of the transfer reaction

which produces incorporation of the cocatalyst in the polymer chain. Although this topic falls outside the scope of the present review, we wish to underline that the above interaction possesses a remarkable potential in the synthesis of polymeric and non-polymeric reactive species. The extreme susceptibility towards alkylation of the 4-position of the phenyl ring in 2,6-disubstituted phenols was demonstrated by Russell[576], who showed that 2,4,6-tri-t-butyl phenol is formed in large amounts when isobutene is polymerised in the presence of 2,6-di-t-butyl phenol. The above transfer reaction is in fact so frequent that it overwhelms propagation by trapping the active species at their monomeric stage.

Sangalov and collaborators[897] recently published an interesting study on the catalytic activity of $EtAlCl_2$-alcohols combinations for the polymerisation of isobutene and the oligomerisation of its dimers.

c) Cocatalysis by Carboxylic Acids

The electrophilic addition of carboxylic acids to olefins can be drastically enhanced by the presence of a Lewis acid. Indeed, for the weakest acids such as acetic acid itself, the addition does not take place in the absence of a catalyst (strong Brønsted acid or Lewis acid) capable of increasing the protonating power of the medium, unless very basic olefins are used[237]. The role of Lewis acids in these esterification reactions must be briefly analysed before dealing with polymerisation systems in which carboxylic acids have been employed as cocatalysts.

The use of boron fluoride[577] or boron fluoride etherate[578] as catalysts for the esterification of olefins by various carboxylic acids dates from the thirties. Since then numerous Lewis acids have been reported to perform the same task[579]. Guenzet and Camps[579] showed that a linear relationship exists between the Hammett acidity function and the concentration of the Lewis acid dissolved in pure carboxylic acid. The effect of the presence of the latter species towards the increase of the acidity of the mixture is very large. Thus, the acidity of a 1 M solution of stannic chloride in glacial acetic acid is slightly higher than that of a solution of sulphuric acid of the same concentration in the same solvent. This is apparently not due to any significant dissociation of the Lewis acid, as shown by cryoscopic results obtained nearly a century ago[580]. Instead, more recent studies[581-583] proved that a strong complex, $SnCl_4 \cdot (CH_3COOH)_2$, is formed which protonates acetic acid in the equilibrium reaction:

$$SnCl_4 \cdot (CH_3COOH)_2 + CH_3COOH \rightleftharpoons SnCl_4 \cdot H(CH_3COO)_2^- \, CH_3COOH_2^+.$$

Guenzet and Camps[579] proposed that the above interaction actually proceeds further and the real acidic species possesses the structure:

$$2\, CH_3COOH_2^+ \, ,$$

and concluded from kinetic evidence that in the esterification of olefins by this entity the protonation of the double bond is the rate-determining step, the carbenium ion intermediate rapidly collapsing with the acetate anion. Zinc, aluminium, titanium and ferric chlorides were also examined in this context and were found to give the same qualitative behaviour[584]. The addition of acetic acid onto some cyclolefins in the presence of stannic chloride or boron fluoride was carefully investigated by the same authors[585], and the following order of ease of esterification was obtained: cycloheptene $\geqslant$ trans-cyclooctene $>$ cyclopentene $>$ cyclohexene. These processes were interpreted according to the mechanism:

$$MtX_n + x\, CH_3COOH \rightarrow MtX_n \cdot (CH_3COOH)_x,$$

$$MtX_n \cdot (CH_3COOH)_x + CH_3COOH \rightarrow MtX_n \cdot H_{x-1}(CH_3COO)_x^- \, CH_3COOH_2^+,$$

$$MtX_n \cdot H_{x-1}(CH_3COO)_x^- \, CH_3COOH_2^+ + \, \overset{\diagdown}{\underset{\diagup}{C}} = \overset{\diagup}{\underset{\diagdown}{C}} \rightleftharpoons I,$$

$$I \rightarrow H-\overset{|}{\underset{|}{C}}-\overset{|}{\underset{|}{C}}{}^{+}MtX_n \cdot H_{x-1}(CH_3COO)_x^- + CH_3COOH, \quad \text{and}$$

$$H-\overset{|}{\underset{|}{C}}-\overset{|}{\underset{|}{C}}{}^{+}MtX_n \cdot H_{x-1}(CH_3COO)_x^- \rightarrow$$

$$H-\overset{|}{\underset{|}{C}}-\overset{|}{\underset{|}{C}}-O-COCH_3 + MtX_n \cdot (CH_3COOH)_x;$$

where I is a complex of unspecified structure, and x = 1 for BF_3, and x = 2 for $SnCl_4$. In these experiments, performed in glacial acetic acid, the measured rate of esterification was found to be proportional to the first power of the olefin concentration and to the 1.5th power of the $SnCl_4$ concentration, but to the second power of the BF_3 concentration. These empirical observation are not in agreement with the proposed mechanism shown above. In the systems involving boron fluoride, we reinterpret the results on the basis of the Ad_E3 mechanism and of the well-known fact that BF_3 with excess acetic acid gives a 2:1 adduct. The selfionisation of this adduct,

$$BF_3 \cdot (CH_3COOH)_2 \rightleftharpoons BF_3 \cdot CH_3COO^- \; CH_3COOH_2^+$$

would provide the acidic and basic moieties for the concerted attack on the double bond to give the intermediate:

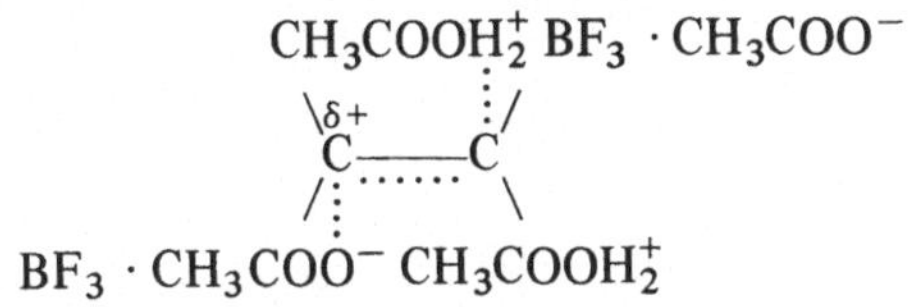

$$
\begin{array}{c}
CH_3COOH_2^+ \; BF_3 \cdot CH_3COO^- \\
\overset{\delta+}{\underset{/ \vdots}{\overset{\backslash}{C}}}\!\!\!\text{------}\!\!\!\overset{\vdots}{\underset{\backslash}{\overset{/}{C}}} \\
BF_3 \cdot CH_3COO^- \; CH_3COOH_2^+
\end{array}
$$

which produces the corresponding acetate. This scheme readily explains the orders in both olefin and boron fluoride. With stannic chloride the order in catalyst is less obvious, but it could be assumed that the reaction goes along the same pathway, and that only the free ions are effective in promoting the esterification, the ion pairs being insufficiently acidic. This is not unreasonable, in view of the weaker acidity of stannic chloride, compared with boron fluoride.

Latrémouille and Eastham[191)] had already studied a similar esterification process. They found that the reaction of butene-2 with acetic acid-boron fluoride catalytic mixtures in ethylene chloride at 20 °C gave a quantitative yield of the corresponding acetate if the molar ratio $[CH_3COOH]/[BF_3]$ was higher than unity. The kinetics of this esterification followed the empirical law:

$$d[ester]/dt = k[CH_3COOH \cdot BF_3]^2[butene],$$

which is entirely compatible with the Ad_E3 mechanism discussed above for the work of Guenzet and Camps. They also noticed that the adduct $BF_3 \cdot (CH_3COOH)_2$ was a far less active catalyst than the 1:1 species. A similar behaviour had been observed by Clayton and Eastham[555] with the 1:1 and 1:2 adducts of boron fluoride and water.

In the above studies the use of an excess carboxylic acid over the Lewis acid determined the specific situation whereby the catalytic species is an adduct between the two acidic components, virtually no free Lewis acid being present. Such situations seems most adequate for maximising esterification and minimising polymerisation. The presence of free Lewis acid in these types of systems tends on the contrary to favour the polymerisation of the olefin, or its isomerisation. This situation has been exploited in a large number of investigations in which carboxylic acids have been used as cocatalysts for the cationic polymerisation of alkenyl monomers. The majority of these reports are qualitative in that they do not examine the kinetics and mechanism of these cocatalytic processes. But a few excellent studies have been carried out dealing with these fundamental aspects, and these are reviewed below.

The decisive contributions made by Plesch to the understanding of the chemistry of cationic polymerisation also includes a pioneering piece of research in this specific area. In 1950, he published a classical paper on the role of trichloroacetic acid as cocatalyst in the polymerisation of isobutene by titanium tetrachloride in hexane at low temperature[586]. An analysis of the dependence of the maximum polymerisation rate upon the concentration of monomer was carried out by us from the plots published by Plesch and it can be concluded that a second power law was approximately followed. The presence of an excess of $TiCl_4$ over the cocatalyst gave a maximum rate proportional to the concentration of the latter. We assume that in these conditions the following mechanism accounts for the experimental observations:

$$TiCl_4 + CCl_3COOH \rightleftharpoons TiCl_4 \cdot CCl_3COOH$$

$$TiCl_4 + \overset{\backslash}{\underset{/}{C}}{=}\overset{/}{\underset{\backslash}{C}} \rightleftharpoons \begin{matrix} \backslash\,/ \\ C \\ \| \\ C \\ /\,\backslash \end{matrix} \rightarrow TiCl_4$$

$$\begin{matrix} \backslash\,/ \\ C \\ \| \\ C \\ /\,\backslash \end{matrix} \rightarrow TiCl_4 + TiCl_4 \cdot CCl_3COOH \longrightarrow \begin{matrix} \backslash^{\delta+} & \overset{\vdots}{\underset{\vdots}{C}}{-}\overset{TiCl_4}{\underset{}{}} \\ C{-}{-}{-}C \\ /\,\vdots & \ddots \backslash \\ & TiCl_4 \cdot CCl_3COO{-}\!\!-H \end{matrix} \longrightarrow$$

$$\longrightarrow \begin{matrix} \backslash^{+} & / \\ C{-}\!\!-C \\ / & |\,\backslash \\ & H \end{matrix} \; TiCl_4 \cdot CCl_3COO^- + TiCl_4.$$

It can be assumed that since $[CCl_3COOH]_0 \ll [TiCl_4]_0$, all the carboxylic acid is complexed with the catalyst and consequently the rate of polymerisation is directly proportional to the concentration of the cocatalyst, i.e. the first equilibrium in the above scheme is very strongly shifted to the right-hand side. An inspection of Plesch's plots shows moreover that at relatively low titanium tetrachloride concentrations, and constant trichloroacetic acid concentration, the maximum polymerisation rate depended on $[TiCl_4]^2$, in agreement with our proposed Ad_E3 mechanism. However, when the

Lewis acid concentration is increased above a certain value, its effect on the rate becomes less marked and finally the rate becomes independent of $[TiCl_4]$ for even higher concentrations. This is also reasonable, since an excess of $TiCl_4$ makes the overall equilibrium

$$TiCl_4 \cdot CCl_3COOH + TiCl_4 + \overset{\diagdown}{\underset{\diagup}{C}}{=}\overset{\diagup}{\underset{\diagdown}{C}} \rightleftharpoons H{-}\overset{|}{\underset{|}{C}}{-}\overset{|}{\underset{|}{C}}{}^+ \ TiCl_4 \cdot CCl_3COO^- + TiCl_4$$

independent of its concentration, the number of active species being controlled by the amount of cocatalyst and olefin present. Previous work by Evans and Meadows[565] on the system isobutene–BF_3–CH_3COOH can probably be interpreted in a similar fashion, but we preferred to focus our attention on Plesch's study because of the valuable kinetic information it provided.

Clayton and Eastham[587] investigated the *cis-trans* isomerisation of butene-2 promoted by the boron fluoride-acetic acid combination at 20 °C in ethylene chloride. With olefin concentrations lower than about 0.5 M and an excess of Lewis acid with respect to the cocatalyst, the initial rate of isomerisation was found to obey the simple kinetic relationship $R_i = [Butene] [BF_3] [BF_3 \cdot CH_3COOH]$. This behaviour suggests the formation of an Ad_E3 intermediate characterised by the conjoint attack of boron fluoride and its adduct on the olefin double bond. At higher butene concentrations, the order in olefin concentration progressively decreased to become practically zero. This observation can be rationalised in terms of virtually complete complexation of the free boron fluoride by the olefin when the concentration of the latter is higher than a certain critical value. Considering the relative efficiency of different cocatalysts in this process, the authors concluded that the experimental ratio of $50:15:1$ for the rate of isomerisation with water, acetic acid and methanol respectively, was due to the different stabilities of the anions derived from the active complex $BF_3 \cdot$ cocatalyst. We prefer to look upon these results within the perspective of our new generalised Ad_E3 mechanism, and attribute the different cocatalytic power to the relative nucleophilicities of the anionic moiety of the $BF_3 \cdot$ cocatalyst adduct.

Isobutene was reported to polymerise rapidly in the presence of boron fluoride and trifluoroacetic acid[191], while only the corresponding ester was obtained in the absence of the Lewis acid (see discussion in Sect. III-B). The fact that trifluoroacetic acid complexes weakly with boron fluoride and thus leaves appreciable amounts of free Lewis acid in solution produced a situation most favourable to polymerisation. The nature of the counterion in this system is a particularly interesting point. Brownstein[588] recently showed that the trifluoroacetate anion gives strong complexes with many metal fluorides, including BF_3 with which the following disproportionation occurs:

$$BF_3 + CF_3COO^- \rightarrow BF_3 \cdot CF_3COO^-$$

$$2 \, BF_3 \cdot CF_3COO^- \rightleftharpoons BF_4^- + BF_2(CF_3COO)_2^-.$$

It is difficult to speculate about the different stabilities of the two anions formed, i.e.

which one would be the suitable companion of the growing carbenium ion in the polymerisation of isobutene and other monomers.

Prosser and Young[589] reported an interesting study of the polymerisation of indene
and acenaphthylene by titanium tetrachloride and trichloroacetic acid in methylene
chloride at $-10\ ^\circ C$. The process was characterised by two distinct stages. During the
first, the rate of polymerisation fell from an initial maximum value to a limiting value
after which the second stage took over to give internal first-order kinetics. The initial
rate was found to be directly proportional to the monomer and the carboxylic acid
concentration, but, at constant Lewis acid Brønsted acid concentration ratio, independent of $TiCl_4$ concentration. The latter observation was obtained with a $[TiCl_4]/$
$[CCl_3COOH]$ ratio of 1.85, although the highest rates were observed for a value close
to unity. It was also reported that a decrease in temperature produced an enhancement
of the yield due to the first stage. Finally, no carbenium ions could be detected spectroscopically during these polymerisations. All the above peculiarities, underlining the
complexity of these systems, seem to indicate that the initiation process is quite different from the general scheme invoked for most of the cocatalytic studies analysed up
to now. We feel that we might be witnessing another example of pseudocationic polymerisation (see Chap. III) induced by the complex acid $TiCl_4 \cdot CCl_3COOH$, i.e. active
species consisting of ester molecules (trichloroacetates) activated by titanium tetrachloride, but not to the extent of giving carbenium ions.

Recent work by Mach and collaborators[590] showed that while stannic chloride
forms complexes with the weaker carboxylic acids as known previously, but not with
stronger ones such as di- and trihalogenated acids, titanium tetrachloride gives rise to
complexes *and* chemical interactions with formation of hydrogen chloride and the
carboxylate $RCOOTiCl_3$. The latter reaction is particularly important with the stronger
carboxylic acids and especially trichloro and trifluoroacetic acid. These results complicate further the interpretation of the cocatalytic mechanism of systems involving such
initiating combinations. Indeed, the formation of hydrogen chloride in equilibrium
with the original components and with the complex gives rise to several possibilities
of initiation pathways. Moreover, the titanium chloride carboxylate must also be considered as potential initiators. Following this specific study of catalyst-cocatalyst interactions, Mach and coworkers published an investigation of the polymerisation of isoprene by $TiCl_4$ with trichloro and trifluoroacetic acids[591] and their *t*-butyl esters as cocatalysts. The behavoiur of these systems was extremely complex and only a superficial
kinetic survey was carried out. The authors concluded that the active species were formed from "oligoester of halogenoacetic acids with $TiCl_4$ and its carboxylates, in which
the ester bond is either considerably polarised or dissociated". In other words it was
supposed that the monomer reacted with the $TiCl_4$ (or carboxylates)-carboxylic acid
complex to give the ester which then initiates polymerisation probably after suffering
ionisation. This scheme is highly speculative in view of the many other possibilities
which could be envisaged in that complex mixture of potential promoters of cationic
polymerisation, and the lack of detailed kinetic and mechanistic data.

d) Cocatalysis by Ethers

The use of ethers as cocatalysts for the cationic polymerisation of alkenyl monomers induced by Lewis acids has received little systematic attention and the mechanism through which these compounds operate is not well understood. The complex diethyl-ether-boron fluoride has been extensively used as a very convenient cationic initiator, but mostly for preparative purposes. As in the case of alcohols and water, ethers are known to act as inhibitors or retarders in the cationic polymerisation of olefins, if used obove cocatalytic levels, because they are more nucleophilic than most π-donor monomers. Imoto and Aoki[592] showed that diethyl ether, tetrahydrofuran, β-chloro-diethyl ether and diethyl thioether are inhibitors for the polymerisation of styrene by the complex $BF_3 \cdot Et_2O$ in benzene at 30 °C, at a concentration lower than that of the catalyst, but high enough (0.5×10^{-3} M) to quench the active species formation for a time. Their action was temporary in that the quenching reaction consumed them, and therefore induction periods were observed, but the DP's of the polystyrenes were independent of the presence of such compounds, as expected from a classical temporary inhibition.

Zlamal and Kazda[593] carried out the best study to our knowledge aimed specifically at understanding the way ethers intervene in a cocatalytic process. They investigated the polymerisation of isobutene catalysed by aluminium chloride in ethyl chloride at −78 °C in the presence of dibutyl ether, anisole and diphenyl ether, using equimolar amounts of catalyst and cocatalyst (1 : 1 adducts). The degree of dissociation of these complexes was also examined by electrical conductivity measurements. Correlations were obtained between the DP's of the polymers and the conductivity of the reacting solutions to show that the length of a chain was inversely proportional to the concentration of anions derived from the initiating complex. The initiation mechanism proposed was based on the postulate that Lewis acid-ether complexes are partly dissociated according to the equation:

$$2 \, ROR' \cdot MtX_n \rightleftharpoons MtX_nRO^- \, MtX_nROR_2'^+,$$

and the positive moiety is responsible for the electrophilic attack on the monomer double bond. In the specific case of the mixed arylalkyl ethers used in this work, it was shown that the alkyl group goes to form the anion and the two aryl groups are present in the cation. A tentative initiation reaction would therefore be:

$$MtX_n \cdot ROR_2'^+ \, MtX_nRO^- + _{/}^{\backslash}C{=}C_{\backslash}^{/} \longrightarrow R'{-}\overset{|}{\underset{|}{C}}{-}\overset{|}{\underset{|}{C}}{}^+ \, MtX_nRO^- + ROR' \cdot MtX_n,$$

while termination would involve the collapse of the active species with formation of a terminal −C−OR ether group. Thus, as in the case of certain systems of water cocatalysis, the ether would be consumed in the overall process leaving uncomplexed or chain-complexed metal halide. More kinetic information would be needed to formulate a detailed mechanism of initiation and in particular the effect of free Lewis acid upon the rate of active species production would provide some important indications about

the transition state formed in the attack of the double bond and the possible intervention of an Ad_E3 reaction pathway.

Giusti and collaborators[594] studied the polymerisation of styrene by $BF_3 \cdot Et_2O$
in ethylene and methylene chloride at 25 °C and the effect of added water to this
system. In this very important paper the initiation steps were discussed in detail and
we report here a summary of the main arguments used and our views on them. The
first point concerns the experiments "in the absence of water". The solvents were re-
fluxed over phosphorus pentoxide and we think that such a drying technique might
not have removed all other cocatalytic impurities completely. Thus, in the experiments
where no water was added a slight doubt remains that initiation might still have been
cocatalysed by traces of other coinitiators. However, if one accepts that the complex
$BF_3 \cdot Et_2O$ is capable of inducing alone the attack on the monomer double bond, the
following kinetic and mechanistic evidence must be taken into account before formulating a suitable reaction scheme for initiation. The initial rate of polymerisation was
found to be directly proportional to the monomer concentration and proportional to
the square of the catalyst concentration. The internal order of these polymerisations
was close to two (rate of monomer consumption) and the second order rate constant
was found to be inversely proportional to the monomer concentration. Isotopically
labelled ethyl groups in the catalyst produced some labelling of the polymer chains.
Added water accelerated the reaction if present in small amounts but slowed it down
if present in larger concentrations. The maximum-efficiency concentration of added
water increased linearly with the initial monomer concentration. In the presence of
added water, the incorporation of labelled fragments in the polymer was drastically
reduced with respect to the amount found in the absence of water.

The authors of this paper proposed an initiation step consisting of the direct attack
of boron fluoride etherate onto the monomer:

$$BF_3 \cdot Et_2O + \overset{\backslash}{\underset{/}{C}}{=}\overset{/}{\underset{\backslash}{C}} \longrightarrow Et{-}\overset{|}{\underset{|}{C}}{-}\overset{|}{\underset{|}{C}}{}^+ \; BF_3 \cdot OEt^-,$$

but gave no definite proof for such a straightforward interaction. The partial incorporation of ethyl groups is not a conclusive piece of evidence in favour of this mechanism
because it could well arise from transfer reactions[595] with the catalyst. These would
be replaced by more efficient reactions with water thus explaining the diminished incorporation in the presence of moisture. Moreover, the above initiation reaction does
not explain the complex kinetic behaviour of the system. Indeed it seems that if a
simple interaction between catalyst and monomer is an inadequate representation of
the process of formation of active species, the alternative Ad_E3-type of mechanism is
not completely satisfactory either. The following scheme:

$$BF_3 \cdot Et_2O + \overset{\backslash}{\underset{/}{C}}{=}\overset{/}{\underset{\backslash}{C}} \rightleftharpoons BF_3 \leftarrow \overset{\overset{\backslash / }{C}}{\underset{\underset{/\ \backslash}{C}}{\|}} + Et_2O$$

$$\begin{array}{c} \diagdown\diagup \\ C \\ \parallel \\ C \\ \diagup\diagdown \end{array} \to BF_3 + BF_3 \cdot Et_2O \rightleftharpoons I$$

where I is a concerted intermediate of the Ad_E3 type, viz.

$$\begin{array}{c}
BF_3 \\
\diagdown^{\delta+}\quad :\diagup \\
C\text{------}C \\
\diagup\!\!\vdots\quad\cdots\cdots\searrow\quad\diagdown \\
BF_3 \leftarrow O\text{---}Et \\
\mid \\
Et
\end{array} \quad ,$$

explains the second order in catalyst, but certainly does not explain the inverse dependence of the second-order rate constant on the initial monomer concentration. The release of a molecule of ether in the first step of this mechanism and the formation of free BF_3 following the formation of active species from the final intermediate are complicating factors which make the interpretation of the behaviour of this system very difficult. Moreover termination with the counterion generates further free Lewis acid:

$$\sim\!\!\!\sim\!\!\overset{\mid}{\underset{\mid}{C}}{}^+ \; BF_3.EtO^- \longrightarrow \sim\!\!\!\sim\!\!\overset{\mid}{\underset{\mid}{C}}\!-O\!-\!Et \; + \; BF_3$$

and the role of uncomplexed boron fluoride in conjuction with its etherate is not known. It would be interesting to carry out further studies on this system to achieve a deeper understanding of the chemistry of initiation with ether cocatalysis.

5. Cocatalysis by Alkyl and Aralkyl Halides and Halogens

a) Introduction

This section deals with the difficult problem of establishing the most plausible initiation mechanism for polymerisation systems in which cocatalysis is ascribed to alkyl or aralkyl halides. This type of cocatalysis is by no means a general phenomenon. We have already mentioned in preceding sections dealing with other types of cocatalysis that, for example, t-butyl chloride is *not* a cocatalyst in the polymerisation of isobutene induced by titanium tetrachloride[397,475] and that 1-phenylethyl chloride is *not* a cocatalyst in conjunction with stannic chloride[389,518,519]. Indeed, the reaction

$$RX + MtX_n \longrightarrow R^+ MtX_{n+1}^-$$

does not occur readily, unless the carbenium ion formed is particularly stable, e.g. Ph_3C^+, or if the Lewis acid is very strong and suitable experimental conditions are employed, as in the work with boron fluoride-alkyl fluorides carried out at low temperature by Olah et al.[596]. Both extreme situations are discussed in the following chapter where initiation by carbocationic salts is reviewed, since they do not fall within the scope of cocatalysis. Extensive studies by Kennedy's group at Akron and by Cesca's and Giusti's groups in Italy have led these authors to conclude that aluminium-based Lewis acids in conjunction with organic halides or halogens give rise to active catalytic systems in which carbenium or halonium ions are generated from the cocatalyst and can initiate the polymerisation of such monomers as isobutene and styrene. We recall here that certain aluminium catalysts used with suitable cocatalysts are among the most efficient initiators for the synthesis of high molecular-weight polymers[597]. Also, as we have shown in Sect. IV-B-4-c), aluminium and alkylaluminium halides can effectively induce the polymerisation of certain olefins via direct initiation.

Our attention in the present section will therefore focus upon aluminium-based catalysts, considering that most of the relevant work has been carried out with them. We hope however that our conclusions will bear a more general validity.

b) The Problem of the True Catalyst

When going from an aluminium trihalide to an aluminium trialkyl, passing through the two mixed alkyl halides, one witnesses a progressive decrease in acidity and a corresponding decrease in activity vis-à-vis initiation of cationic polymerisation. Thus, aluminium trichloride and tribromide as well as ethyl aluminium dichloride are strong Lewis acids capable, as we have seen in detail, to induce the polymerisation of olefins without the aid of a cocatalyst. Both the chloride[469] and the bromide[486,598−600] give well-defined complexes with their respective methyl halides. It is difficult to compare the relative acidity (and activity) of $AlCl_3$ and $AlEtCl_2$ because of the former's lack of, or very limited, solubility in most of the solvents used in cationic polymerisation, which makes proper assessments of relative initiating power impossible. It is well established however that dialkyl aluminium halides and trialkyl aluminium derivatives are inactive towards isobutene polymerisation[601,602] in the absence of a cocatalyst, and are therefore to be considered as weaker Lewis acids, at least for cationic polymerisation. It has also been claimed[603] that chloroalkyl aluminium compounds can initiate the polymerisation of styrene only if traces of water or hydrogen chloride are added.

Another important aspect of the chemistry of aluminium alkyls and alkyl halides is the readiness with which they can exchange halogen atoms for alkyl groups, both among themselves or with organic and inorganic halides. This behaviour is encountered in the basic interactions occurring in the initiation processes of Ziegler-Natta polymerisations, e.g. with the classical catalytic combination $Et_3Al−TiCl_4$. Also, the facile exchange is well illustrated by the fact that the addition of an aluminium alkyl halide to another of different composition rapidly brings about an equilibrium mixture reflecting the scrambling of both groups. This reaction is the basis of the preparation of such compounds as $EtAlCl_2$, ethylaluminium sesquichloride and Et_2AlCl from Et_3Al, $AlCl_3$ and appropriate mixtures of the above compounds. The reaction of halogens with alkylaluminium halides proceeds on an entirely similar pathway[604−606]:

$$R_y AlX_{3-y} + X_2 \rightarrow R_{y-1} AlX_{4-y} + RX \cdots \xrightarrow{X_2} AlX_3 + yRX.$$

Given these premises, we can now analyse the essence of the work published by Kennedy and collaborators and by Cesca's and Giusti's groups in order to try and establish the real nature of the catalyst(s) (Lewis acid) in their systems. Since the peculiarities of these polymerisations have been repeatedly summarised in several extensive articles and monographs by Kennedy and coworkers[607,608], another detailed enumeration of all the features in this review would be superfluous. Only the most salient points of this rather complicated work will be recalled.

The basis of the initiation mechanism in systems involving R_3Al or R_2AlX as catalysts, $R'X$ as "cocatalyst" and MeX as solvent has been proposed to be constituted by the following reactions:

(i) *complexation:* the solvent penetrates in the bridged dimeric structure of the catalyst, breaks it and forms a 1 : 1 complex with the latter, viz.,

$$(\text{Catalyst})_2 + 2\,\text{MeX} \rightleftharpoons 2\,\text{Catalyst} \leftarrow \text{MeX};$$

(ii) *displacement:* the cocatalyst replaces the solvent in the complex with the catalyst,

$$\text{Catalyst} \leftarrow \text{MeX} + R'X \rightleftharpoons \text{Catalyst} \leftarrow R'X + \text{MeX};$$

(iii) *ionisation:* the catalyst-cocatalyst complex is partly ionised,

$$\text{Catalyst} \leftarrow R'X \rightleftharpoons \text{Catalyst}X^- \; R'^+.$$

Ionisation is followed by different types of interactions, depending upon the presence in the system of a monomer (e.g., isobutene), of a non polymerisable olefin (e.g., 2,4,4-trimethylpentene-2), or of no further reagent. In the latter instance the ion pair produced in reaction (iii) collapses by an alkylation reaction involving the migration of an alkyl group onto the cation, leaving a more halogenated aluminium alkyl, e.g.:

$$Me_3AlCl^- \, t\text{-}Bu^+ \rightarrow Me_2AlCl + (CH_3)_4C.$$

In the presence of a polymerisable olefin, initiation takes place, through the addition of the cation R'^+ to the monomer double bond to give the active species, e.g.:

$$t\text{-}Bu^+ \, Et_2AlCl_2^- + \;{}^{\backslash}_{/}C{=}C^{/}_{\backslash} \rightarrow t\text{-}Bu{-}\overset{|}{\underset{|}{C}}{-}\overset{|}{\underset{|}{C}}{}^+ \, Et_2AlCl_2^-.$$

Finally, in the presence of a non polymerisable olefin the above initiation reaction is not followed by chain growth, but only by elimination (transfer) or alkylation (termination) reactions involving an "active" ion pair incapable of propagating.

Two other important reactions must be considered in these complex systems. The first has been termed "hydridation" and occurs when the alkyl substituent(s) on the aluminium possess beta hydrogen atoms. Thus, for example:

$$RCl + Et_3Al \rightleftharpoons R^+ Et_3AlCl^- \rightarrow RH + C_2H_4 + Et_2AlCl.$$

The second takes place if the organic halide contains aromatic rings; this is a classical Friedel-Crafts electrophilic substitution, e.g.[619]:

$$C_6H_5-CH_2Cl + Et_2AlCl \rightleftharpoons C_6H_5-CH_2^+ Et_2AlCl_2^- \xrightarrow{+ C_6H_5-CH_2Cl}$$

$$\rightarrow C_6H_5-CH_2-C_6H_4-CH_2Cl + Et_2AlCl + HCl$$

$$Et_2AlCl + HCl \rightarrow EtAlCl_2 + C_2H_6.$$

Obviously, this process can continue and the final products will be "polybenzylic" hydrocarbons.

The above mechanism includes various reactions leading to an increase in the halogenation of the catalyst, i.e. to an enhancement of the initiating power (Lewis acidity) of the system. Kennedy and collaborators recognised this fact and considered that apart from the original initiation pathway arising from the ionisation of the cocatalyst by the *added* catalyst, secondary initiation processes should occur as a result of the formation of more halogenated aluminium derivatives. Thus, for example, if diethylaluminium chloride is used as catalyst, ethylaluminium dichloride will be generated and will be able to initiate more effectively than the original catalyst. Moreover, aluminium trichloride will be formed following this second initiation process and this third catalyst will also actively promote polymerisation. However, having discussed these possible secondary initiation processes, Kennedy and coworkers concluded that the original pathway involving the added catalyst was the dominant route to active species.

It is our purpose to reexamine all available evidence in this context and to show that Kennedy's conclusions are not entirely satisfactory. Kennedy and Gillham[609] investigated the relative cocatalytic efficiency of a series of alkyl and aralkyl halides in the polymerisation of isobutene by diethylaluminium chloride. By arranging the presumed carbenium ions formed from the cocatalyst in order of stability, they found that the efficiency went through a maximum for the *t*-butyl and the 1-phenylethyl ions. The authors argued that this behaviour demonstrated that the cocatalytic efficiency was not simply related to the ease of formation of further chlorinated aluminium derivatives and that therefore the initiation must have primarily arisen from the interaction of the starting catalyst with the cocatalyst (heterolytic carbon-halogen bond cleavage). We observe, however, that the parameter used as a measure of cocatalytic efficiency was a complex one which included the overall polymerisation process, i.e. initiation, propagation and termination reactions, and therefore it seems hazardous to draw conclusions on the relative power of initiation on the basis of such global data, since different halides might well affect differently the course and the rate of propagation and termination. Given this basic flaw, we will only discuss our views on the mode of initiation with the most important halides used, without referring to their supposed relative activities.

Trityl chloride was reported to be an effective cocatalyst. This observation cannot be rationalised in terms of initiation through ionisation because it is well known that the trityl cation is unable to initiate the polymerisation of isobutene. The fact that this monomer was polymerised indicate that other reactions must have led to the formation of active species. In particular, we propose that the exchange reaction

$$Ph_3CCl + Et_2AlCl \rightarrow Ph_3CEt + EtAlCl_2$$

must be regarded as the principal cause of initiation since it produces what we consider to be the real catalyst, ethylaluminiumdichloride. The same considerations most probably apply to diphenylmethyl chloride cocatalysis, given the unlikelihood of initiation by the corresponding cation. Concerning benzyl and 1-phenylethyl chlorides, both halogen-alkyl exchange with the catalyst and electrophilic substitution on the ring must be considered as important reactions giving rise to $EtAlCl_2$, and we prefer to envisage these interactions as the origin of the activity of the systems (through catalysis by $EtAlCl_2$), rather than the ionisation of the halide. Finally, when t-butyl chloride was used, its "maximum" efficiency is compatible with a fast and efficient halogen exchange with the catalyst as shown by the ease of alkylation[610] and by the fact, reported by Cesca et al.[611] that at $-45\ ^\circ C$ in methylene chloride t-butyl chloride reacts rapidly and quantitatively with Et_2AlCl and all the chlorine of the former ends up bound to the aluminium.

We conclude therefore that the real catalyst in the above systems must have been a mixture of ethylaluminium dichloride and aluminium trichloride (formed as a result of further exchange of the former with the cocatalyst). However, in order to substantiate this claim, several other pieces of evidence must be critically discussed:

— It has been found that the Me_2AlCl-t-BuCl catalytic system generally leads to higher polymer yields than $MeAlCl_2$ used alone. The latter initiator does not seem to require any cocatalyst for the polymerisation of isobutene, but gives incomplete conversions. At first sight these observations could be considered as a strong argument against our interpretation, since one would expect $MeAlCl_2$ to be equally or more active than its supposedly precursor combination Me_2AlCl-t-BuCl. However, we believe that in situ generation of the dichloride results in a higher catalytic activity than addition of its dimer solution to the monomer in a single "shot".

— In a similar vein it has been argued that in the system Me_3Al-t-BuCl the amounts of cocatalyst used are so low that they could hardly be considered sufficient to produce the amounts of di- and trichloride necessary to induce appreciable polymerisation; yet, the above system gives high polymer yields. This argument again implies that the highly chlorinated species produced during the interactions of Me_3Al with the co-catalyst are in the same aggregation state as when they are used as normal additives in solution. Our contention is instead that nascent $MeAlCl_2$ and $AlCl_3$ are monomeric and therefore probably much more reactive than their dimeric solutions.

— It has been pointed out that the initiating potential of the system Me_3Al-t-BuCl is high in solvent methyl chloride, but nil in n-pentane. On the other hand, $AlCl_3$ and alkylaluminium dichlorides can induce the polymerisation of isobutene in both solvents. These facts do not disprove our interpretation, since it can be readily argued that the chlorine-methyl exchanges between catalyst and cocatalyst are extremely slow in a medium of such a low dielectric constant as n-pentane and that the absence of polymerisation arises precisely from the lack of formation of di- and trichloride.

— The fact that t-butyl bromide gives higher molecular weights than t-butyl chloride when these cocatalysts are used with Et_2AlCl has been used as supporting evidence for the mechanisms of ionisation invoked by Kennedy's school. However, we must emphasize that changes in DP do not necessarily correlate with changes in the rate or efficiency of initiation. In this specific instance we believe that they were due to the formation of aluminium mixed halides when t-BuBr was used. It has in fact been shown on many occasions that mixed halides behave quite differently (and are generally more active) than homogeneous ones.

— Yet another of Kennedy's arguments against catalysis by di- and trichlorinated aluminium compounds formed in situ runs as follows. While the "molecular weight activation energy" for polyisobutenes produced in methyl chloride by the initiator systems trialkylaluminium-t-butyl chloride or monochlorodialkylaluminium-t-butyl chloride is independent of the nature of the alkyl substituent for isobutyl, ethyl and methyl, a much higher value of that parameter is obtained when $AlCl_3$ or $EtAlCl_2$ are used as catalysts *without* t-butyl chloride[609,610,612]. First of all, the identity of DP activation energy for the cocatalysed systems cannot be explained so readily, if one assumes with Kennedy and coworkers[612] that these polymerisations proceed through ion-paired active species. On the contrary one would expect the counterion to influence all reactions affecting the molecular weight of the product. Secondly, it is not correct to compare a group of systems all involving t-butyl chloride with two others in which this cocatalyst was not added. It would be interesting indeed to study the behaviour of $AlCl_3$ and $EtAlCl_2$ in the presence of added t-BuCl. We consider that the "molecular weight activation energy" is the same in all cocatalysed systems because the same transfer reactions occur, most probably transfer with the alkyl halide. Given the above considerations and the additional fact that the study of the DP's in those systems appeared to be difficult due to their sensitivity to impurities (see important scatter in the experimental results), we consider the initial argument as inconclusive. Note moreover that Kennedy and Rengachary[610] showed that t-BuCl does act as a transfer agent.

— The production of graft copolymers from an initial polymer containing carbon-chlorine bonds, Et_2AlCl and a monomer sensitive to cationic polymerisation seemed until recently the best indirect proof of Kennedy's initiation scheme:

$$Et_2AlCl + Polymer\text{-}Cl \rightleftharpoons Polymer^+ \, Et_2AlCl_2^-,$$

$$Polymer^+ + nM \rightarrow Polymer\text{-}(M)_n^+ \text{ (graft)}.$$

It has however been shown recently in Sigwalt's laboratory[613] that grafting can take place on the unsaturations of a polymer. Thus, one could envisage that grafting *onto* the polymer occurs rather than *from* it. Also, the fact that a polymer bearing tertiary chlorine atoms could be grafted can be rationalised in terms of alkylation of these sites with formation of further chlorinated catalyst (just as in the model experiments performed by Kennedy's school[610]), and subsequent grafting onto the polymer at unsaturated moieties catalysed by the newly formed di- or trichloro aluminium compounds, or grafting from the C–Cl bonds but again catalysed by the new species. The degradation of chlorinated butyl rubber by $AlCl_3$ and $EtAlCl_2$, and the unimportance of such reaction when trialkylaluminium compounds are used, can also be explained by an initial alkylation reaction in the case of the latter catalyst, degradation becoming significant on the double bonds only when most of the allylic chlorine atoms of the original polymer have been consumed. See also Sect. X–C.

— The synthesis of block copolymers from a chlorine-terminated polymer, Et_2AlCl and a cationically sensitive monomer was achieved by Kennedy and collaborators[614,850] and apparently constitutes a strong indication for the validity of their initiation mechanism. The blocking of styrene on polyisobutene with chlorine end groups and of isobutene on polystyrene with chlorine end groups was highly efficient[850]. This is apparently in contradiction with our interpretation, because if we assume the occurrence of the exchange reaction

$$\text{\textasciitilde\textasciitilde M}_1^+ \text{ A}^- + \text{\textasciitilde\textasciitilde M}_2\text{Cl} \longrightarrow \text{\textasciitilde\textasciitilde M}_1\text{Cl} + \text{\textasciitilde\textasciitilde M}_2^+ \text{ A}^-,$$

one would expect about 50% homopolymerisation, which is not observed in practice. We think that, as in the previously discussed systems, ethylaluminium dichloride and probably aluminium trichloride were formed by chlorine-alkyl exchange with the polymer. These stronger Lewis acids then catalysed the electrophilic aromatic substitution onto the rings of polystyrene, thus causing *grafting* onto styrene-containing chains and giving little homopolymerisation. We believe that in Kennedy's experiments it could well be that each initial polymer molecule is terminated by a chlorine atom, but that no proof is given that these macromolecules are the cationogenic species. It would be interesting in this context to carry out a polymerisation of isobutene by Et AlCl$_2$ in the presence of normal polystyrene (i.e. one not bearing chlorine). According to our interpretation the result would be a grafted copolymer. Note that this proposal is not entirely speculative, since apart from the already quoted grafting experiments onto unsaturated sites[613], Polton has achieved the grafting onto a polymer not bearing halogens or unsaturations using Kennedy's techniques[615].

To summarise, Kennedy's initiation theory based on the heterolytic cleavage of the carbon-halogen bond promoted by trialkyl or dialkylaluminium monohalides is not fully established, since the experimental evidence in its favour can all be interpreted adequately by assuming that the real catalysts are the stronger acids produced from further halogenation of the aluminium compounds. Only kinetic experiments, performed in accurately controlled conditions, can give a clear answer to this problem. A recent paper in this vein was published[619] and we believe that, contrary to the authors' conclusions, the results described in it corroborate our hypotheses. In that study the reaction of *p*-methylbenzyl chloride with triethyl aluminium was investigated in methyl chloride at low temperature. Three products were obtained viz. *p*-methylpropylbenzene, *p*-xylene and oligomeric polybenzyls, arising from alkylation, hydration and polybenzylation, respectively. The authors claimed that the initial step responsible for those subsequent interactions was the following equilibrium:

$$CH_3C_6H_4CH_2Cl + Et_3Al \rightleftharpoons CH_3C_6H_4\overset{+}{C}H_2 \ Et_3AlCl^-.$$

It was noticed that the rate of disappearance of the aralkyl chloride strongly depended on the relative amount of Et_3Al present, being the lower the higher the ratio catalyst/

substrate. It was argued that this was due to the fact that with a large excess of Et_3Al the initiation equilibrium above was so strongly shifted to the right that little unionised chloride was left for polybenzylation. In other words, it was concluded that a very high concentration of ion pairs must have been present under these conditions. Yet, quenching the reacting solutions with methanol did not produce any detectable amount of the corresponding *p*-methylbenzyl methyl ether and to explain this the authors advocated a very small equilibrium concentration of carbenium ion pairs, in contradiction to their previous conclusion. We think that this serious contradiction is yet another indication of the unlikelihood of the above ionisation process. Indeed, the decrease in substrate rate of disappearance with increase in Et_3Al concentration is better explained by assuming chlorine-ethyl exchange between the two reactants, i.e., the basic interaction of our alternative mechanism. Obviously when less Et_3Al was available this exchange could go further and lead to the formation of higher amounts of $EtAlCl_2$, the more active catalyst. With a large excess of Et_3Al the main product of the exchange would be Et_2AlCl, and this weaker Lewis acid would consume the substrate much more slowly.

The work of Cesca's and Giusti's group[409,493,602,606,616−618] concentrates on the use of chlorine or iodine as cocatalysts in the polymerisation of isobutene in methylene or methyl chloride at −45 °C catalysed by diethylaluminium chloride. These authors noticed that the addition of chlorine to the quiescent catalyst-monomer solution produced a sudden rise in conductivity, and polymerisation slowly began. Analysis of the time-conversion curves revealed that the rate of monomer consumption increased steadily at the beginning, indicating S-type non-stationary conditions. The rise in conductivity followed a similar behaviour. A kinetic analysis showed that above about 40% conversion the rate of polymerisation was directly proportional to the monomer concentration. The maximum rate was found to depend on the first power of both the catalyst and the cocatalyst concentrations[602,493]. The following initiation scheme was formulated[493]:

$$Cl_2 + Et_2AlCl \rightleftharpoons Cl^+ + Et_2AlCl_2^- \rightleftharpoons EtCl + EtAlCl_2,$$

and the chloronium ion was assumed to be the species responsible for the attack on the olefinic double bond to give the active carbocationic species. We feel however that the real catalyst is the ethylaluminium dichloride (and perhaps aluminium trichloride resulting from further reaction with the cocatalyst). The autoacceleration and the observed kinetics support this point of view. Moreover, it was also found that addition of a second portion of monomer at the end of the first polymerisation resulted in a fast polymerisation *without* acceleration. The authors concluded that more active catalytic entities must have formed during or after the first polymerisation. It is surprising that having stated that obvious conclusion they did not see that those more efficient initiators are the same which produced the acceleration in the first reaction, since they were being formed from the interaction of the original catalyst and the cocatalyst. To us these experiments leave little doubt about the fact that $EtAlCl_2$ and $AlCl_3$ were the real catalysts in those systems.

c) The Problem of the True Cocatalyst

If one accepts that the combination of an alkylaluminium or alkylaluminium halide
with an organic halide gives rise to a cationogenic reaction and that the carbenium ions
thus produced add onto the olefinic double bond of the monomer to generate the active
species, then the problem of cocatalysis seems settled and in a very different way than
in the systems analysed in all the previous sections. However, the real situation is far
from being so straightforward. Kennedy and Rengachary[610] reported in their study
of model systems that when the unpolymerisable dimer of isobutene 2,4,4-trimethyl-
pentene-1 was mixed with the catalytic pair Me_3Al-t-BuCl in methyl chloride at
−78 °C, very little alkylation took place if some care had been taken to exclude moisture and other impurities from the reaction medium, the only product detected being
neopentane. The authors estimated that the residual moisture in these experiments
amounted to about a quarter of the t-butyl chloride added. We find that this amount
is very high in terms of cocatalysis and are surprised that the reaction did not proceed
more efficiently. Unfortunately the experimental procedures were not described in
detail, but perhaps the low yields obtained might have been due to the use of calcium
hydride as the drying agent for the solvent. We have in fact experienced that unless
extreme care is taken in the distillation of highly volatile compounds from their suspension with CaH_2, some fine dust of this solid (and/or of calcium hydroxide) is carried over with the distillate causing inhibition or retardation of cationic polymerisation
and related base-sensitive reactions. In conclusion, we believe that the system described
by Kennedy and Rengachary either contained in reality less water than estimated by
the authors or basic impurities limited its activity. The second aspect of this study deals
with the effect of added water to the system, resulting in appreciable conversions into
products arising from the interaction of the t-butyl cation with the olefin. These results
were rationalised by suggesting that in "dry" systems alkylation of t-butyl chloride by
the organometallic catalyst to give neopentane predominates, while in "wet" ones the
t-butyl cation is formed through chloride ion migration to an active cationic species
arising from ordinary water cocatalysis. Of course these conclusions must await confirmation, i.e. some experiments performed in really dry conditions. As far as we are
aware, no such study has yet been conducted. Only then the validity of the claimed
ordinary cocatalysis by water with alkylaluminium compounds can be properly assessed.

It has already been mentioned that Kennedy's work showed that t-butyl chloride
and trimethylaluminium do not interact in n-pentane, although some residual moisture
must have been present in these non-rigorously dried media. The reaction occurs as
soon as some methyl chloride is added to the system and this phenomenon has been
attributed to the breaking down of the trimethylaluminium dimers and formation of
an active methyl chloride-trimethylaluminium complex. Thus, the role of water in
these interactions seems negligible or absent.

Some interesting information concerning the possible intervention of cocatalytic
impurities in the mechanism of isobutene polymerisation by diethylaluminium chloride
and chlorine was gathered in the work of the Italian groups[602]. These authors remarked that when they introduced the diethylaluminium chloride in a chlorinated
solvent like methyl chloride, the electrical conductivity of the solution increased appreciably within a few seconds from mixing. This sudden conductivity jump was irre-

producible and could not be correlated with the concentration of added catalyst. More-over, further introductions of the organoaluminium compound gave much smaller con-ductivity increases and an improved purification procedure of the solvent resulted in a decrease of the magnitude of the jumps. It was therefore concluded that this behaviour was due to adventitious impurities capable of reacting with the catalyst to give ionic species. The important aspect of these observation was that whenever the conductivity jump was high (high level of impurities), a slow polymerisation of isobutene was de-tected in the *absence* of added chlorine. Thus, the ionogenic impurities (water?) acted as cocatalyst with Et_2AlCl. The problem remains of establishing the mechanism of this cocatalytic action, i.e. do the impurities facilitate the chlorine-alkyl exchanges so as to promote the formation of more acidic catalysts (Et_2AlCl_2 and $AlCl_3$), or do they act in a conventional fashion with the weak Lewis acid Et_2AlCl? No adequate answer can be given to this question, since so little is known about the catalytic properties of the re-action products of alkylaluminium halides or trialkylaluminiums with such protogenic impurities as water. Indeed, while it is well established that traces of water will be rap-idly scavenged by these compounds to form hydroxyalkylaluminium derivatives, the fate and "strength" of the latter in polymerisation reactions is not clear. If on the other hand the impurity is a halide which can readily exchange the halogen atom for an alkyl group, obviously its "cocatalytic" role will simply be to increase the acidity of the cat-alyst, as we have seen extensively in the preceding section. This is also the case for such impurities as traces of hydrogen chloride. In other words, in the latter situation, the process of initiation cannot be defined as cocatalytic in the classical sense of the term.

We can conclude that the lack of specific studies dealing with the exact role of added protogenic substances in these systems precludes for the moment any meaning-ful mechanistic conclusions about these presumed cocatalytic processes[603,861,862].

d) The Problem of the True Initiation Mechanism

The general initiation scheme proposed by Kennedy and coworkers and sketched out at the beginning of this discussion in Sect. b), predicts a direct cationogenic interaction between catalyst and added halide. Before examining the merits and demerits of this scheme, it is instructive to look briefly into the field of racemisation of asymmetric halides promoted by Lewis acids, since these studies reflect a situation similar to that under scrutiny.

The racemisation of 1-phenylethylchloride catalysed by various Lewis acids in diethyl ether at 25 °C was studied by Evans and Satchell[620]. Gallium, aluminium and boron chlorides as well as boron fluoride induced a clean first-order loss of optical activity and the rate of this process was found to be directly proportional to the cat-alyst concentration. The effect of added water was investigated quantitatively in the case of gallium chloride and the retarding role of this compound followed the kinetic expression:

$$k_{obs} = k_1 ([GaCl_3]_0 - [H_2O]).$$

This result would suggest that racemisation does not involve cocatalysis by water, but

arises instead from the direct ionisation of the halide induced by the Lewis acid. However, the fact that these experiments were conducted in ether introduces a complication, since it must be remembered that the metal halides used can readily complex with
this solvent and the real catalysts were therefore the Lewis acid etherates. In the section
devoted to cocatalysis by ethers we pointed out that the ether molecule bound to the
catalyst, e.g. boron fluoride, plays a determining role in the initiation mechanism and
that added water has a retarding effect on this process at concentrations higher than a
low critical value[594]. Given the possible cocatalytic intervention of the solvent in the
racemisation, any conclusion about direct ionisation must obviously be taken with
caution. Evans and Satchell[621] also studied the racemisation of 1-phenylethylchloride
by stannic chloride in benzene around room temperature and the results obtained in
this second investigation are not marred by possible solvent cocatalysis. These authors
found that the reaction was first order in catalyst and that water was apparently not involved in the process, since racemisation proceeded in carefully dried media and was
slowed down by small quantities of added moisture. Hydrogen chloride alone was inactive towards racemisation, a surprising result which is in contrast with previous observations by Pocker[515] who had claimed that this acid could induce the ionisation
of alkyl and aralkyl halides in benzene. Evans and Satchell noticed on the other hand
that hydrogen chloride added to stannic chloride increased the rate of racemisation.
The two ionisation reactions below were proposed to explain the results:

$$Ph(CH_3)CHCl + SnCl_4 \rightleftharpoons Ph(CH_3)\overset{+}{C}H\ SnCl_5^-,$$

$$Ph(CH_3)CHCl + SnCl_4 + HCl \rightleftharpoons Ph(CH_3)\overset{+}{C}HCl\ SnCl_5^- \cdot HCl.$$

The second equation means in fact that the exchange reaction is slightly faster when
the hydrogen dihalide anion is complexed with stannic chloride, viz. $HCl_2^- \cdot SnCl_4$.
It could be argued that the above study always involved some traces of hydrogen
chloride which was therefore the necessary cocatalyst for the racemisation, i.e. that
ionisation of the organic halide was preceded by some dehydrochlorination of the
halide. Thus, we think that the fact that racemisation is observed with an organic
halide in the presence of a Lewis acid does not necessarily imply that this halide is a
cationogen for initiation of polymerisation. It must also be emphasised that even if
an ionic intermediate were formed by direct interaction of the halide with the Lewis
acid in the absence of a cocatalyst, the lifetime of this intermediate might be too low
to allow the attack onto a monomer (initiation of polymerisation), but sufficient to
cause inversion of configuration. These considerations are supported by the apparently
contrasting observations that while $SnCl_4$ can racemise optically active 1-phenylethyl
chloride, the combination of these two compounds is not effective in initiating the polymerisation of styrene through carbenium ion formation[389,519]. Other evidence pointing to the lack of cocatalytic power of organic halides in the presence of Lewis acids
has been given in this chapter and it all points to the essential suggestion that any carbenium ion intermediate resulting from the interactions of these two families of compounds is too short-lived to initiate polymerisation.

The above considerations can be complemented by a reminder of the work of
Fairbrother[622,623], Willard[624,625] and others[626,627] on the isotopic halogen exchange

between metal halides and organic halides. The occurrence of such reaction must necessarily imply the formation of carbenium ion intermediates, but the lifetime of these species must be extremely short, since many of these systems are known to be inactive in the initiation of cationic polymerisation in the absence of a suitable cocatalyst like water or hydrogen chloride.

Turning now to trialkylaluminiums and dialkylaluminium halides, we encounter an apparently different phenomenology suggesting at first sight a qualitative change in mechanistic behaviour. Thus, Kennedy et al.[628] observed that the alkylation of (−)-1-phenylethyl chloride by Et_3Al in EtCl gave an essentially racemic product. The readiness with which this system and many others of the same type, i.e. those composed of R_3Al or R_2AlX as catalysts and alkyl or aralkyl halides as cocatalysts in halogenated solvents, can also initiate the polymerisation of isobutene and other olefins would suggest that the carbenium ion intermediates must be sufficiently long lived to be able to add to the olefinic double bond with a frequency which competes favourably with that of ion-pair collapse. This is the basis of Kennedy's mechanism repeatedly expounded in his publications and summarised in Sect. b). This drastic change of features with respect to other Lewis acids, leaves us quite perplex and we find it difficult to rationalise how such weak Lewis acids as R_3Al and R_2AlX could possess the capacity of inducing the heterolytic bond fission of organic halides and moreover produce the necessary environment (counterion) capable of stabilising the carbenium ion formed for a time long enough for initiation to take place, while strong Lewis acids as $TiCl_4$ cannot achieve the same result. We have already examined in detail all the indirect evidence offered by Kennedy's school to support its mechanism and tried to show that in fact none of the arguments put forward points unequivocally to that mechanism to the exclusion of others. Indeed, our contention that all the major features observed could be explained by assuming halogen enrichment of the catalytic aluminium compound prior to real initiation seems equally tenable and perhaps more realistic in view of the higher acidity of the catalysts formed in situ and their possibly higher reactivity in the nascent state. Moreover, we showed that in certain specific instances the direct ionisation mechanism seems unacceptable: thus, the interaction of trityl chloride with diethylaluminium chloride might well give rise to the corresponding trityl cation (although this was not proved, and it would be very interesting to check this point spectroscopically), but even if this first step occurred it seems unreasonable to postulate that this ion can initiate the polymerisation of isobutene and give high yields of polymer when it is well established that such a stable entity is incapable of accomplishing that feat.

In an attempt to elaborate an alternative initiation mechanism for these systems which would adequately explain all the major experimental observations without involving fundamental contradictions with generally accepted facts, we undertook the task of examining in great detail the voluminous literature published on the subject by Kennedy's laboratory. The subject is very complex and the vast amount of information collected covers a lot of ground in an extensive manner, but seldom goes sufficiently deep to permit the unravelling of fundamental questions. A step-by-step commentary of the many series of experiments reported in the last decade would make this discussion far too long and we have therefore limited ourselves to the series of points raised in Sect. b) and to some further observations below. We hope that this procedure will not seem partial in the choice of the topics to be dealt with and we emphasize that the unfamiliar reader should of course become acquainted with the original work before assessing the validity of our comments.

The complexity of these systems becomes apparent when one considers that: (i) the catalysts used have a strong tendency to exist in a dimeric form; (ii) certain solvents (particularly methyl halides) can effectively break that association to give 1 : 1 monomeric complexes; (iii) certain monomers, e.g. isobutene, also form complexes with the catalysts, but can be displaced by certain halogenated solvents; (iv) addition of the "cocatalytic" halide produces a displacement of the solvent from its complex with the catalyst, prior to ionisation; (v) the "cocatalytic" halide, e.g. t-butyl chloride, does not displace the monomer from its complex with the catalyst in a neutral solvent, e.g. n-pentane; (vi) the presumed ionisation of the added halide can be accompanied by formation of halonium ions, particularly in the case of bromides and iodides, e.g. $Me_3C-\overset{+}{I}-Me$; (vii) elimination (transfer) and alkylation (termination) reactions are common to all the systems, but two more specific interactions also intervene in a substantial way with certain catalysts or cocatalysts, namely "hydridation" and polybenzylation; (viii) halide enrichment of the catalyst giving stronger Lewis acids results from many of the above reactions and therefore the activity of the systems changes continuously with time; (ix) the newly formed catalysts can act as initiators in the monomeric state or form complexes with the various components of the polymerising mixture, including their less-halogenated precursors.

In view of the maze of possible physical and chemical phenomena arising from an initially "simple" four-component system, it is not surprising if a massive qualitative and occasionally semiquantitative approach has not provided in our view a totally satisfactory elucidation of the major features of the initiation process. We feel that some interesting indications could be obtained more readily from a systematic kinetic study of a given system in rigorously controlled experimental conditions (let us not forget that one should have added to the above list the further complicating role of reactive impurities such as residual moisture). Moreover, certain mechanistic experiments accompanying such a kinetic study could adequately complement it. Thus, for example, we have already mentioned that it would be instructive to follow spectroscopically the reaction of trityl or diphenylmethyl halides with one of the typical catalysts used in these systems to see if the corresponding carbenium ions can be detected, and if so what is their lifetime and what degree of ionisation is achieved as a function of the nature and the relative amount of catalyst used. Another interesting experiment could consist in using an optically active cocatalyst in the presence of isobutene, using different media such as pentane and methyl chloride, and a given typical catalyst. The purpose of this investigation would be to assess the conditions under which racemisation takes place and whether it is accompanied or not by polymerisation. In other words, it would provide precious information on the properties of the carbenium ion intermediates postulated for these reactions.

If we now turn to the second family of systems under discussion, namely that involving halogens as cocatalysts, and principally chlorine, we find again that the evidence in favour of initiation by halonium ions such as Cl^+ produced in the interaction with the catalyst is not entirely satisfactory. Indeed, the arguments based on the formation of certain chlorinated products in studies conducted with the model compound 2,4,4-tri-methylpentene-1[606,493] are not conclusive, since the same results, in lower yields, were observed with chlorine added without the catalyst. Thus, all that can be said is that the chlorination of the olefin proceeds at a higher rate when the aluminium compound is present. Our alternative interpretation based on formation of catalysts richer in chlorine finds some supporting evidence in an experiment reported by Di Maina et al.[409] in which addition of chlorine to a partially polymerised, quiescent mixture of isobutene and ethylaluminium dichloride in methyl chloride at −45 °C gave no further polymerisation. We see no reasons why chloronium ions could not have formed in this system by the reaction of the halogen with the catalyst, and the fact that the polymerisation was not reactivated indicates that initiation is not due to chloronium ions. We have already discussed the role of ethylaluminium dichloride in the direct initiation of isobutene polymerisation (see Sect. IV-B-4-c) and the reasons for limited conversions; obviously, the addition of chlorine did not alter the equilibrium between accumulated anions and non-dissociated $EtAlCl_2$.

Our appraisal of the work with halogens is similar to that given above for organic halides, i.e. the systems are rather complex and one would need more kinetically-oriented studies to gain a better insight into the mechanism, together with some discriminating specific experiments. For example, the chloronium ion theory could be checked with other Lewis acids, incapable of exchanging chloride ions and unable to give direct initiation.

Our general, but tentative, conclusions on the initiation mechanism with both types of systems can be summarised as follows. The relatively weak Lewis acids of general formulae R_3Al and R_2AlX, incapable of promoting cationic polymerisation without the help of a cocatalyst, are transformed more or less rapidly into stronger more halogenated derivatives of the type $RAlX_2$ and AlX_3 in the presence of organic halides or halogens. The in situ formation of these more powerful catalysts is particularly conducive to initiation of cationic polymerisation, more so than addition of their solution to the monomer. This initiation probably takes place by various routes, and it is impossible at this stage of knowledge to establish which one, if any, predominates. Direct initiation must be considered as one of these routes, since it is known that both dialkylaluminium halides and aluminium trihalides can achieve such task. Cocatalytic initation promoted by residual moisture and perhaps by traces of hydrogen halides formed in the complex set of reactions arising from these systems, is also possible; the role of water is quite clear with aluminium trihalides, but less so in the case of alkylaluminium halides which are known to react rapidly and completely with this compound. A third type of initiation can finally be envisaged, i.e. that involving ionisation of the added halide (for those systems in which such addition is carried out) by heterolytic fission of the carbon-halogen bond induced by the Lewis acids present in the solution. This last route is probably valid only with certain halides and its occurrence needs further study to be confirmed. The specific mechanism of the third type of initiation is in fact not clear and the intervention of the organic halide might well take place in a concerted fashion to assist the polarisation of the monomer double bond fundamentally induced by the Lewis acid molecules. In this scheme, similar to the Ad_E3 mechanism, ionisation of the halide would therefore not be required, but rather its strong polarisation followed by the addition of the alkyl or aralykyl group to the β-carbon of the olefin. Thus, tentatively:

$$\begin{array}{c} \backslash\ \diagup \\ \overset{\displaystyle C}{\underset{\displaystyle C}{\parallel}} \\ \diagup\ \backslash \end{array} \rightarrow AlWYZ + \overset{\delta+}{R}\ \overset{\delta-}{X}\ AlWYZ \longrightarrow \quad \begin{array}{c} AlWYZ \\ \backslash^{\delta+}\ \vdots \diagup \\ C\text{-----}C \\ \diagup\ \vdots \qquad \searrow \\ \overset{\delta-}{X}\text{-----}\overset{\delta+}{R} \\ \\ AlWYZ \end{array} \quad ,$$

where AlWYZ is the aluminium Lewis acid in its monomeric or dimeric form and RX the alkyl or aralkyl halide added as cocatalyst. If initiation with RX is at all possible, we prefer to envisage it through the above concerted intermediate rather than through ionisation to give R^+ (free or paired), since this proposal seems more in tune with the general mode of cocatalysis with other substances which we have discussed throughout this chapter, and since Cesca et al.[629] showed that the complex between aluminium chloride and t-butyl chloride is not ionised in detectable amounts.

We fully recognise the highly speculative character of these last remarks and offer them to the reader as a (perhaps premature) new way of looking at the role of organic halides in initiation of cationic polymerisations promoted by aluminium-based Lewis acids.

In the case of cocatalysis by chlorine and iodine, the above conclusions should apply equally well for the part concerning the first two initiation routes. It remains to be established if a third route, namely that involving the addition of a chloronium ion to the monomer double bond, is also operative. We feel that present knowledge does not allow a satisfactory answer to that point.

To summarise our views on this topic we can say that the Kennedy's mechanism is probably operative, but the overwhelming process is the one originating from the formation of highly chlorinated aluminium catalysts formed in halogen-alkyl exchange reactions and possessing a much more pronounced acidity than the Lewis acids initially added.

e) Other Lewis Acids

Pecka et al.[630] investigated the effect of some organic halides on the polymerisation of isobutene catalysed by titanium tetrahalides. They found that the addition of t-butyl fluoride and benzyl fluoride to heptane solutions of isobutene and $TiCl_4$, $TiBr_4$ or TiI_4 led to very fast polymerisations. The effect of n-butyl fluoride was less marked. A careful examination of the $TiCl_4-C_6H_5CH_2F$ system led the authors to conclude that the organic fluorides do not add as cationogens, but rather exchange halogens with the catalysts to give more powerful mixed titanium halides which are probably stronger Lewis acids capable of promoting direct initiation in a very effective way. This type of halogen exchange has recently been observed also between aluminium chloride and alkyl fluorides[631]. The above investigation seems to support in a slightly different context our main ideas about the principal role of organic halides in Kennedy's systems discussed in the previous sections.

Kennedy et al.[557] recently reported a survey on the role of numerous alkyl and aralkyl chlorides in the polymerisation of isobutene by boron fluoride, boron chloride and boron bromide. The experiments with BF_3 showed that in methylene chloride at -53 °C, with a monomer concentration of 0.89 M, about 10^{-3} M of Lewis acid alone produced 15% yield in 30 min (apparently only one control run); additions of about 10^{-3} M of such chlorides as allyl, methallyl, p-methylbenzyl and 1-phenylethyl gave approximately double yields, while t-butyl chloride enhanced the yield to 42%. The relevance of these results is extremely difficult to assess, In fact, no details were given of the preparation of BF_3 solutions in methylene chloride and of how the concentration of this Lewis acid was determined. Polymerisations were carried out in test tubes with teflon stoppers in a nitrogen-purged stainless steel enclosure. Given the very high sensitivity of this catalyst to cocatalysis by water, we feel that variable levels of residual moisture could well account for the reported changes in yield from one experiment to the other. The presumed action of the halides, thought by the authors to generate the corresponding carbenium ions by loss of Cl^- to the catalyst, and thus produce further initiation apart from that due to the BF_3 alone, remains entirely to be proven.

When BCl_3 was used, some model studies were conducted before the actual polymerisations. These consisted in mixing high concentrations of the Lewis acid with the alkyl or aralkyl chloride and observing the behaviour of the system either visually or by proton magnetic resonance. In some of these experiments a third reactant was also added, viz. 2,2,4-trimethylpentene-1, also in high concentrations. Aliphatic tertiary chlorides did apparently not interact with BCl_3, i.e. no carbenium ions were formed as inferred from the lack of reaction with the isobutene dimer, nor dehydrochlorination took place. Allyl and methallyl chlorides gave small amounts of gel of unknown composition within about a day at -25 °C, and the solutions turned yellow upon the mixing of the reactants, but the lack of isomerisation and of reaction with the isobutene dimer was again taken as evidence for the lack of production of carbenium ions. A different situation was encountered with 3-chlorobutene-1 and 1-chlorobutene-2: both halides were found to isomerise to give the same mixture of 34% of the former and 66% of the latter. The formation of allylic cations $CH_3-\overset{+}{\overbrace{CH \cdots CH \cdots CH_2}}$ in these systems was confirmed by the occurrence of "allylation" of the isobutene dimer (product detected by PMR). 1-chloro-3-methyl-butene-2 also allylated the dimer in the presence of boron chloride, presumably through the carbenium ion derived from chloride ion abstraction. Finally, p-methylbenzyl chloride gave polybenzyl oligomers and diphenyl and triphenylmethyl chlorides gave the corresponding carbenium ions (the latter conclusion was derived only from "visual observation", i.e. colour detection). The above experiments seem to show that a strong Lewis acid such as BCl_3 can induce the heterolytic bond cleavage of certain chlorides, more specifically those which give the most stable carbenium ions. Following these model experiments, Kennedy and collaborators examined the initiating power of these systems on isobutene. They found that only 1-chloro-3-methylbutene-2 and cumyl chloride (the latter was not used in model runs) gave reasonably high yields of polymer, while other chlorides, and in particular t-butyl chloride and trityl chloride, were ineffective as cocatalysts. It was concluded that initiation was due to the attack of the formed carbenium ions onto the monomer double bond. We find that this conclusions is unwarranted by the paucity of quantitative and systematic work. First of all there is a lack of correlation between the results obtained with

model compounds and in polymerisation runs. Thus, 3-chlorobutene-1 was found to isomerise and allylate the isobutene dimer efficiently, but its cocatalytic effect in the polymerisation of isobutene was minimal. This fact is hard to explain, and was indeed not
discussed by the authors, because it is in open contrast with the behaviour of 1-chloro-3-
methylbutene-2 which performed well in both sets of experiments. Nothing very definite can be drawn from a panoramic series of tests in which only the polymerisation
yields are measured and nothing reported concerning reproducibility: note for example
that with 3-chlorobutene-1 two polymerisations were reported but, unaccountably
when more cocatalyst was used the yield fell to the same level as in the blank (BCl_3
without organic halide). This observation possibly reflects the range of experimental
error, i.e. the reproducibility of experimental conditions, and in particular of the
level of residual moisture, and not an effect imputable to the added chloride. The high
activity of cumyl chloride as cationogen was recently demonstrated by Kennedy's
group. Elegant experiments were performed with dicumyl chloride and BCl_3 to show
that bifunctional poly(isobutenes) could be synthesised (see Notes added in proof).
There seems to be no doubt that initiation in that system took place through chloride-
ion abstraction from the organic halide by BCl_3 to give a carbocation capable of inducing the polymerisation of isobutene. The success obtained by Kennedy's school
with the cumyl structure represents a happy medium which places the carbenium ion
formed between those too reactive to survive long enough to initiate and those too
stable to attack the monomer.

Very recently, Kennedy et al.[632] asserted to have prepared a polyisobutene with
unsaturated head groups derived from initiation promoted by BCl_3 and 1-chloro-3-
methylbutene-2. It was postulated, as in the previous work, that the allylic cation derived from the cocatalyst attacked the monomer and that thereafter no transfer reactions would take place, but only termination with the BCl_4^- counterion, leading to
the formation of carbon-chlorine end groups. The presence of approximately one terminal double bond per polymer molecule was taken as supporting evidence for the
initiation mechanism. This work was then continued with the aim of synthesising a
block copolymer, making use of the chlorine end groups as macrococatalytic sites for
the second initiation, this one catalysed by diethylaluminium chloride. The operation
was carried out using α-methylstyrene as comonomer and the products claimed to be
essentially pure diblock copolymers on the basis of DP, GPC and composition data.
Several points must be raised about this investigation. The first one concerns the total
absence of transfer reactions with BCl_4^-: this assertion is not backed by experimental
evidence and the authors extrapolated it from their previous results with the hypothetical counterion BCl_3OH^- (water cocatalysis)[556], since in their work with organic chloride cocatalysis[557] they did not measure DP's. If transfer cannot be ruled out, the
occurrence of one terminal unsaturation per chain is not surprising and does not
necessarily prove initiation by the unsaturated cationogen. The second point concerns
the yield of polyisobutene obtained in the first polymerisation (23–25%) compared to
that reported for the same system in the previous publication (45%, single experiment).
The only sensible difference of conditions between the two runs is the concentration of
cocatalyst and we again find (see above) the surprising result that an increase in cocatalyst concentration reduces the yield of polymer. This fact is difficult to explain if
the initiation theory based on ionisation of the cocatalyst is accepted. Finally, the

arguments put forward for the mechanism and purity of sequential polymerisation are tenuous. There is no proof that the polyisobutene chains contained one terminal chlorine each; indeed the blocking could also have been the result of a reaction *onto* the unsaturated end groups, rather than *from* the ionised polymer. As for the lack of homopolymers in the products, the calculation performed by the authors to demonstrate it applies to one of the two copolymerisation reported, but not to the other.

D. Conclusions

The interactions of Lewis acids with alkenyl monomers in the presence or absence of cocatalytic substances can lead to a variety of initiation pathways whose probable mechanisms have been discussed throughout this chapter. The following general scheme summarises the alternative routes which bring about the formation of active species:

$$H-\overset{|}{\underset{|}{C}}-\overset{|}{\underset{|}{C}}{}^{+}\ MtX_nB^-
\qquad\qquad\qquad\qquad
H-\overset{|}{\underset{|}{C}}-\overset{|}{\underset{|}{C}}{}^{+}\ MtX_nB^- + MtX_n$$

$$\underset{/}{\overset{\backslash}{C}}=\underset{\backslash}{\overset{/}{C}}\Big\uparrow \qquad\qquad\qquad\qquad\qquad\qquad \Big\uparrow$$

$$MtX_n\cdot HB \qquad\qquad\qquad Ad_E3\ \ \text{Intermediate}$$

$$HB\ \Big\updownarrow \qquad \underset{/}{\overset{\backslash}{C}}=\underset{\backslash}{\overset{/}{C}} \qquad \Big\updownarrow\ MtX_n\cdot HB$$

$$Mt_2X^-_{2n+1} \underset{MtX^-_{n+1}}{\overset{\longrightarrow}{\rlap{\ \ }\longleftarrow}} MtX_n \overset{\longrightarrow}{\longleftarrow} \underset{/\ \backslash}{\overset{\backslash\ /}{\underset{C}{\overset{C}{\|}}}} \rightarrow MtX_n \longrightarrow MtX^-_n-\overset{|}{\underset{|}{C}}-\overset{|}{\underset{|}{C}}{}^+$$

$$MtX_n\ \Big\updownarrow$$

$$MtX^+_{n-1}\ MtX^-_{n+1}$$

$$\Big\downarrow \underset{/}{\overset{\backslash}{C}}=\underset{\backslash}{\overset{/}{C}}$$

$$MtX_{n-1}-\overset{|}{\underset{|}{C}}-\overset{|}{\underset{|}{C}}{}^+\ MtX^-_{n+1}$$

where HB indicates a cocatalyst molecule.

The occurrence or predominance of any one specific mechanism depends of course on the Lewis acid and the monomer used and on the adventitious or controlled presence of a cocatalyst. It is obvious however that initiation can take place simultaneously through more than one of the possible mechanisms envisaged above. Indeed, the often complicated kinetic behaviour encountered in these studies probably arises from such a situation.

The above scheme does not include any free ion-ion pair equilibria. Too little is known about the state of chain carriers in most of these systems and we must content ourselves with supposing that the active species are an equilibrium mixture of ions in different states of aggregation.

Because olefins are soft bases and most Friedel-Crafts halides are hard acids, the primary interaction between these two types of compounds must be regarded as a weak one, the outcome of which is generally limited to the equilibrium formation of the relatively feeble π-complex. Only with the softer of the strong Lewis acids would one expect this interaction to proceed further and give direct Hunter-Yohé initiation: titanium tetrachloride seems to comply with such a requirement, as suggested by the experimental evidence discussed in Sect. IV-B-4-b).

Self-dissociation of the Lewis acid produces a situation more conducive to direct initiation, since the cationic moiety formed from the metal halide can attack the monomer more effectively and add onto its double bond to give the corresponding carbocation.

Initiation through cocatalysis takes place more efficiently than direct initiation in the same conditions. The presence in a system of small amounts of residual cocatalytic impurities can therefore mask (at least in the first phase of the polymerisation process) any contribution to initiation arising from direct formation of active species. Only if the cocatalyst is consumed in a termination reaction, will the possible occurrence of a concomitant direct initiation process become manifest after the removal of all the cocatalyst from the reaction medium. These considerations have been abundantly dealt with in the previous sections of this chapter and their validity is well established. As for the mechanism of cocatalytic initiation, our attempt to draw a unifying principle based on the formation of an Ad_E3-type intermediate finds a good deal of experimental support from the old and recent literature, particularly with systems involving cocatalysis by water and by organic protogenic substances. To complete the parallel with classical mechanisms dealing with the addition of Brønsted acids to olefins, it can be said that if cocatalysis is adequately rationalised by the Ad_E3 scheme, direct initiation is equivalent to the Ad_E2 one. Indeed, it is interesting that the latter process seems to occur preferably with the stronger Brønsted acids, just as direct initiation takes place with the stronger Lewis acids.

An inspection of the general initiation scheme sketched above clearly indicates that the kinetics of chain carriers formation can take a multiplicity of forms characterised by different combinations of the actual values of the exponents in the general rate law

$$R_i = k_i \left[\begin{matrix} \diagdown \\ \diagup \end{matrix} C=C \begin{matrix} \diagup \\ \diagdown \end{matrix} \right]^n [MtX_n]^m [HB]^p.$$

In fact, the values of n, m, and p will depend on the initiation pathway prevailing in the system, the values of the equilibrium constants for the relevant interactions, the aggregation state of the ionic species (intermediates and chain carriers), and of course the rate-determining step of the process. Many specific examples have been analysed throughout this chapter, and attempts made to correlate the empirical kinetics reported in the literature with a specific initiation pathway.

Cationic polymerisations initiated by Lewis acids have been studied for several decades, but it is really only in the last few years that the details of the chemistry involved in the generation of chain carriers is beginning to become well understood. Our present proposals and interpretations were developed in an attempt to shed more light on this basic problem and we trust that they will prove of some relevance.

V. Initiation by Carbenium Salts and Related Species

This chapter deals with those cationic polymerisations in which the initiators contain an electrophilic organic moiety capable of attacking the vinylic double bond of the monomer to produce an active species. Included in this family are such compounds as stable carbenium and acylium salts, less stable ones which are usually prepared in the presence of monomer to avoid extensive decomposition before initiation, and ester molecules possessing a high degree of polarisation. Cation-radical salts will also be considered, but oxonium salts will not, because they are too stable to initiate the polymerisation of vinyl monomers.

As pointed out in Chap. II, the study of these systems began receiving considerable attention just over ten years ago. It was realised then that much fundamental information could be obtained about the initiation and propagation reactions if the complex phenomenology characteristic of polymerisations promoted by Brønsted and Lewis acids was replaced by the much simpler mechanism through which these new initiators operated, or were thought to operate. The discussion below offers an assessment of the achievements attained in this field, but also a critique of the weak points of some of the published investigations.

A. Stable Carbenium Salts

As we have mentioned in Chap. II, it is possible to prepare stable, fully ionised crystalline salts of carbocations provided the positive charge is sufficiently delocalised over the cationic moiety and the nucleophilicity of the anion is sufficiently low. Within this family of compounds, we will limit our attention to two major groups of salts with which most of the fundamental studies on cationic polymerisation of olefins has been carried out: the triphenylmethylium (trityl) and the cycloheptatrienylium (tropylium) salts. It has been a common practice to assume that if these salts are fully ionised in the solid state, they will also be completely dissociated in solution. Thus, the equilibria

$$RB \rightleftharpoons R^+B^- \rightleftharpoons R^+ + B^-,$$

B^- being the anion of a strong Brønsted acid (ClO_4^-, HSO_4^-, $CF_3SO_3^-$, etc.), and

$$RX + MtX_n \rightleftharpoons R^+MtX_{n+1}^- \rightleftharpoons R^+ + MtX_{n+1}^-,$$

where MtX_n is a "strong" Lewis acid ($SnCl_4$, $SbCl_5$, PF_5, etc.), should involve a negligible contribution from the unionised species on the left-hand side. For a series of trityl salts, this condition has been verified experimentally[633–635] and in fact the measurement of the apparent extinction coefficients at 413 and 435 nm (the two characteristic maxima of the paired or free Ph_3C^+ in the visible spectrum) provides an accurate indication of the extent of ionisation for these salts since the real values of ϵ have now been established at $40,000 \pm 2,000$[635,636] in several solvents. These values have tended to increase in recent years from earlier figures. This is probably due to improved purification and drying techniques applied to the solvents. As far as we know, these trityl salts are the only group of compounds used as initiators in cationic polymerisation for which total dissociation into ions has been proved experimentally, at least in methylene chloride, 1,2-dichloroethane and nitromethane. The tropylium salts with similar anions should behave likewise, given the exceptional stability of the carbocation. Of course, with anions of weaker Brønsted acids[637,638] and with weaker Lewis acids[639,640] the above equilibria include important contributions from the non-ionised species.

As for the equilibria between ion pairs and free ions in solutions of these salts, we refer to the values collected by Ledwith and Sherrington in their recent review[641]. Their Tables 2 and 3 show that except for the work of Lichtin and Pappas, carried out in liquid sulphur dioxide, the values of the dissociation constant for reliable studies in chlorinated hydrocarbons are quite independent of the counterion. Thus, $K_d = (1 - 2) \times 10^{-4}$ M at 25°C for the trityl salts, and $K_d = 0.5 \times 10^{-4}$ M at 0°C for the tropylium salts. Note that those Tables contain some misprints and the original references should be consulted for the correct values. A recent study by Gogolczyk et al.[635] is particularly interesting in this context because while it confirms the above value of K_d for trityl salts in dichloromethane, it also reports for the first time values of the dissociation constant in nitromethane and in mixtures of the two solvents. The value in pure nitromethane is more than a hundred times higher than that in dichloromethane, as expected from the considerable increase in polarity. Obviously then, the proportion of free ions in CH_3NO_2 is very high and at salt concentrations below 10^{-4} M they will be the predominant species. It must be emphasised that these ionic salts are only sparingly soluble in solvents of low dielectric constant such as carbon tetrachloride and therefore polymerisations cannot be carried out in these media.

Having established some general criteria concerning the state of these salts in the solvents suitable for cationic polymerisation, it is advisable briefly to survey the most relevant unwanted reactions of these species which could arise in the course of their use as initiators.

1. Reactions with Water and Other Nucleophiles

a) Trityl Salts

Brauman and Archie[642] used NMR spectroscopy to study the reaction of trityl fluoroborate with D_2O in acetonitrile with fairly high salt concentrations. They established the occurrence of a series of equilibria leading to the formation of the carbinol:

$$R^+B^- + D_2O \overset{K}{\rightleftharpoons} ROD_2^+B^- \rightleftharpoons ROD + DB.$$

They could not obtain a value of K, but concluded that the second order rate constant for the hydration reaction must be greater than 10^3 M^{-1} sec^{-1}. Slomkowski et al.[643] studied the same reaction by visible spectroscopy using various trityl salts in low concentrations in dichloromethane and derived the following values of K/M^{-1}: 60 at 25°C and 1.2 x 10^5 at -73°C, for Ph$_3$C$^+$ 10^{-4} M and H$_2$O 10^{-3} M. Apart from the ingenious extension of these results to the determination of the concentration of residual water in aprotic solvents[153], this work clearly shows the extreme sensitivity of the trityl ion to moisture, particularly at the lower temperatures. This point is to be kept in mind when setting up a polymerisation involving trityl salts as initiators. Rigorous drying of both solvent and monomer is a fundamental prerequisite for any meaningful quantitative result.

Reactions of trityl salts with other strong nucleophiles such as alcohols, amines etc., are also very fast and efficient as recently shown by the excellent work of Dorfman's group[111]. Recent work has also shown the high reactivity of this salt towards ethers and acetals[644,645].

b) Tropylium Salts

The fact that the ultraviolet spectrum of the tropylium ion is the same in 98% sulphuric acid and 0.1 N hydrochloric acid is a clear indication of the remarkable stability of this carbenium ion towards nucleophilic attack. The extinction coefficient of this ion in water is however somewhat lower than that calculated in acid solutions[646]. This indicates that the tropylium ion must be in equilibrium with its alcohol:

$$C_7H_7^+ + H_2O \rightleftharpoons C_7H_7OH_2^+.$$

It is also known that this ion is not stable in dilute solutions in ethanol[646]. Other interactions of the tropylium ion with nucleophilic agents have been recently reviewed by Harmon[646].

If it can be safely assumed that residual moisture in systems involving tropylium salts as initiators for the cationic polymerisation of olefins or vinyl ethers is not detrimental to their activity, the same might not be necessarily true for the chain carriers formed from them. It is likely that the carbenium ions derived from the monomers are much more reactive towards water than these initiators. Thus, even in these circumstances, it is advisable to carry out a thorough drying prodecure to avoid termination reactions.

2. Photolysis

a) Trityl Salts

After Dauben's observation that trityl perchlorate produced 9-phenylfluorene upon daylight exposure[647], van Tamelen and Cole[648] systematically studied the photolysis of the trityl ion in various acidic media with visible light from a medium-pressure mercury lamp. Depending on the acidity and the composition of the media in which these reactions were conducted, different products and product distributions were observed. The essential point about the primary processes involved was that the triplet Ph_3C^+ reacted with its ground state precursor to give trityl radical cations. The subsequent reactions depended strongly on the medium, and various products such as 9-phenylfluoren-9-ol, triphenylmethane, 1,1,1,2-tetraphenylethane, and hexaphenyl hydrocarbons and carbinols were characterised. While the mechanistic details of these photolyses are outside the scope of the present review, it is important to stress the sensitivity of trityl salts to visible light. Obviously, when using these compounds, both in the solid state and in solution, care must be taken to avoid over exposure to laboratory light and ideally their handling should be carried out in the dark.

b) Tropylium Salts

Many tropylium salts are coloured[646] although the carbenium ion only absorbs at 217 and 273 nm. The photolysis of colourless tropylium fluoroborate in 5% sulphuric acid with a medium-pressure mercury lamp was studied by van Tamelen et al.[649]. Valence-bond isomerisation of excited tropylium ions yielded bicyclic products. Although no study of the photolysis of coloured tropylium salts has been reported, it seems advisable to recommend the same precautions as with trityl salts in order to avoid undesirable photochemical losses or formation of products which might be detrimental to the normal development of the polymerisation.

3. Other Relevant Reactions

a) Trityl Salts

The trityl ion has been shown to be an effective hydride-ion abstractor in reactions with cyclic ethers[645 − 650] and polynuclear hydrocarbons[651]. In this context Olah and Svoboda[652] have shown that the trityl ion can exhibit ambidental reactivity in hydrogen transfer processes, the two sites accepting the incoming H^- being the aliphatic carbon atom and the para positions of the rings. These observations clearly confirm the extensive delocalisation of the positive charge in this stable ion. Heck et al.[653] had also reported that the trityl ion reacted with acetals through one of its para positions, instead of through the aliphatic atom.

The above reactions should be kept in mind when analysing the mechanistic behaviour of trityl salts in cationic polymerisation.

b) Tropylium Salts

The higher stability of these ionic species compared to trityl ions is directly proved by the fact that tropylium salts can by prepared by hydrogen abstraction from cycloheptatriene induced by trityl salts[646]. An excellent account of the reactions of the tropylium ion has been published by Harmon[646].

4. Polymerisation Systems

Ledwith and Sherrington[641] have already discussed some of the basic aspects of polymerisation processes involving stable carbenium salts as initiators and reviewed the literature in this field up to 1974. Since then, some interesting and pertinent material has
been published which makes it necessary to reassess the interpretation of the older work.
We therefore cover this whole area of cationic polymerisation from its beginning.

Depending on its nucleophilicity, when an alkenyl monomer is added to a solution
containing a stable carbenium salt, different situations can arise, namely:

- fast and complete initiation characterised by the total disappearance of the catalyst's
  carbenium ions upon mixing and the formation of an equivalent amount of active
  species;
- slow initiation continuing during most or even all of the polymerisation process;
- no appreciable interaction between monomer and initiator i.e., no polymerisation.

The first limiting situation is obviously an ideal one since it leads to the possibility of
determining propagation rate constants through relatively simple kinetic studies. Although various authors have claimed that some systems display this type of behaviour,
little evidence was in fact given to substantiate this claim. The rapid consumption of the
initiator was in fact inferred from indirect kinetic measurements, but not proved, as will
be shown in the specific sections below. Any study aimed at establishing a reliable understanding of the nature and rate of the initiation reaction in these systems, must be
carried out using techniques capable of specifically revealing the fate of the stable carbenium ions, such as ultraviolet and visible spectroscopy. It is moreover essential to demonstrate that the disappearance of these ions is accompanied by the formation of active
species since the first does not necessarily prove the second. Only recently, has this rigorous approach been applied and it has been shown that most systems do not behave in
such a straightforward manner as just described.

The second, intermediate, situation corresponding to concurrent initiation and
propagation, covers of course a wide variety of possibilities depending on the relative
rate of these two reactions. It represents by far the most common occurrence among
the systems which produce polymerisation.

As for the third limiting situation, it is characteristic of systems involving monomers of low nucleophilicity such as alkenes incapable of reacting with stable carbenium
ions.

The chemical details of the initiation reaction can vary from one system to another.
Thus the carbenium ion can add on to the monomer double bond in a classical electrophilic reaction or it can abstract a hydride ion or an electron from the monomer. The
first process seems to be the most common according to the experimental evidence gathered to date. Whatever the nature of this interaction, it remains to be established if the
active species formed from them are also ionic and if so, if they are predominantly free
ions or ion pairs. This is undoubtedly the most delicate problem implicit in studies of
this type. It has been often assumed that since the initiator was essentially composed of
free ions in the low concentrations used it would follow that the same would apply to
the chain carriers derived from it. However, it has become apparent that this assumption
is generally unjustified since the stability of the carbenium ions formed from the mono-

mers can be considerably lower than that of the initiating species. Thus, the extent of dissociation of the former is not necessarily as high as that of the latter.

This introduction offers warning against unsubstantiated oversimplifications in the drawing of kinetic and mechanistic conclusions from a study which has not made use of all possible experimental resources. We show below, that unfortunately these systems are not as straightforward as it was suggested in some of the earlier investigations.

a) Vinyl Ethers

The reaction of vinyl ethers with tropylium salts was investigated by Volpin's school in different media[654, 655]. There seems to be no doubt that the tropylium ion readily adds on to the double bond at the CH_2 site. Bawn et al.[656] found aryl groups in poly(vinyl ethers) prepared with trityl salts and no evidence for the formation of triphenylmethane among the products. It seems therefore, that initiation in these systems takes place according to a simple electrophilic addition:

$$R^+ + CH_2 = CHOR' \rightarrow RCH_2\overset{+}{C}HOR',$$

although other pathways cannot be excluded.

The first detailed study of a cationic polymerisation of vinyl ethers induced by stable carbenium salts was reported in 1971 by Bawn et al.[656]. Isobutyl vinyl ether was polymerised with trityl and tropylium hexachloroantimonates and trityl fluoroborate. From calorimetric measurements of the rate of polymerisation, it was concluded that all the initiator used was consumed soon after mixing and the assumption was made that an equal number of active species was formed in this fast initiation reaction. Propagation rate constants were thus obtained and attributed to the action of free ions. It was also claimed that no significant termination took place during the polymerisation since successive monomer additions produced polymerisations having the same propagation rate constant. Later work performed in the same laboratory on other vinyl ethers [657] led the authors to similar conclusions except that the possible existence of ion pairs alongside free ions was envisaged for the chain carriers. In these investigations no attempt was made to prove mechanistically the simple initiation pattern proposed. Chung et al.[658] repeated some of the experiments of the Liverpool school using more rigorous purification and drying techniques. Following the usual assumptions, they obtained k_p values 2 to 3 times higher than those reported previously and concluded that propagation was essentially due to free ions since these rate constants agreed with those measured in radiation-induced polymerisations of the same monomers. However, again no attempt was made to study the process of initiation.

Subira et al.[659] recently carried out a very detailed and thorough study of the polymerisation of isobutyl vinyl ether initiated by trityl hexachloroantimonate in methylene chloride at various temperatures. The major difference between this investigation and the previous ones consisted in the fact that for the first time the rate of initiation was measured spectroscopically at the same time as the rate of monomer consumption (calorimetry). These determinations clearly showed that the behaviour of the system was by no means as simple as postulated by previous workers. Initiation was in fact found to be fairly slow to the point that below 0°C not all the catalyst was consumed by the end of the polymerisation. Also, second monomer addition experiments showed that some termination took place during polymerisation. Thus, the living character of these processes proposed by the Liverpool school was disproved. A specific kinetic analysis was performed by the French authors to prove that the oversimplified reaction scheme previously assumed to hold for these systems was in fact erroneous and the origin of the mistake arose from the lack of a direct study of the initiation process. The values of k_i (5.4 M^{-1} sec^{-1}) and k_p (7 x 10^3 M^{-1} sec^{-1}) at 0°C calculated by Subira et al. clearly show that initiation is relatively slow with respect to propagation. This work thus substantially changed the picture by an even more comprehensive and rigorous attack.

Two major problems remain open concerning the nature of some of the reactions involved in these systems, namely the mechanism of the unimolecular termination reaction detected by Subira et al. and the structure of the active species, i.e. their state of dissociation. An answer to the first of these problems might come from considering the following reaction:

$$\sim\!\!\sim\!\!\overset{|}{\underset{|}{C}}{}^{+}\ MtX_{n+1}^{-}\ \longrightarrow\ \sim\!\!\sim\!\!\overset{|}{\underset{|}{C}}X\ +\ MtX_n\ \bullet$$

A systematic search for halogen in the polymers could give an indication about the likelihood of such termination. As for the second problem, it seems now well established that both free ions and ion pairs propagate the growth of polymer chains. This conclusion is borne out by the observation made by Stannett et al.[660] that free ions constiture only 20 to 50% of the total amount of chain carriers in these systems. This would imply that the calculated propagation rate constants represent composite parameters and that probably the specific k_p values for free ions are as much as 5 times higher.

b) Cyclopentadiene and Other Dienes

The study of the polymerisation of cyclopentadiene by stable carbenium salts has been the exclusive domain of Sigwalt and his group. In 1967, they reported the first observations on a system involving trityl hexachloroantimonate in methylene chloride between – 70 and 20°C[661]. Because of the interesting features of this polymerisation, the work was pursued by kinetic and mechanistic investigations. In 1969[662], it was shown that initiation took place by direct addition and that transfer and termination were unimportant particularly in the first stages of the process. The kinetics of initiation were followed by visible spectroscopy and provided further evidence for a one-to-one reaction. Thus,

$$Ph_3C^+\ SbCl_6^-\ +\ \langle\!\!\bigcirc\!\!\rangle\ \overset{k_i}{\longrightarrow}\ Ph_3C\!-\!\langle\!\!\overset{+}{\bigcirc}\!\!\rangle\ SbCl_6^-\ ,$$

E_i = 9.2 kcal mole^{-1}, A_i = 5 x 10^6 M^{-1} sec^{-1}. The third study on this system[663] dealt with further kinetic and mechanistic investigations. The combination of spectroscopy (initiation), dilatometry (propagation) and DP determination at different yields allowed the formulation of sound overall mechanism. At low conversions no transfer could be detected and termination was important only at the higher temperatures. It was also observed that both k_p and k_t had apparent negative activation energies. This was rationalised in terms of the existence of free ions and solvent-separated ion pairs in equilibrium with tight ion pairs, the latter being almost inactive towards propagation. A decrease in temperature would provoke an overal shift of these equilibria towards an increase in the concentration of free ions, and consequently a corresponding increase in the observed value of the propagation rate constant.

The above three publications constitute an excellent example of what can be achieved with this type of initiators by working rigorously and using several complementary techniques to gain a comprehensive understanding of all aspects of a system. Of particular interest is the observation that at low temperature cyclopentadiene gives rise to a simple polymerisation characterised by concurrent initiation and propagation but without chain-breaking reactions. This behaviour was recently exploited by Vairon and Villesange[636,898] to carry out an ingenious series of experiments. The dicationic salt $SbCl_6^-(Ph)_2\overset{+}{C}(C_6H_4)CH_2CH_2(C_6H_4)\overset{+}{C}(Ph)_2SbCl_6^-$ was prepared and used as an initiator for the polymerisation of cyclopentadiene at $0°C$ in methylene chloride. The products obtained below about 20% conversion followed closely the expected DP's for a process without chain-breaking reactions, i.e., $DP_n = ([M]_0 - [M]_t)/0.5[C]_0$. Each polymer molecule contained six phenyl groups (NMR), as expected. These preliminary results allowed an extension of the original idea leading to the preparation of nearly monodisperse block copolymers in the following way. A monodispersed polymer of α-methylstyrene was prepared by anionic polymerisation; this was functionalised at both ends by introducing $Ph_3CSbCl_6^-$-type end groups; cyclopentadiene was then polymerised from this bifunctional cationic polymer to produce a block copolymer $(CPD)_n$——$(MST)_m$——$(CPD)_n$. We consider this combination of anionic and cationic living copolymerisation a most pleasing achievement.

Within the context of their investigation Vairon and Villesange observed that the interaction of cyclopentadiene with trityl hexachloroantimonate in carbon tetrachloride or benzene produced the rapid disappearance of the characteristic yellow colour of the trityl ion but no polymerisation. This interesting phenomenon of *false* initiation has also been reported in some systems involving styrene and will be discussed when we deal with this monomer.

Isoprene has been polymerised by both tropylium and trityl salts in nitrobenzene by Gaylord and Švestka[664] who also found that less polar media were ineffective for this reaction. The complex behaviour of these systems did not allow any fundamental conclusion to be reached on the mechanism of initiation except that some incorporation of trityl groups (but not of tropylium groups) was detected in the polymers. Interestingly, isoprene seems to display a limiting nucleophilicity and sit at the boundary between monomers polymerisable and non-polymerisable by these stable carbenium salts.

The only other dienic monomer studied in this context is *spiro*[2,4]hepta-4, 6-diene. Its polymerisation by various trityl salts was investigated by Kunitake et al.[665] in methylene chloride-toluene mixtures at $-76°C$. This study was entirely devoted to the alternative 1,2 and 1,4 propagation modes and no attention was given to the mechanism of initiation or other fundamental problems.

c) Indene

Sauvet et al.[661] were the first to show that indene can be polymerised by trityl salts. A few years later, Eckard et al.[666] studied its polymerisation by tropylium hexachloroantimonate and showed that only with very high salt concentrations was it possible to obtain complete conversion. Chain termination was found to be an important reaction probably because of the formation of stable, non-propagating carbenium ions as end groups. No attempt was made to characterise an addition product between monomer and initiator. The development of a red colour during the polymerisation was also reported and attributed to a stable tropylium-indene carbenium ion. This interpretation did not take into account a simpler alternative explanation which Prosser and Young[667] put forward a year later as a feature common to all cationic polymerisations involving indene. It consists in the formation of an allylic-type carbenium ion formed in a termination reaction between an active species and an unsaturated polymer molecule (hydride-ion transfer). This type of termination is now well known in the cationic polymerisation of several monomers[45, 281].

Subira et al.[668] studied the polymerisation of indene by trityl pentachlorostannate in methylene chloride. Again this group carried out a deeper investigation of the process by examining the kinetics of initiation before dealing with the polymerisation reaction. A simple kinetic law implying a one-to-one reaction was obtained with $k_i = 3.7 \times 10^{-2}$ M^{-1} min^{-1} at 20°C and $E_i = 18.4$ kcal mole^{-1}. Initiation was slow at low temperature and propagation extremely sluggish, but no real termination reaction was detected since all polymerisations eventually went to completion. The authors' suggestion that a growing chain could expel a proton to give stannic chloride and hydrogen chloride seems a plausible one to explain the observed behaviour of this system. It is a pity that no experiments were performed to prove that initiation took place via addition of the trityl ion to the monomer double bond.

d) p-Methoxystyrene and Anethole

Sauvet et al.[661] first reported that p-methoxystyrene could be effectively polymerised by trityl salts in methylene chloride, and this was later confirmed by Burns et al.[669]. It was again Sigwalt's cationic group[670] which carried out an accurate kinetic and mechanistic investigation using trityl hexachloroantimonate in methylene chloride. With the dual spectroscopic-calorimetric technique they followed both initiation and propagation. The kinetics of initiation was shown to involve one molecule of monomer and one of catalyst, $k_i = 0.1$ M^{-1} sec^{-1} at −2°C and $E_i = 13.3$ kcal mole^{-1}. Incorporation of trityl groups into the polymer gave the proof that the initiation reaction consisted of the electrophilic addition of the initiator cation to the monomer double bond. Time-conversion curves were found to be S-shaped indicating that initiation was not fast. In fact, a considerable amount of catalyst was still present at the end of the polymerisations. A complete kinetic treatment allowed the calculation of the propagation, the monomer transfer and the very low termination rate constants. Again, as in the case of cyclopentadiene, a negative activation energy for propagation was obtained and to explain this two alternative mechanisms were considered. The first was based on an increase in the

proportion of free ions at low temperature while the second invoked the formation of
a complex between the monomer and the active species before the actual propagation
step took place. The latter mechanism was proposed because a similar interaction had
been observed spectroscopically in the dimerisation of 1,1-diphenylethylene (see Sect.
III-E-12-b). Very recent work by the same group[671] has shown that both free ions and
ion pairs are responsible for propagation in this system, and kinetic measurements in
the presence and in the absence of a common-anion salt have resulted in the determination of the propagation rate constant for both species. Surprisingly, k_p^+ was found to be
only about four times larger than $k_p^\pm$. In view of these findings, the negative activation
energy for propagation is probably due to the shifting of the equilibrium between ion
pairs and free ions towards the latter as the temperature is lowered.

While the above investigation was being completed, Goka and Sherrington[672] published a study on the polymerisation of *p*-methoxystyrene by tropylium hexachloroantimonate in methylene chloride. These authors showed unequivocally that the initiation reaction proceeded through the addition of the tropylium ion to the monomer
double bond. They were able in fact to isolate and characterise this addition product.
The kinetics of polymerisation were followed calorimetrically and attempts to follow
the initiation reaction by NMR spectroscopy failed. The absence of appreciable termination was claimed on the basis that addition of a second monomer sample at the end of
a polymerisation resulted in a second polymerisation with approximately the same propagation rate constant. The authors concluded that after a short acceleration period one
could equate the concentration of active species with that of the added initiator. However, the arguments they proposed to support this assumption are not convincing and
the fact that the observed propagation rate constants were directly proportional to the
initial monomer concentration shows on the contrary, as pointed out by Cotrel et al.
[671], that initiation continues throughout the polymerisation. Also, the relatively low
values of k_p obtained in this work, compared with those reported by Cotrel et al.,
clearly indicate that the concentration of active species was grossly overestimated. Other
shortcomings of this work include the lack of any detailed study on termination processes, and the extremely high concentrations of common-anion salt compared with the
amount of initiator used. It is not surprising that the rate suppressions observed were
so large in view of the extensive ionic aggregation provoked by the presence of such
high quantities of added salt.

Very recently, Rooney[673] reported on the polymerisation of anethol by trityl hexafluoroantimonate in methylene chloride. He showed that this monomer is much less reactive than *p*-methoxystyrene both in terms of initiation and propagation. This is obviously due to the β-methyl substitution. We found this study very diffucult to accept for
the following reasons. The initiation rate is assumed to be monomolecular in both reactants, but this kinetic law was not proved experimentally, thus the value of k_i are simply
a rough indication of the slowness of the reaction but do not bear any quantitative significance. It is stated that the maximum rate of polymerisation is directly proportional to
the square of the initial monomer concentration and three experimental points are given
in a log-log plot to prove this (Fig. 6 Ref.[673]); yet in Fig. 2a three polymerisation curves
are shown (same temperature, same initial catalyst concentration) in which the claimed
dependence is certainly not verified. The kinetic arguments employed to arrive at values
of k_p are weak and again one can only give these values a qualitative significance, i.e.

that suggesting that propagation is sluggish. The connection between termination and the peak at 501 nm is not proved and its supposed demonstration in Fig. 8 is at least shaky given the freedom with which straight lines were drawn among the experimental points. Equally unconvincing is the mechanistic test for initiation because it does not specify how the "products" were isolated and their characterisation was inadequate. On the whole therefore, this work is rough, but it certainly shows that secondary reaction producing branching and termination are important drawbacks of this monomer.

e) N-vinylcarbazole and N-ethyl-3-vinylcarbazole

Although the polymerisation of N-vinylcarbazole by stable carbenium salts was first reported in 1964[674], the details of this study appeared several years later[675]. Tropylium hexachloroantimonate and perchlorate were used as initiators and the rates of polymerisation determined calorimetrically. While the kinetics of initiation were not followed, a model reaction was performed to establish that the tropylium ion adds on to the monomer double bond to give chain carriers. The assumption was made that initiation was "instantaneous under most conditions". The values of the propagation rate constant were thus obtained by supposing that the concentration of active species was equal to that of the catalyst added. We have already commented on the danger of such an assumption. In this context it can only be concluded that the order of magnitude of k_p was probably correct but the lack of complementary studies on the rate of initiation with respect to the rate of propagation, makes any further assessment premature. The high values of k_p obtained – about 10^5 M^{-1} sec^{-1} – are consistent with the well-known reactivity of this monomer in cationic polymerisation. Again, the assumption that these values were to be attributed essentially to free ions was not corroborated by any experimental backing. Finally, the claimed lack of termination was equally unsubstantiated.

Rooney has recently revived work on this monomer in an investigation of its polymerisation by trityl hexafluoroantimonate[676]. He used a spectroscopic stop-flow apparatus to follow initiation and an adiabatic calorimeter to measure rates of polymerisation. Propagation was shown to compete effectively with initiation to the point that some initiator was often present at the end of the polymerisations. These observations cast some doubts on the assumption made in the paper by the Liverpool school discussed above. A kinetic analysis of the initiation reaction showed it to be bimolecular, with a rate constant of about 130 M^{-1} sec^{-1} at 20°C. The determination of the propagation rate constant was less straightforward despite the fact that further monomer-addition experiments seemed to rule out any appreciable termination. The k_p values fluctuated considerably as the initial catalyst concentration was varied, a fact which induced Rooney to propose that the empirical constant was a composite function of k_p^+ and $k_p^\pm$. Experiments with a common-anion salt supported this proposal and their kinetic treatment led to the individual values of $k_p^+ = 6 \times 10^5$ M^{-1} sec^{-1} and $k_p^\pm = 5 \times 10^4$ M^{-1} sec^{-1}. It is difficult to assess the reliability of these values in view of the following statement by the author: "the reaction at a 5×10^{-5} M concentration of initiator, thought to proceed exclusively through paired ions . . .". This statement is certainly incorrect as far as the initiator is concerned for which the proportion of ion pairs for a concentration 5×10^{-5} M at 20°C is only about 20% in methylene chloride[635]. However, the experiments with added common-anion salt clearly showed that in those conditions a large excess of salt did not retard the polymerisation. It could be argued that this dilemma might be resolved since most probably the dissociation constant for the propagating species is considerably smaller than that of the initiator, i.e. that at that concentration while the initiator is essentially present in the form of free ions, the chain carriers are almost entirely paired. This important problem will be discussed in detail in the conclusions to this section.

In a second paper devoted to the polymerisation of N-vinylcarbazole, Rooney[677] used various tropylium salts as initiators. Only calorimetric measurements of the rate of monomer consumption were made. The treatment of the data was again based on the assumption that complete initiation had been achieved shortly after mixing the reactants in methylene chloride. The author claimed that tropylium salts were more effective initiators than trityl salts. As we will discuss later, this unsupported statement does not seem reasonable. Values of the propagation rate constant at $-40°C$ were calculated and a discussion was given on the origin of their dependence upon the type of counterion, the initiator concentration and the additions of a common-anion salt. However, in view of the paucity of mechanistic experiments, this discussion seems speculative and rather gratuitous. Particularly disturbing is the statement that free ions and ion pairs possess the same k_p in the case of SbF_6^-, considering that the same author had calculated a difference of one order of magnitude in the reactivity of the two species in the previous paper (see above). This paper does stand as a preliminary approach to these systems but hardly sustains the mechanistic speculations. We fail to understand how the tropylium ion which is considerably more stable than the trityl one, should be more effective in promoting the polymerisation of N-vinylcarbazole and give rapid and complete initiation. Since the previous study with trityl hexafluoroantimonate had shown conclusively that initiation was slow compared with propagation[676], only experimental determinations of the rate of initiation with the tropylium salts would satisfactorily settle this point.

Rooney et al.[678] published a study on the possibility of synthesising block copolymers of vinyl ethers with N-vinylcarbazole. These experiments were based on the assumption that no termination takes place in the polymerisation of vinyl ethers with trityl salts; that transfer can be minimised in that polymerisation; and that N-vinylcarbazole polymerisation can be initiated by a polymerised solution obtained as above. The first assumption was verified kinetically by conducting a double polymerisation and measuring the two apparent propagation rate constants. However, the DP's of the first and second polymer were not measured. The second assumption was not verified but it was decided that in order to minimise solvent transfer it would be advisable to work with high monomer concentrations. It is difficult to understand how this measure could in fact minimise transfer since the monomer transfer constant was reported to be about five times larger than the solvent transfer constant so that a large increase in the monomer concentration would certainly not help in reducing the importance of transfer reactions. The third assumption was verified experimentally but no study was conducted of the polymerisation of N-vinylcarbazole initiated by the trityl salt used, to check how important transfer reactions were in this homopolymerisation. The major flaw of these experiments resides in the fact that under the conditions chosen for the copolymerisation runs, it was not shown that the DP of the poly(vinyl ethers) increased during the polymerisation or simply with a second addition of the same monomer. It is surprising that this classical experiment was not performed. Also, if block copolymerisation had been achieved, the copolymer should have had a higher DP than the first block of poly(vinyl ether). This point was not verified either. In fact, the only evidence offered in favour of the formation of block copolymers came from fractionation experiments and measurements of intrinsic viscosities as a function of temperature. These tests are certainly not conclusive, particularly in view of the importance of monomer transfer in these polymerisations. Therefore, the claim purported by this paper remains to be verified by more adequate tests.

The extremly high susceptibility of N-vinylcarbazole towards cationic polymerisation is well illustrated by the ease and rapidity with which this monomer is converted into polymer even at low temperatures under the action of very small concentrations of stable carbenium salts. We have already mentioned that oxonium salts are inactive for the polymerisation of olefins. Turchi et al.[679] tried triethyloxonium hexafluorophosphate as a possible initiator for the polymerisation of this monomer and proved in a series of elegant experiments that this salt is not an initiator. Polymerisation was observed but caused by traces of PF_5 present in the system. The importance of this paper stems from its conclusive evidence showing that even the most basic among the vinylic monomers is insensitive to oxonium salts.

An extremely promising monomer, N-ethyl-3-vinylcarbazole, has recently been polymerised with tropylium salts[96]. This polymerisation seems to proceed without transfer and termination reactions judging from the linear and almost theoretical increase of the DP's of the products with [M]/[I] ratio. It is a pity that this interesting analysis was not corroborated by multiple monomer-addition experiments to prove that the same rate was obtained after each addition (no termination), and that the DP increased after each addition (no transfer either). In this investigation, the initiation process was not studied but it was assumed to be fast and complete. From this assumption and the usual kinetic treatment, the value of $k_p = 2 \times 10^4 \ M^{-1} \ sec^{-1}$ at $0°C$ was computed and attributed to free ions propagation. A visible absorption band at 468 nm detected during the polymerisation was attributed to the monomer carbenium ion and the authors found that its maximum optical density (at t = 0) correlated linearly with the free ion concentrations of the initiator. In apparent contradiction with this finding, the addition of appreciable concentrations of common-anion salts, which should have suppressed the concentration of free ions considerably, did not alter the above correlation. Obviously, more work is necessary to prove the origin of that band before it can be assigned unequivocally. As the authors have pointed out, this monomer has remarkable potentials and future studies might well prove that it can give a truly living cationic polymerisation. The absence of branching will have to be proved before such conclusion is reached.

f) o-Divinylbenzene

The work of Aso et al.[680] on this monomer has been discussed at length by Ledwith and Sherrington in their review[641]. This investigation does not contribute substantially to the basic understanding of the mechanism of initiation by stable carbenium salts and no further comments are therefore necessary in the present context.

g) α-Methylstyrene

The polymerisation of this monomer by stable carbenium salts was first reported by
Sauvet et al.[661] who only remarked that at −70°C in methylene chloride, about 30%
yield was obtained in two hours with a 10^{-3} M concentration of trityl hexachloroantimonate. In the same year, Higashimura et al.[681] measured the rate of interaction between trityl pentachlorostannate and α-methylstyrene at 30–60°C in ethylene chloride
and mixtures of this solvent with benzene. They monitored the disappearance of the
characteristic visible bands of the trityl ion as the reaction proceeded. Good first order
plots were obtained and the external order in monomer was found to be unity so that
kinetically the reaction was bimolecular with a k_i of 1.3 M^{-1} min^{-1} at 30°C in pure ethylene chloride, and E_i = 5.0 kcal mole^{-1}. The product of this reaction was not characterised so that it is not known whether the trityl ion added on to the monomer double
bond or abstracted a hydride ion from it. These authors claimed that the initiator was
in equilibrium with its precursors in the media they used and that for this reason the extinction coefficients they measured were much lower than the one typically expected
for the trityl ion. This argument has been proved wrong both by Subira et al.[668] and by
Johnson and Pearce[682] who showed that trityl pentachlorostannate was entirely dissociated in methylene chloride, with an extinction coefficient at 410 of (3.4–3.5) x 10^4
M^{-1} cm^{-1}, i.e., the normal value. The origin of the low value found by Higashimura et al.
is unclear but could be due to the high moisture level in their reaction systems, quoted
to be 1 to 2 mM. In a recent publication Yamamoto and Higashimura[683] again quote
an apparent molar extinction coefficient of only 17,000 for that salt without commenting on the fact that it is about half the normal value. In any case, the value of the initiation rate constant calculated in both these papers must of course carry a systematic
error due to the incorrect value of the extinction coefficient used. No study of the polymerisation of α-methylstyrene was carried out by these authors.

Kunitake's group has reported on two occasions observations about the polymerisation of this monomer by a whole family of trityl salts at −78°C in solvents of various polarities. Polymerisations were reported to be slow, particularly in media of low
dielectric constant. These studies were not concerned with the mechanism of initiation
or with the kinetics of polymerisation but rather with the influence of the counterion
on the steric structure of the polymers obtained in conditions where ion pairs predominate over free ions. No further studies have been published on this monomer.

In conclusion, although it is known that trityl salts can initiate the polymerisation
of α-methylstyrene and that the polymerisation is rather slow, no systematic study has
yet been carried out on these systems and the chemistry of initiation is still unknown.

h) Styrene

Reports on the possibility of inducing the polymerisation of styrene by trityl salts are
among the oldest in this field[684]. However, surprisingly enough these processes have
not yet been fully explored. Styrene is rather reluctant to react with the trityl ion and
does not react at all with the tropylium ion. The latter observation has been reported
by Ledwith and Sherrington[641] who describe an experiment in which *p*-methoxystyre-

ne was polymerised by a trityl salt in *solvent* styrene, the product being a homopolymer of the former monomer. Unfortunately, these authors did not quote the source of this work.

Sambhi and Treloar[685] published in 1965 a brief study of the polymerisation of styrene by a mixture of trityl chloride and mercuric chloride in ethylene chloride at 30°C. The following mechanism was proposed on the basis of their experimental results:

$$Ph_3CCl + HgCl_2 \overset{K}{\rightleftharpoons} Ph_3C^+HgCl_3^-, \quad K = 1.31 \ M^{-1};$$

$$-d[Ph_3CCl]/dt = k_i[St] [Ph_3CCl] [HgCl_2], \quad k_i = 3.7 \times 10^{-3} \ M^{-2} \ min^{-1};$$

and

$$-d[St]/dt = k'[St]^2 [Ph_3CCl] [HgCl_2], \quad k' = 0.33 \ M^{-3} \ min^{-1}.$$

A termination reaction consisting in the migration of a chloride ion to the active species was also claimed,

$$\sim\sim\sim CH_2CHPh^+ \ HgCl_3^- \xrightarrow{k_t} \sim\sim\sim CH_2CHPhCl + HgCl_2 \ .$$

On the basis of a presumed stationary state, k' was equated to $k_i k_p/k_t$. No mention was made of S-shaped time-conversion curves despite the fact that this is a general feature in systems involving styrene and trityl salts[669,681,686].

In 1967, Higashimura et al.[681] studied the polymerisation of styrene by trityl pentachlorostannate. We have already commented in the previous section about the error made by these authors in evaluating the extinction coefficient of the salt, and on its possible source. The same reproach applies to this system studied at the same time. The kinetics of the interaction between the salt and the monomer gave simple first orders and a value of $k_i = 5.3 \times 10^{-2} \ M^{-1} \ min^{-1}$ in ethylene chloride at 30°C was calculated ($E_i = 6.7$ kcal mole^{-1}). The chemistry of this simple one-to-one reaction was not investigated but the authors claimed that the trityl ion added onto the monomer double bond to give the corresponding styryl carbenium ion. The polymerisation reaction was only very briefly studied. S-shaped rate curves were reported and it was stated that the maximum rate of polymerisation was proportional to the square of the initial monomer concentration but independent of the initiator concentration. On the whole, this study, though frequently quoted in the past, does not provide any fundamental information about the mechanism of these processes, except for the (wrong) values of the initiation rate constant.

Sambhi returned to the $HgCl_2 - Ph_3CCl$ system in 1970[687]. For this investigation of the polymerisation of styrene, both the decay of trityl ion (413 nm) and that of styrene (291.5 nm) were followed spectroscopically in ethylene chloride at 30°C. No special precautions were reported to exclude adventitious moisture from the reaction media. A plot of trityl ion concentration vs. the product $[Ph_3CCl] \times [HgCl_2]$ was linear indicating that ion pairs were the predominant species. The rest of the kinetic treatment was entirely similar to that given in the previous paper by Sambhi and Treloar discussed above, except that the values of the various constants were slightly changed. The only new evidence mentioned in this new publication was:

— $HgCl_2$ does not seem to form complexes with styrene since addition of this Lewis acid to a styrene solution did not cause any change in the latter's ultraviolet spectrum.
— Polymerisations proceeded without an induction period. The monomer concentration vs. time plots indicated extremly slow reactions with half lives of the order of 50 h, but unfortunately the conditions of these experiments were not specified.
— Polymers were found to contain chlorine (0.9–1.5%), but no correlation was established between this interesting parameter and other reaction conditions.
— An internal order of 2 was claimed for the rate of monomer consumption, but apparently this was only verified up to about 30% conversion.

We feel that the experimental evidence gathered in this study is insufficient to justify the detailed and all-embracing mechanistic interpretation proposed by the author. Thus, for example, no serious attempt was made to detect the possible presence of trityl groups in the polymers, i.e., no proof was obtained concerning the proposed chemistry of initiation. The existence of termination reactions was invoked, but not demonstrated. Finally, the second-order behaviour of the polymerisations rested on shaky grounds given the inadequate kinetic analysis of the experimental data.

The extreme slowness of these polymerisations makes one wonder if the trityl ion (thought to be present exclusively as the pair $Ph_3C^+HgCl_3^-$ in the above two papers, but see below for a reassessment of this conclusion) was the real initiator and not the weak Lewis acid used to form it. No "blank" experiments were reported with mercuric chloride alone as catalyst, to verify the absence of polymerisation at the concentrations used for the study with trityl chloride. One could well envisage that in the presence of traces of moisture of hydrogen chloride (note that $HgCl_2$ was used without purification) this Lewis acid, present in concentrations of about 10^{-2} M, could have catalysed the polymerisation of styrene with half lives of tens of hours.

A recent study by Bos and Treloar[640] on the interaction of mercuric chloride with trityl chloride in ethylene chloride showed that the trityl trichloromercurate is formed as a mixture of free ions and ion pairs, roughly in equal amounts when the concentrations of the precursors were similar to those used in the polymerisation studies reported above. This conclusion makes the interpretation of Sambhi's results even more difficult, because the extremely low polymerisation rates he observed are incompatible with initiation by free trityl ions. In fact, the work of Johnson and Pearce discussed below, clearly showed that initiation is fast with styrene when one uses a trityl salt which is appreciably dissociated into free ions. This paradox raises a new question, namely that concerning the role of the counterion in these processes. It is conceivable that with $HgCl_3^-$ transfer of a chloride ion is so favoured that appreciable quantities of l-phenylethyl chloride or homologous oligomers could be formed and not be detected as "polymer". Sambhi's few polymerisation curves do not allow to assess this point.

Pepper and collaborators[669] examined the polymerisation of styrene initiated by various trityl salts in chlorinated hydrocarbons. Induction periods were observed and initiation was so slow compared with propagation that the characteristic colour of the trityl ion persisted during the whole of the polymerisation processes, except with trityl hexafluoroantimonate. With some salts limited yields were reported. Unfortunately, this exploratory piece of work was never followed up.

The latest and undoubtedly the most serious attempt to rationalise the behaviour of styrene in this context comes from Johnson and Pearce[682]. They used trityl hexafluoroantimonate and hexachloroantimonate as initiators and conducted their interesting study in methylene chloride at room temperature. Rigourous high vacuum handling was employed for the measurement of electrical conductivity, visible spectra and volume contraction (dilatometry). With trityl hexachloroantimonate concentrations of $10^{-5} -$ 10^{-4} M (mostly free ions) addition of excess styrene produced a steady decrease in both the optical density at 413 nm and the conductivity which eventually reached values close to 0. A kinetic analysis of these reactions gave evidence for a straightforwad one-to-one process, $k_i = 5.3 \times 10^{-2}$ M^{-1} min^{-1}, i.e., a slow initiation. The decrease in conductivity indicates that the species formed in these interactions were not free ions. The surprising fact was that no detectable polymerisation occurred during the lifetime of the initiator, i.e., no active species were produced in these reactions which however consumed all the trityl salt added. Only with much higher salt concentrations was polymerisation observed and this became fast and complete with $Ph_3C^+SbCl_6^-$ 10^{-1} M. All the polymerisation rate curves were S-shaped. In order to characterise the products of the reactions with low salt concentrations, an experiment was performed where successive portions of styrene were added. Cyclic and linear styrene dimers were isolated together with very small amounts of trityl substituted dimers. Thus, for the first time, some evidence was obtained for the direct addition of the trityl ion to styrene, but the very low proportion of this compound with respect to the dimers clearly indicates that transfer reactions were very important. What is not clear is why, following the addition reaction, polymerisation or oligomerisation does not proceed steadily. Obviously, an important termination reaction must be taking place. One such possibility is a chloride ion transfer from the anion to the chain carrier:

$$\sim\sim\sim CH_2\overset{+}{C}HPh \ SbCl_6^- \longrightarrow \sim\sim\sim CH_2CHPhCl + SbCl_5 \ ,$$

the amount of antimony pentachloride formed being too low for direct or cocatalysed reinitiation. As Plesch suggested in the discussion of this paper[688], a material balance on the anion could prove very useful to elucidate the reasons for such an important termination reaction.

Contrary to former claims that small amounts of moisture did not influence the course these of reactions[681], Johnson and Pearce noted that traces of residual water were most detrimental to the stability of the catalysts, confirming previous findings of Slomkowski and Penczek[153]. Once again the importance of working under the most stringent experimental conditions becomes evident. It is not surprising that the best and most reliable studies in this specific area, where often submillimolar concentrations of initiator are used, are those conducted under high vacuum and with rigorously purified reagents.

The second part of Johnson and Pearce's investigation dealt with trityl hexafluoroantimonate. This initiator was found to be extremely active for the polymerisation of styrene, as already reported by Pepper and coworkers[669]. Monomer conversions reached 100% before complete bleaching of the trityl salt, and as polymerisations proceeded the conductivity decreased at the same time as D_{435}. Thus, the active species were certainly not free ions. After a brief acceleration, the polymerisation rate was shown to be first order in both monomer and initiator. Assuming that the number of active species at the time the rate of polymerisation had reached a maximum, was equal to the amount of initiator used, the authors calculated an approximate value for the propagation rate constant, 7.5×10^3 M^{-1} min^{-1} at 0°C. However, this treatment is incorrect since initiation is far from complete at the beginning of the polymerisation as recognised by the authors themselves, so that probably k_p is considerably higher. The remarkable difference in behaviour between the two salts used in terms of their activity towards the polymerisation of styrene could be due to the fact that with the SbF$_6^-$ no termination by fluoride ion transfer is possible, and thus the active species can continue propagating. The nature of these active species is unclear. As we have already remarked, they cannot be free ions, and it would be interesting to search in the ultraviolet region corresponding to the absorption of the polystyryl carbenium ion (around 330 nm) to see if any band arose following the disappearance of the peaks in the visible region due to the trityl ions.

We can summarise the situation with styrene by a few observations. Obviously this monomer is not easily attacked by the trityl ion and the most favourable case encountered is that in which trityl hexafluoroantimonate it used. Termination reactions seem to plague most of these polymerisations and we believe that halogen-ion transfer from the anion to the growing species soon after initiation is responsible for the loss of activity of these systems. Again, the SbF$_6^-$ counterion is an exception in that it does not seem to undergo this termination reaction. Except for the last study discussed, little reliable evidence is available to date to elucidate these difficult problems. Much fundamental information could be obtained by continuing on the line followed by Johnson and Pearce.

i) 1,1-Diphenylethylene

Recent unpublished work by Sauvet[500] has shown that trityl hexachloroantimonate can induce the dimerisation of 1,1-diphenylethylene. However, this process requires high concentrations of initiator probably because the interaction of the trityl ion with the monomer is sterically hindered. Most of the dimerisation proceeds by proton transfer once initiation has taken place. It is therefore difficult to look for evidence on the chemistry of the initial step since most of the product is the linear unsaturated dimer.

j) Isobutene

While this nomomer cannot be polymerised either by trityl or by tropylium salts, Plesch and Pask[689] were able to promote its polymerisation by using diphenylethylium hexafluoroantimonate. This recent discovery opens up new interesting potentials in the field of carbenium-salt initiated polymerisation of aliphatic olefins.

k) Conclusions

This relatively new area of cationic polymerisation has undoubtedly provided new fruitful ways of exploring and understanding the fundamental aspects related to the mechanism of initiation and the structure of the chain carriers formed thereby. Progress has been remarkably rapid thanks to the fact that some of these systems proved quite manageable particularly in the hands of experienced and exacting groups. This research has provided precious information about the chemistry of initiation and quantitative data concerning its kinetics and that of propagation. We would like to dwell a little longer on the problems and complications which still face polymer chemists engaged in this type of research.

The mode of attack of stable carbenium ions on to alkenyl monomers has been proved in some cases to proceed via direct electrophilic addition. However, for certain monomers, e.g., styrene and α-methylstyrene, the mechanism of this reaction has not yet been established unambiguously. We believe that studies on model compounds such as substituted styrenes incapable of polymerising because of steric impediment could provide an answer to this important question.

The difference in behaviour observed experimentally between tropylium and trityl salts clearly confirms the higher stability of the former cation towards nucleophilic agents. Thus, the trityl ion, despite its bulkiness, is reactive enough to initiate the polymerisation of styrene, while the tropylium ion is inert toward that monomer. No initiation rates have ever been measured with tropylium salts mainly because spectroscopic studies are not easily carried out with a cation absorbing around 280 nm. However, in view of its lower reactivity than the trityl ion, and since most initiation reactions with the latter have been proved to be relatively slow with respect to propagation, it is most unlikely that the claims for fast and complete initiation put forward in several systems involving tropylium salts are correct. The original optimistic approach towards these systems, particularly of the Liverpool school, to the effect that stable carbenium salts produce almost instantaneous initiation and therefore give rise to simple kinetic patterns from which the propagation rate constant can easily be calculated, must now be viewed as an oversimplification of the issue. Many investigations in recent years have shown that such an assumption cannot be made and that initiation rates must actually be measured.

One of the most important problems still lingering in this field, is the determination of the degree of dissociation of the propagation species, i.e. the relative amount of ion pairs and free ions formed in the initiation reaction. Much of the work published has tacitly or explicitly assumed that if the concentration of the initiator was such that it was essentially present as free ions, by extension also the chain carriers would be in that form. However, this reasoning is not valid and one could easily argue for the opposite, that is, since most carbenium ions derived from alkenyl monomers are less stable than the carbenium ions used as initiators, one would expect for the former a higher degree of association into ion pairs. Some evidence to this effect has recently been published. We have already discussed Rooney's results on the system N-vinylcarbazole-trityl hexachloroantimonate where, while the initiator was totally dissociated into free ions, the absence of any effect on the polymerisation rate when appreciable quantities of a common-anion salt was added to the system, clearly showed that the propagating species were predominantly ion pairs. An interesting article by Sherrington[690] has recently approached this problem indirectly through the study of the dissociation constants of various ammonium salts. This author showed that while the dissociation constant of quaternary salts was virtually independent of the chain length of the alkyl substituent on the nitrogen atom and moderately sensitive to the presence of nucleophilic substances added to the methylene chloride solutions, binary salts gave much lower dissociation constants than their quaternary counterparts and moreover, these constants were quite sensitive to the presence of small amounts of added donors. As Sherrington appropriately pointed out, the latter salts bear a much closer resemblance to the carbenium ions formed from vinylic monomers than do the quaternary salts. Thus, after addition of the free ion from the initiator to the monomer, the degree of association into ion pairs of the new carbenium ion formed will strongly depend on the structure of the latter, but if this is a secondary species, it is likely to be high or very high. The reaction of styrene with trityl salts is a dramatic example of this situation since the disappearance of the trityl ion in the initiation reaction is accompanied by a corresponding fall in electrical conductivity[682]. With more nucleophilic monomers the change in the degree of association might not be so drastic, as illustrated by the study of Littlejohn et al.[691] who investigated the extent of dissociation of ethoxy methyl carbenium hexachloroantimonate in methylene chloride as a model for the propagating species arising from vinyl ethers. The dissociation constant obtained for that salt was 3 to 4 times lower than that of the corresponding trityl one. This implies that with the concentrations of the latter initiator, normally used for inducing the cationic polymerisation of vinyl ethers, one would pass from a totally dissociated catalyst to chain carriers with 20 to 50% of free ions only. The repercussions on the meaning of the values of k_p obtained experimentally are obvious. If k_p^+ is much higher than $k_p^\pm$, it can be assumed that the largest contribution to propagation comes from free ions and therefore a reduction in their concentration with respect to that of the initiator due to appreciable pairing of the chain carriers, will introduce important errors in the kinetic calculation, the errors being the larger the less nucleophilic the monomer. Although it is difficult to assess relative values of the propagation rate constants for free ions and ion pairs, the apparent negative activation energies of propagation obtained by Sigwalt's group with some of these systems[663, 671] clearly indicate that free ions are considerably more active than ion pairs, apart from proving that both species are in fact formed from a totally dissociated initiator. We feel that much of the work dealing with these complex aspects reflects the poor understanding as yet available on the subject, compared with the situation in anionic polymerisation. Nevertheless, the advances made in the last few years are hopeful indications that future searching studies will contribute to the unravelling of these questions.

A problem which has received little attention and yet seems of great importance to us, is that of the role of the counterion in these processes. We have already proposed that the failure of certain systems to polymerise steadily after initiation could be due to termination reactions involving the transfer of a halide ion from the anion to the growing carbocation. Little experimental work has been done to search for halogen in the polymers and virtually none in the direction of following the fate of the anions. While tests to investigate the former aspect are simple and straightforward, it is obviously less easy to find reliable techniques for the determination of anion concentrations during a polymerisation. Some specific solutions to this problem are however already available, e.g., SbCl$_6^-$ has a strong absorption at 270 nm[115] and its concentration can therefore be followed spectroscopically as has been done in cyclic ethers polymerisation. Recent work by Heublein[692] in this area was briefly communicated at the Tashkent meeting but must await full publication before it can be assessed.

The major drawback concerning the use of these stable carbenium salts as initiators for the polymerisation of vinylic monomers, is of course their inability to attack certain olefins such as isobutene. The search for alternative catalysts within this family has resulted in the remarkable success obtained by Plesch already quoted. Hopefully more such new initiators will become available after this breakthrough.

Although initial expectations of finding among these systems the first living cationic polymerisation have been proved too optimistic by subsequent more rigorous work, the investigation of the polymerisation of 3-vinyl-N-ethylcarbazole[96] appears to be much more promising in this respect.

B. Polymerisation by Carbenium Salts or Esters Prepared in Situ

In 1960, Olah and coworkers[596] reported a study on the polymerisation of olefins by various carbenium salts prepared at low temperature just before the addition of the monomer. Alkyl tetrafluoroborates in particular were shown to be fairly active initiators, but the molecular weight of the products were low. The above salts are not stable under normal conditions, but can be handled in an inert atomsphere at low temperature and their reactivity is such that they can attack aliphatic olefins such as butene and isobutene. However, the very instability of these initiators is an obvious stumbling block against fundamental studies.

Kennedy and collaborators have studied the polymerisation of olefinic monomers induced by catalytic combinations consisting of mixtures of organic chlorides and aluminium alkyls. Although the authors claimed that these mixtures gave rise to carbenium salts capable of initiating vinyl polymerisation, we believe that the process leading to active species does not necessarily imply the formation of transient carbenium ions derived from the catalyst pair. For this reason, these systems are not discussed in this section but rather in Chapter IV together with other aluminium halide initiated polymerisations.

Cesca et al.[629] isolated a complex between aluminium chloride and t-butyl chloride, stable at low temperature and showed by NMR spectroscopy that a considerable degree of polarisation characterised its structure, viz.,

$$(CH_3)_3\overset{\delta+}{C}Cl \rightarrow \overset{\delta-}{Al}Cl_3.$$

This crystalline complex was found to initiate the polymerisation of isobutene at −78°C more effectively than an equivalent amount of AlCl$_3$ alone or with t-butyl chloride added to the mixture. This was taken as a proof that the activity of the complex pointed to its structure approaching that of a carbenium salt. This work was never continued with other monomers to demonstrate that the t-butyl cation was the initiating species.

Gandini and Plesch[127, 336] prepared 1-phenylethyl perchlorate from silver perchlorate and 1-phenylethyl bromide in a methylene chloride solution containing styrene to provide evidence for their pseudocationic theory. The reaction produced poly(styrene) at the same rate as one with an equivalent amount of perchloric acid, except for a short acceleration period due to the metathetic interaction between the two catalytic components. The necessity of preparing the ester in the presence of excess monomer arose from the instability of that initiator on its own.

Similar experiments were carried out a few years later by Kagiya et al.[128] In this interesting study, a series of alkyl and aryl perchlorates were prepared in situ in bulk styrene and their relative activity as initiators assessed at 0°C; t-butyl perchlorate was found to be by far the most active catalyst and trityl perchlorate the least effective. No termination was detected in these processes. Other esters were also tried and their activity followed approximately the strength of the parent acid. Although these systems were too complicated to allow detailed mechanistic studies, mainly because the synthesis of the initiator proceeded alongside the polymerisation itself, they certainly provided further support for the pseudocationic theory, considering the very low polarity of the medium.

Sangalov et al.,[693] prepared arenonium ions in the form of Gustavson complexes starting from various substituted benzenes, hydrogen chloride and aluminium chloride. These complexes were shown to be active initiators for the polymerisation of styrene and isobutene at −30 and −78°C in methylene chloride. No kinetic investigation was carried out on these systems. The authors claimed that initiation took place by protonation of the olefin and that incorporation of aromatic groups from the cataylst in the products was due to transfer reactions.

C. Polymerisation by Stable Halides and Esters

In 1962 Tokura and Kawahara[221] reported that styrene could be polymerised by various alkyl and aryl chlorides in sulphur dioxide. The solvent did not participate in the polymerisation as shown by the absence of polysulphone copolymers among the products. Benzyl, 1-phenylethyl, propyl and butyl chlorides were all modestly active initiators in these experiments. Typically about 10% polystyrene, DP = 100–300, was obtained in 150 minutes with 0.37 M of initiator at 25°C. Under these same conditions, both trityl chloride and hydrogen chloride failed to give any polymer. We do not know if these results have ever been repeated and confirmed by other workers. If they are genuine and do not arise from some unknown artifact, they could be interpreted in terms of specific solvation of the chlorides by the solvent to give a C—Cl polarisation sufficient to induce the pseudo cationic polymerisation of styrene. However, more work would be necessary to confirm these proposals.

N-Vinylcarbazole has been polymerised by a whole series of organic halides by Otsu and coworkers[694, 695]. In view of the extreme sensitivity of this monomer to traces of acidic impurities and to charge-transfer initiation, most of these results could arise from such causes and not from direct initiation by the halide. The observed accelerating effect of light in some of these reactions seems to confirm this supposition.

Hernàndez and Gandini[696] have studied the polymerisation of N-vinylcarbazole by diethyl sulphate in ethylene chloride using high-vacuum techniques. Extreme precautions were taken to purify both catalyst and solvent in order to remove all acidic impurities. It was ascertained that the solvent was totally inert towards the monomer and that the catalyst contained less than 10^{-7} M of sulphuric acid and/or monoethyl sulphate. Polymerisations carried out in vacuo, in rigorously dry conditions, proceeded rapidly but to incomplete yields. Typically, a solution 0.8 M in catalyst and 0.4 M in monomer produced about 40% polymerisation within a few minutes and thereafter propagation ceased (room temperature, DP higher than 100). Experiments conducted in the presence of added moisture or in open systems gave no polymerisation. If one excludes that trace amounts of acidic impurities in the catalyst could have been the cause of these polymerisations, a consideration which is all the more acceptable in view of the limited yields and the negative effect of water (hydrolysis of the catalyst produces both monoethyl sulphate and sulphuric acid), the only other logical interpretation of the observed behaviour is direct ester initiation according to the following interaction:

$$Et_2SO_4 + CH_2=CHCarb \rightleftharpoons C_3H_7 - \overset{+}{C}HCarb \; EtSO_4^-$$

The above equilibrium would obviously lie strongly to the lefthand side. The presence of small amounts of water would kill the carbenium ions formed in the initiation reaction and no polymerisation would thus occur in a wet system. As for the limited yields following rapid initial polymerisation, the authors postulated that the polymer might from a complex with the catalyst and so remove it from the above equilibrium. A firm conclusion on the real nature of the initiation process would however require further fundamental work on this system.

D. Polymerisation by Acylium Salts and Acyl Esters

The preparation, characterisation and reactivity of acylium ions and acyl complexes and esters have been extensively investigated, particularly in the last two decades. Their use as initiators in the cationic polymerisation of alkenyl monomers is however a more recent extension of these studies, except for some old exploratory work.

Three main routes are available for obtaining these species:
— the interaction of Lewis acids with acyl halides;
— the metathetic reaction between silver salts and acyl halides; and
— the dissolution of anhydrides in very strong acidic media.

While the first two allow the isolation of the reaction products, the third does not. When a Lewis acid is mixed with an acyl halide, a donor-acceptor complex RCOX . . . MtX$_n$, or full ionisation by halide ion transfer, to give $RCO^+MtX_{n+1}^-$, or both, take place. The extent of ionisation depends mostly on the nature and strength of the Lewis acid used[697–705]. Thus, for example, acetyl halides react with stannic chloride and titanium tetrachloride to give mostly the coordination complex while with antimony pentachloride, pentafluoride and boron fluoride they give the corresponding acylium salts. Many of these complexes and salts have been fully characterised and an excellent review on acylium ions recently published[61]. The preparation via metathesis is a very convenient method, and it allows studies of the stability of the resulting salt following its synthesis. It has been shown that some of these salts exist in equilibrium with their covalent precursors [706]. This observation is particularly relevant to cationic polymerisation since it implies that if one of these salts is used as initiator it is essential to know the extent of its ionisation in the medium used before conducting any quantitative study of the fundamental processes.

The acylation capacity of these species is well documented.

The classical work of Burton and collaborators[120] and more recently that of Commeyras et al.[707] on aromatics has contributed notably to the understanding of the nature of these mechanisms. The reaction of acylium salts with olefins has been studied by Lyubinskaya et al.[708] at low temperature and showed to give β, γ-unsaturated ketones or stable five-membered carboxonium salts.

1. Acetyl Perchlorate

Burton and Praill were the first to prepare this species both by the interaction of perchloric acid with acetic anhydride and the metathetic reaction of silver perchlorate with acetyl chloride[709, 120], in a comprehensive study of acylation reactions. They did not attempt a characterisation of the compound and assumed that in acid solutions it existed in a fully ionised form. A few years later, Jander and Surawski[710] followed the formation of acetyl perchlorate in acetic anhydride by measuring the electrical conductivity changes which took place when acetyl bromide was added to silver perchlorate or vice versa. In both titrations an inflection point was observed for a mixture of equimolar quantities of the two reactants indicating the formation of acetyl perchlorate. Moreover, it was clearly shown that this compound was at least partly ionised since its conductivity was higher than that of acetyl bromide. Avedikian and Commeyras[711] characterised acetyl perchlorate by infrared and Raman spectroscopy both in acetic anhydride and carbon tetrachloride. They concluded that some ionisation was present but could not assess its extent. Molecular acetyl perchlorate was also detected. As far as we are aware, no other study of the structure and extent of dissociation of acetyl perchlorate in solution has been conducted.

Longworth and Plesch[712] were the first to report that aromatic olefins and dienes could be polymerised by acetyl perchlorate in chlorinated hydrocarbons but did not extend their study beyond a qualitative report. It was not until 1971 that this initiator was revived in the context of the cationic polymerisation of alkenyl monomers. Masuda and Higashimura[713] described the polymerisation of styrene by acetyl perchlorate in methylene chloride at 0°C. No kinetic study was performed, the main aim of this investigation being to show that the polymer formed had a bimodal distribution, in contrast with similar polymers obtained with Lewis acid initiators. Addition of sufficient amounts of n-Bu$_4$N$^+$ClO$_4^-$ suppressed the high-DP peak leaving the other unaffected. A decrease in the polarity of the medium also reduced the contribution of the former peak. The conclusion was reached that in this polymerisation system two different active species operated independently. In a subsequent paper[714] the same authors discussed the effect of various tetrabutylammonium salts on the rate of polymerisation and the DP's in various solvents. The decrease in rate was found to be very pronounced with the iodide and the tetrafluoroborate and much less important with the perchlorate. Before proceeding to analyse other papers by the Kyoto school it is convenient to make a few observations. The preparation of acetyl perchlorate was very briefly discussed by these authors in another context[715]. It was apparently conducted in methylene chloride alone despite the fact that silver perchlorate is insoluble in this solvent. The authors used a two-to-one excess of acetyl chloride probably in order to assure that most of the silver perchlorate would react, but this procedure obviously leaves at the end a mixture of both acetyl chloride and acetyl perchlorate. The determination of the perchlorate concentration was conducted by adding an excess of trityl alcohol and assuming a one-to-one reaction between the two species to give the trityl ion. This reaction has never been investigated quantitatively and its analytical value is by no means ascertained. No attempt was made in these studies or in any of the later ones to establish the stability and the nature of acetyl perchlorate solutions. In methylene chloride one would not expect any interaction of these species with the solvent but nothing is known about the intrinsic lifetime of acetyl perchlorate. As to the ionic or covalent character of this compound, the authors simply refrained from commenting or from attempting any specific physico-chemical determination. It is surprising that when they used benzene and nitrobenzene as solvents they did not worry about acylation processes on the aromatic nucleus. Finally, the residual concentration of water in the media used for polymerisation was stated to be $(2-3) \times 10^{-4}$ M, i.e., of the same order of magnitude of acetyl perchlorate concentrations used in experiments at $0°C$. The degree of hydrolysis of the initiator could have therefore been quite high which implies that in practice a mixture of acetyl perchlorate, acetic acid and perchloric acid (plus hydrochloric acid coming from the hydrolysis of the excess acetyl chloride) was present in these polymerising solutions. This state of affairs is unfavourable for studies on a new catalyst.

The detrimental effect of tetrabutylammonium iodide and tetrafluoroborate on the rate of polymerisation of styrene in the presence of acetyl perchlorate can easily be explained by an anion-exchange reaction with the propagating ester to give a much less active chain carrier.

A subsequent more important paper by Higashimura and Kishiro[716] dealt with the system styrene-acetyl perchlorate from the point of view of the kinetics of polymerisation and DP distribution. At $0°C$ a simple kinetic pattern was observed resembling closely that characteristic of the classical system styrene-perchloric acid (see Sect. III-E-13-a). The polymers displayed bimodal distributions and the relative importance of the two peaks was studied as a function of temperature, added water, dielectric constant of the medium, catalyst concentration, and common anion concentration. Two alternative possibilities must be considered in the overall interpretation of these results:

— as we have mentioned above, the residual moisture concentrations in these systems could have been sufficient to produce near complete hydrolysis of the catalyst. The authors did not investigate the possible presence of free perchloric acid. It could well be that in fact these polymerisations were exact replicas of perchloric acid-styrene ones. All the features observed would agree entirely with such a possibility. We calculated a propagation rate constant from the authors' results at $0°C$ and obtained approximately 1.4 M^{-1} sec^{-1} which is about one half of the value for the same situation but with perchloric acid[127]. It is difficult to say whether this one half relationship is fortuitous or if it bears a mechanistic significance. It could be that perchloric acid formed in the hydrolysis of acetyl perchlorate reacted with the acetic acid

$$CH_3COOH + 2\,HClO_4 \rightleftharpoons CH_3OOH_2^+\,H(ClO_4)_2^-,$$

and thus its activity as initiator was one half of that for a system where only perchloric acid is present. These speculations might be unfounded, but as long as no information is available on the ease of hydrolysis of acetyl perchlorate the presence of perchloric acid in these wet systems must be seriously taken into account.

— If one supposes that the real initiator is acetyl perchlorate, then it must be accepted that this compound attacks styrene in a fast initiation reaction and thereafter the situation is in any case totally equivalent to that of a polymerisation induced by perchloric acid. Should this be so , the explanation given by Hamman et al.,[337] concerning the formation of an active π-complex between perchloric acid and styrene (see Sect. III-E-13-a) is untrustworthy since acetyl perchlorate, which cannot give rise to any such complex, produces exactly the same phenomenology as perchloric acid in the polymerisation of styrene.

Most probably, the real situation in these systems is one of compromise between the two possibilities discussed above, i.e., the initiator is a mixture of perchloric acid and acetyl perchlorate. It seems obvious to us that the mechanism of these polymerisations is pseudocationic and the minor effect of added perchlorate salts is simply due to homoconjugation of perchloric acid present at the beginning of the reaction or liberated in spontaneous transfer processes, and *not* to common-ion effects suppressing the concentration of free ionic chain carriers.

Further work by Higashimura's group on the polymerisation of styrene and styrene derivatives by acetyl perchlorate[359 – 361, 717, 718] does not provide any substantial new evidence in favour or against the arguments discussed above. Again, the low catalyst concentrations used in the presence of wet media cast serious doubts about the real nature of the initiator, and the studies involving common-anion salts ignore the basic phenomenon of homoconjugation with the acid.

The significance of bimodal distributions in polymers obtained from cationic processes has already been discussed in Section III-F and no new elements arise here.

Higashimura and Nishi[316, 719] have recently studied the dimerisation of styrene and α-methylstyrene by acetyl perchlorate. An appropriate choice of temperature and solvent allowed the optimisation of linear dimers formation. These authors postulate a transition state for the formation of the linear dimers which resembled closely the spontaneous-transfer mechanism proposed thirteen years earlier by Gandini and Plesch[127] in the pseudocationic polymerisation of styrene by perchloric acid.

Kohjiya and Yamashita[720] investigated the polymerisation of cyclopentadiene by acetyl perchlorate in toluene-methylene chloride mixtures. They found a kinetic behaviour entirely similar to that of the system cyclopentadiene-perchloric acid, which they had studied previously. In fact, in the same solvent mixture and at the same temperature the time-conversion curves were practically identical. For a criticism of the method used for calculating a "propagation" rate constant, see Sect. III-E-13-c. The striking resemblance of phenomenology between this work and that with perchloric acid[296] seems to suggest that the real catalyst might have been in both cases the acid arising in the former from the hydrolysis of acetyl perchlorate by residual moisture. These authors adopted uncritically both the method of preparation, and the analytical technique for the determination, of acetyl perchlorate elaborated by Higashimura's group (see above discussion). No attempt was made to establish the extent of ionic dissociation of this compound in the solvents used, nor to determine its sensitivity to adventitious moisture.

Acetyl perchlorate has also been used as an initiator in the cationic polymerisation of substituted butadienes[721, 722] in methylene chloride and toluene (possible acylation of this solvent by the catalyst was ignored). These studies do not advance the limited knowledge on the properties of this catalyst in solution and on its mode of initiation.

It is disappointing to see that all the work carried out with acetyl perchlorate to the present, has failed to provide any real fundamental information about the way this catalyst operates. It seems in fact that its use was simply decided in order to avoid the more laborious preparation of solutions of anhydrous perchloric acid. In this respect, it might have served a purpose. We think that acetyl perchlorate would be a very interesting catalyst to investigate under strictly controlled experimental conditions.

2. Other Acyl Initiators

Biswas and Kabir[723] reported that benzoyl chloride is an initiator for the polymerisation of vinyl ethers and N-vinylcarbazole. Given the high nucleophilicity of these monomers the above observation is not surprising. The authors envisaged an initiation scheme involving the self-ionisation of the catalyst,

$$2\,C_6H_5COCl \rightleftharpoons C_6H_5CO^+ + C_6H_5COCl_2^-,$$

followed by attack of the benzoylium ion on the monomer double bond. These conclusions were based on kinetic measurements. No attempt was made to search for catalyst fragments in the polymer.

Acylium hexafluoroantimonates have been successfully used as initiators in the polymerisation of tetrahydrofuran[724, 725], but little work has been undertaken to our knowledge to test their effectiveness for the polymerisation of olefins. Plesch and Pask[689] discovered recently that some of these salts can induce the polymerisation of isobutene. This area of research should provide much useful and fundamental information.

E. Initiation by Radical-Cation Salts

This section is devoted to a discussion of those cationic polymerisations in which the initiator is a radical cation prepared electrochemically before the addition of the monomer. The formation of radical cations in certain cationic polymerisation processes was recognised in the early sixties. The systems in which these transients are supposed or have been proved to intervene will be discussed in later sections. Here, we are concerned exclusively with the use of relatively stable radical-cation salts as promoters of polymerisation.

About ten years ago, various workers first reported the anodic production of radical cations which showed appreciable stability in certain solvents[726–728]. Mengoli and Vidotto[729] recognised that these species could be used as initiators for cationic polymerisation. They prepared the cation radical of 9,10-diphenylanthracene in acetonitrile and nitrobenzene by the anodic oxydation of the parent hydrocarbon in the presence of tetrabutylammonium perchlorate as supporting electrolyte. The half life of these species was about one hour in the experimental conditions used, but addition of styrene and n-butylvinylether to these solutions produced a much faster decay of the radical cations and polymerisation of these monomers. The initiation reactions regenerated the parent hydrocarbon in yields of 50% or more, some incorporation of 9,10-diphenylanthracene groups as polymer end groups being observed. With n-butylvinylether, one half of the cation radicals present reappeared as the parent hydrocarbon and the authors interpreted this result as evidence for an initiation process involving the addition of the cation radical onto the monomer followed by the reaction of this adduct with another cation radical to give a dicationic species (chain carrier) and a molecule of the parent hydrocarbon. With styrene, the above initiation process was also invoked but with it another interaction seemed to take place, namely, an electron transfer from the monomer to the radical cation to give the parent hydrocarbon and the styrene radical cation. The second order rate constants for both initiation processes determined by measuring the rate of decay of the original radical cations, were also reported in this paper. As expected from the large difference in nucleophilicity, initiation with the vinyl ether was about 500 times faster than with styrene in the same experimental conditions. Mengoli and Vidotto[730] extended their original study to various styrene derivatives and showed that their reactivity in the initiation reaction followed closely their relative basicity. This interaction mainly occurred through an electron transfer mechanism. Polymerisations were fairly slow owing to the important role of side reactions causing the loss of activity of the chain carriers. It was assumed that the monomer radical-cations would couple to give dications soon after their formation. The latter species were claimed to be the chain carriers in these processes, although no attempt was made to identify them during the polymerisation. The perylene radical cation was also prepared and used in the polymerisation of α-methylstyrene.

Funt and collaborators[731–734] conducted similar studies a few years later. In the first paper of this series, Funt and Verigin showed that the radical cations of 9,10-diphenylanthracene were fairly stable in rigorously dried acetonitrile. Addition of styrene or substituted styrenes resulted in relatively fast initiation processes which were attributed to an electron transfer process giving the monomer radical cation in an equilibrium reaction. The latter species added one monomer molecule irreversibly to give the dimer radical cation and thereafter polymerisation proceeded. In a second paper the polymerisation of styrene was studied in methylene chloride with the same initiator. A preliminary kinetic study of this process was carried out in the presence of tetrabutylammonium perchlorate as background electrolyte. A value for the propagation rate constant at 20°C was calculated following some fairly drastic and unproven assumptions: 0.25 M^{-1} sec^{-1}. This value seems far too low, possibly because the concentration of active species was overestimated in the authors' calculations. This work clearly proved that the real initiator was indeed the anodically prepared radical cations and not perchloric acid resulting from the simultaneous oxidation of the perchlorate anion. In the third paper from this group, styrene and isobutyl vinyl ether were polymerised in various media by a series of radical cations obtained from polynuclear hydrocarbons in electrochemical processes conducted in the presence of the usual supporting electrolyte. In the discussion concerning the mode of initiation, two processes were considered, viz., electron transfer or addition. The authors claimed that their kinetic data supported the second mechanism. The final paper in this series described a rotating ring-disk electrode investigation of the initiation reaction between various stable radical cations and styrene and isobutyl vinyl ether. First order dependence was obtained for both reactants, confirming the original findings of Mengoli and Vidotto[729]. No indication was obtained in this study of the chemical details of these interactions.

Recently, Mengoli and Vidotto[735] restudied these systems and obtained evidence for a direct addition reaction in the initiation of the polymerisation of isobutylvinyl ether by the radical cation perchlorate of 9,10-diphenylanthracene. The kinetics of styrene polymerisations by the above initiator in nitrobenzene at 10°C were also studied. Important termination reactions giving indanyl-type cations were detected and attributed to the perchlorate anion. This system was in fact found to display many similarities with that involving the same monomer and perchloric acid[45, 127].

The above investigations clearly show that despite their relative stabilities, electrochemically prepared radical-cations give rise to fairly complicated phenomenologies when they are used as initiators for cationic polymerisations. While the initiation rate constants reported are probably correct, the chemistry of these processes both with respect to the initial step and to the ensuing reactions of the monomer radical cations, is not fully understood. As for the nature and relative concentration of the chain carriers in these systems, more work would need to be done before any firm conclusion can be attained. All these problems stem at least in part from the fact that the radical cations described are only *relatively* stable and do suffer some annihilation reactions.

Oberrauch et al.[736] recently pointed out some of the drawbacks of the in situ generation of cation radicals and described an interesting piece of work where the perylene radical-cation perchlorate was prepared in the solid state by an anodic electrocrystallisation technique. This stable salt was then isolated and characterised prior to its use as an initiator for the cationic polymerisation of various monomers. It was found that N-vinylcarbazole and vinyl ethers polymerised rapidly in nitroethane, styrene and derivatives less efficiently; surprisingly, even isobutene could be effectively oligomerised. All the polymers were found to exhibit an absorption band in the visible compatible with the presence of perylene groups attached to their chains. This strong indication in favour of an initiation mechanism involving addition of the radical cation to the monomer does not of course rule out, as the authors pointed out, the possible occurrence of simultaneous electron-transfer initiation. In this preliminary study no conclusions could be drawn on the nature of the active species, i.e., whether there were carbenium ions (mono or dications), esters, or cation radicals. However, the achievements of this investigation, consisting in having successfully isolated and used a stable cation-radical salt, will undoubtedly promote further work in this field. This should eventually lead to a better understanding of the fundamentals of these system. The perylene cation-radical perchlorate could in fact be the first of a family of interesting new stable initiators for cationic polymerisation. The recent report by Fritz et al.[737] on the anodic synthesis and characterisation of stable crystalline naphthalene radical-cation salts with various anions, shows that other potential initiators of this type are becoming available.

VI. Initiation by Bare Cations

The term "bare" has been chosen here to define carbenium ions or radical cations which are formed without the corresponding anion. The difference in reactivity between these species and the free ions present in cationic systems involving fully dissociated entities such as carbenium ion salts, resides of course in the absence of all interactions involving the anion, namely ion pairing, collapse to give a covalent bond, etc. Most techniques available for the production of bare cations involve the expulsion of an electron from a monomer molecule, i.e. physically-induced initiation. The present chapter deals with these techniques and their major contributions to the understanding of the chemistry and properties of active species in cationic polymerisation.

A. Initiation by Radiolysis

The investigation of radiation-induced cationic polymerisations in the last few decades has been decisive in bringing about a better knowledge of the activity of free carbenium ions in propagation reactions. In fact, the first reliable values of propagation rate constants for these species were obtained in this field. These data are still used as reference points in any study where the presence of free ionic chain carriers is presumed and kinetic studies allow the determination of an apparent k_p. Many excellent reviews have covered the advances in this area up to very recently[113, 738–744]. Particularly in the field of continuous irradiation, there is little that we can add in the present context without repeating the very appropriate discussions of such experts as Williams, Metz, Chapiro, Hayashi etc. We simply wish to remind the reader that if cationic polymerisation in general requires careful and painstaking experimental work, radiation-induced studies are even more exacting precisely because the bare cations are the most reactive species encountered in this field and their sensitivity to adventitious impurities (moisture in particular) imposed the most stringent measures to eliminate them. Many examples of this high-standard of purification and drying will be found in the literature quoted in the above reviews. We can just add a recent contribution by Stannett's school[745, 746] on the radiation-induced polymerisation of vinyl ethers.

The primary process of initiation in this type of polymerisations consists in the abstraction of an electron from a monomer molecule with a consequent formation of the corresponding radical cation. A detailed knowledge of the nature of these species and of their subsequent reactions leading eventually to chain carriers implies the use of fast-detecting techniques because of the short lifetime of the intermediates. Continuous irradiation, moreover does not usually provide a sufficient steady-state concentration of any of these entities to allow measurements other than electrical conductivity, a rather undiscriminating tool. For these reasons, pulse radiolysis coupled with fast spectroscopic analysis, has been the major source of information in this context. While all the studies by this technique have been adequately discussed in the reviews quoted above, some recent advances will be outlined below. Tabata and collaborators have investigated the initial processes in the radiation-induced cationic polymerisation of styrene, α-methylstyrene and N-vinylcarbazole[747−751]. They followed the spectra of various intermediates on a microsecond time scale and obtained some extremely valuable information. Thus, for example, pulse radiolysis of styrene solutions in *iso*pentane/butyl chloride at −165 °C gives rise to a complex absorption spectrum made up of several components. The time-resolved history of these peaks allowed an interpretation of their origin. The monomer radical cation was identified as the species giving maxima at 350 and 630 nm. The peaks at 600 and 1,600 nm were assigned to the associated dimer radical cation, i.e. a complex between the monomer radical cation and a monomer molecule. The dimer radical cation absorbed at 450 nm and the growing bare carbenium ion gave a maximum at 340 nm. A mechanism compatible with these fast kinetic and spectroscopic observations was proposed. Similar studies have been conducted in the last few years by Hayashi et al.[752] and by Brede and coworkers[753−756].

The importance of these studies extends beyond the specific realm of radiation-induced polymerisation since many of these species are formed or thought to be formed in other types of cationic polymerisation. The development of even faster irradiating instruments such as the Febetron, which allows nanosecond pulses, will provide an even deeper insight into these processes.

B. Initiation by Photochemical Ionisation

The only work carried out in this area concerns the polymerisation of isobutene. In a very interesting experiment, Vermeil et al.[757] induced the photoionisation of isobutene by irradiating its solutions in isopentane and ethyl chloride at low temperature with 123.6 and 147 nm photons (at higher wavelengths photoionisation was not achieved). Low yields of polymer were obtained as a consequence of the formation of isobutene cation radicals. Quantum yields of photoionisation of about 0.01 were obtained. A similar investigation was conducted by Schlag and Sparapany[758], but in their experiment isobutene was photoionised in the gas phase and the cations produced accelerated towards a negatively charged electrode placed under liquid isobutene kept at low temperature. This injection of ions into the pure liquid monomer induced its polymerisation. A more detailed description of such as investigation was published two years later by Sparapany[759]. This author demonstrated the cationic nature of the process and showed that one could initiate the polymerisation of isobutene also in the gas phase and in various solvents. Yields were low but proportional to the duration of irradiation and to its intensity. Very high molecular weights were obtained. The following reactions were proposed to account for the formation of the *t*-butyl cation:

$$CH_2 = C(CH_3)_2 + h\nu \longrightarrow \dot{C}H_2 - \overset{+}{C}(CH_3)_2 + e^-$$

$$\dot{C}H_2 - \overset{+}{C}(CH_3)_2 + C_4H_8 \longrightarrow C_4H_9^+ + C_4H_7\cdot \quad .$$

While the radical derived from isobutene was shown not to contribute to the polymerisation, the real initiating species was obviously the cation and the oligomeric species formed from it before reaching the liquid phase. It is regrettable that this study was not continued towards a higher degree of quantitative sophistication so that the rate constants of propagation and transfer could be determined. This system was reinvestigated by Viswanathan and Kevan[760, 761] and strangely these authors reported predominant formation of isobutene dimers and trimers in the same operating conditions in which Sparapany had obtained high molecular-weight polymers. The authors did not offer any comment about this major discrepancy and discussed their work in terms of the predominance of transfer reactions. We believe that some important impurities caused this anomalous behaviour and that the results of Viswanathan and Kevan were therefore incorrect. t-Butyl cation injection into liquid isobutene at low temperature must be seen as an efficient way of initiating the cationic polymerisation of this monomer from the point of view of the high DP's obtained.

C. Initiation by Field Ionisation and Field Emission

Various groups have in recent years studied the possibility of inducing the cationic polymerisation of undiluted liquid monomers by the effect of high-intensity electric fields. Two basic principles have been applied to obtain bare cation radicals as primary intermediates in these processes:

— Field emission, where an injection of electrons from a negatively-charged sharp electrode into the monomer causes the expulsion of an electron from its molecule to give the corresponding radical cation,

$$e^- + M \longrightarrow \overset{+}{M}\cdot + 2\,e^-.$$

— Field ionisation where a positively-charged tipped or bladed electrode produces a local field which is sufficiently strong to abstract an electron from the monomer,

$$M \longrightarrow \overset{+}{M}\cdot + e^-.$$

In 1970, Lambla et al.[762] published the first report on this type of polymerisation and specified that with a potential difference of about 5 kv between the submerged electrodes, separated by a distance of 5 mm, the cationic polymerisation of liquid styrene took place both with the tipped electrode charged positively or negatively. They also underlined the necessity of purifying the monomer as thoroughly as possible in order to obtain good and reproducible yields. In 1971, Brendlé[763] also described the polymerisation of liquid styrene under similar conditions and discussed the effect of such parameters as the shape of the injecting electrode, its polarity and its previous treatment on the efficiency of the process. The dramatic role of impurities and residual moisture was emphasised. Lambla's group later reported an excellent study on the polymerisation of styrene by field ionisation initiation[764, 765]. These authors analysed the molecular-weight distribution of the polymers obtained as a function of the polymerisation temperature. While below room temperature only one peak was observed in the GPC spectrum, at higher temperatures a bimodal distribution was obtained and the new high molecular-weight peak increased in importance as the temperature was raised. The similarity of this behaviour with that encountered by Ueno et al.[767] in the radiation-induced polymerisation of styrene, was taken as preliminary evidence in favour of the formation of the monomer radical cation in the primary initiation step. It was postulated that these cation radicals would grow to oligomeric size and then split their double functionality into a growing radical and a growing carbenium ion by reaction with the monomer. From that moment two parallel propagation processes would therefore proceed independently and, given the different activation energy between them, they would produce changing proportions of polymer as the temperature varied. The progressive onset of the high DP peak as the temperature was raised indicated the corresponding increase in importance of the radical contribution to propagation. These conclusions were confirmed by specific inhibition experiments: added water eliminated the low DP product (cationic), while added diphenyl-1,1-picryl-2-hydrazyl inhibited the polymerisation altogether.

Brendlé and Ilvoas-Fremond[767, 768] published in the same period a series of considerations regarding the theory of these new processes and the difficulties inherent in a quantitative treatment of the field distribution between the electrodes in a liquid submitted to the specific turbulence provoked by the injection of a continuous charge. The experimental part of this investigation, both with field ionisation and field emission, confirmed the mixed radical/cationic nature of these polymerisations. Particular emphasis was placed on the fact that since cationic propagation was carried out by bare species, its contribution could only be fully detected under the most stringent experimental conditions.

The third group which exploited these techniques is that of Schmidt and Schnabel[769−772]. While in their original paper these authors erroneously assumed that the polymerisation of styrene by field emission was anionic, in later contributions they recognized that both field emission and field ionisation lead to cationic chain carriers. This was confirmed by experiments in which amines and water were shown to be inhibitors of these processes. The existence of a parallel radical propagation was also proved in copolymerisations involving monomers which cannot polymerise cationically.

The fundamental qualitative and quantitative considerations put forward in the course of all these studies were tested experimentally in the last work of this series by Brendlé and Ilvoas[773]. Field-ionisation polymerisations were conducted both in high-vacuum glass-metal devices with rigorously purified and dried monomers or with less stringent operational procedures. Styrene, α-methylstyrene and iso-butyl vinyl ether were studied. It was shown that residual moisture was consumed during polymerisation, in contrast to what happens in radiation-induced polymerisations. Thus, repeated cycles of charge injections in semiopen systems produced progressively higher polymerisation rates ultimately approaching the values observed under ultra-dry conditions. The authors developed from these observations a new technique for the determination of very low levels of water concentrations in monomers. A complex physical model was elaborated in order to evaluate the propagation rate constant in these experiments. The values obtained were about one order of magnitude higher than those previously encountered in radiation-induced polymerisations. While such upper limits for k_p must await confirmation owing to the important assumptions made in their calculation, this work provided some interesting new insight into the behaviour of naked ions in pure monomers. In particular, Brendlé and Ilvoas showed that when ultra-dry systems are used, the lifetime of a kinetic chain is governed only by the deactivation on the negatively charged electrode, i.e., by the time of migration of the ions from one electrode to the other.

It seems to us that the chemistry of these rather special interactions is well worth further study and not only in the specific context of cationic polymerisation. For example, simple molecules such as benzene and acetonitrile produce low yields of "polymerisation products" through hitherto unexplored mechanisms,[774].

D. Nuclear-Chemical Initiation

An entirely new approach to initiation in cationic polymerisation was recently reported in a short communication by Akulow et al.[775, 776]. These authors used low concentrations of tritium to induce the polymerisation of liquid isobutene and styrene at low temperature. The β-decay of tritium

$$^3H_2 \longrightarrow HeT^+ + \beta^- + \tilde{\nu}$$

provides a convenient source of low-energy bare cations. These can attack the monomer in a simple initiation process which can be dosed uniformly throughout the liquid or the solid phase. The first results arising from the application of this technique are extremely encouraging: very high DP's were obtained from both monomers (in fact for styrene the highest DP ever recorded in a cationic polymerisation), indicating that side reactions did not seriously affect the chain carriers. These authors also succeeded in initiating the polymerisation of isobutene by the bare butyl cation produced in the β-decay of tritiated isobutane.

This novel technique promises to be very useful for fundamental studies in cationic polymerisation since it provides a very convenient tool for the quantitative appraisal of such parameters as the rate of cation production, the degree of cation incorporation in the polymer, etc. Akulov and collaborators defined very appropriately HeT^+ as "a Brønsted acid of maximum strength". What makes this species so particularly attractive is not only its protonating power but the fact that it does not possess a counterion which as we have pointed out throughout this review, is more often than not a source of complications and possible detrimental reactions. The high DP's obtained with styrene clearly show what can be gained by the absence of a counterion, particularly if one compares these results with the DP's of polystyrenes prepared with even such strong acids as trifluoromethanesulphonic acid. Obviously the growth of a chain is controlled by the nucleophilicity of the anion which seems to be the main promoter of transfer and termination reactions.

VII. Electrochemical Initiation

In previous chapters we already discussed some aspects of electroinitiation of cationic polymerisation, namely the anodic production of relatively stable radical-cation salts used in situ as initiators and the application of high-electric fields to promote the formation of highly active radical-cations in liquid monomers. In this chapter we will consider the more traditional type of electroinitiated polymerisation in which the electrolysis of a solution containing monomer and a supporting electrolyte produces the cationic polymerisation of the former. Of course, radical and anionic polymerisation can also be initiated by this technique, but these processes are outside the scope of the present review.

The pioneering work of Breitenbach's group[777] established for the first time in an unequivocal manner the cationic nature of the electroinitiated polymerisations of styrene, N-vinylcarbazole and vinyl ethers with perchlorate supporting electrolytes. Since then, a considerable amount of work has been devoted to this topic and various reviews have summed up the major achievements of this type of studies[778-780]. In view of this, we will limit our discussion to the fundamental aspects most relevant to the present context, and more particularly to an analysis of recent work.

A. Styrene

In 1970, Funt and Blain[781] published a thorough study of the electroinitiated polymerisation of styrene in the presence of tetrabutylammonium perchlorate. The kinetics of these reactions were followed in methylene chloride with a cell provided with a sintered-glass membrane and a sampling device at the anode compartment. S-shaped time-conversion curves were obtained indicating the accumulation of chain carriers during the electrolysis. The results were treated assuming that the concentration of active species was directly proportional to the time-integrated current flow, and that no termination occurred. A k_p of $1.5 \, M^{-1} \, sec^{-1}$ at $25 \, °C$ and an E_p of 8 kcal mole^{-1} were obtained. Both these results are lower than those obtained by Gandini and Plesch for the system perchloric acid-styrene-methylene chloride[127]. The major difference between these two systems is the presence in the former of the supporting electrolyte, a compound whose anion could easily be the cause of the lower rates observed due to its possible homoconjugation with perchloric acid formed in transfer reactions. The initiation mechanism of this electrolytic polymerisation was postulated to originate from the anodic reaction

$$ClO_4^- \longrightarrow ClO_4^{\cdot} + e^-$$

followed by the attack of the perchlorate radical on the monomer:

$$ClO_4^{\cdot} + CH_2 = CHPh \longrightarrow \overset{\cdot}{C}H_2 - \overset{+}{C}HPh \; ClO_4^-.$$

The formation of styrene radical cations was supported by the observation of a green colour at the inner region of the anode. Funt and Blain argued that these primary initiating species would probably soon couple through their radical ends and therefore the polymerisation would proceed mostly through a cationic propagation. The similarity of the situation achieved after initiation with that of the system styrene-perchloric acid is obvious.

Tidswell and Doughty[782] electropolymerised styrene in sulpholane with sodium tetrafluoroborate and concluded that the species responsible for initiation was boron fluoride formed at the anode. Cocatalysis by hydrogen fluoride or water was invoked. Mengoli and Vidotto[783] studied the electroinitiated cationic polymerisation of styrene in nitrobenzene at $-10°C$ using tetrabutylammonium perchlorate as background electrolyte. All polymerisations followed first order kinetics and an analysis of the results at different amounts of current showed that a direct proportionality existed between the concentration of active species and the current effective in initiating the polymerisation. The authors calculated values of the propagation rate constant assuming a kinetic scheme identical to that previously elaborated by Funt and Blain (see above). The values were considerably lower than those obtained by Funt and Blain in methylene chloride and a convincing reason for this difference was not given by the authors. We believe that again the role of the common-anion salt used as supporting electrolyte was particularly detrimental to the efficiency of propagation, because of the homoconjugation with perchloric acid formed in the polymerisation. An alternative explanation for the slowness of this process could be that perchloric acid is a much less effective catalyst in nitrobenzene due to specific interactions with this solvent, as already observed by Pepper and Reilly[335]. Mengoli and Vidotto did not propose any mechanism for the initiation of this polymerisation and simply assumed that the active species must have been "cationic".

Pistoia[784–786] investigated the electroinitiated polymerisation of styrene in propylene carbonate-lithium perchlorate solutions at $25°C$. Mechanistic evidence was obtained for the formation of perchloric acid at the anode and the cationic nature of the process thus proved. The kinetic analysis yielded a k_p value of $0.5 \; M^{-1} sec^{-1}$. Although no comparisons can be made between this result and previous ones in other solvents, the presence of lithium perchlorate was here a source of homoconjugation for the acid produced and thus the cause of considerable deactivation of its initiating power. As in previous cases, this was not recognised by the author. A similar study by Pistoia and Scrosati in dimethylsulphate gave an insoluble polymer at the anode and the nature or the initiator was not elucidated, but it did not seem to be perchloric acid. The cationic properties of this process was however proved[786].

Ghose and Bhadani[787] reported the electroinitiated polymerisation of styrene with tetraethylammonium hexachloroantimonate in nitrobenzene. The anolyte was found to become acidic during the electrolysis and the authors proposed that following the discharge of $SbCl_6^-$ to give the corresponding radical, the latter abstracted a hydrogen atom from one of the components of the reaction mixture and gave $HSbCl_6$ which was the initiating species for this cationic polymerisation.

Akbulut et al.[788] polymerised styrene by a direct electron transfer initiation carried out at the anodic peak potential of the monomer. These authors claimed the following mechanism involving styrene adsorbed on the anode surface:

$$M_{ads} \longrightarrow M_{ads}^+ + e^-$$

$$M_{ads}^+ + M \longrightarrow M–M_{ads}^+ .$$

Following this initiation scheme propagation was thought to be cationic. From a complex kinetic pattern a value of $k_p = 1.2 \times 10^{-2}$ M^{-1} sec^{-1} at 20°C was calculated. This extremely low value for a propagation rate constant supposedly belonging to a dicationic process clearly indicates that the concentration of active species was grossly overestimated in this work.

B. Vinyl Ethers

Funt and Blain[789] reported the electropolymerisation of isobutyl vinyl ether in methylene chloride with various tetrabutylammonium salts and showed the unmistakably cationic character of the chain carriers in these processes. Similar studies by Mengoli and Vidotto[790−792] extended the range of monomers and solvents with sodium tetraphenyl borate as background electrolyte. The anodic initiation mechanism was postulated as a two-electron process:

$$Ph_4B^- \longrightarrow Ph_2B^+ + Ph_2 + 2 e^- ,$$

followed by the attack of the cation formed onto the monomer,

$$Ph_2B^+ + M \longrightarrow Ph_2B^. + M^+ .$$

The authors pointed out that the diphenylboronium ions has similar characteristics to the trityl ion and is therefore a good initiator for the polymerisation of vinyl ethers. The kinetic investigation of these systems did not allow the determination of an absolute rate constant of propagation.

Cerrai et al.[793, 794] carried out an interesting study of the electropolymerisation of isobutyl vinyl ether with tetrabutylammonium triiodide in methyl chloride. Form the electrolysis of the supporting salt alone and the kinetics of the anodic initiated polymerisation it was clearly proved that molecular iodine was the catalyst responsible for the process. Recently, Breitenbach et al.[795] published an investigation of the anodic initiators cationic polymerisation of isobutyl vinyl ether. The background electrolyte was tetraethylammonium perchlorate and the solvent a mixture of acetonitrile and diisopropyl ether. This process exhibited all the typical characteristics of a cationic polymerisation, including the retarding effect of added moisture. The experimental conditions chosen for this work led the authors to propose an initiation mechanism based on the direct oxidation of the monomer at the anode, followed by reactions leading to dicationic active species.

C. Indene

Bhadani's group[796-798] has studied the electropolymerisation of indene with perchlorate and hexachloroantimonate salts. In the first paper of this series the authors used methyl chloride as solvent. Despite markedly S-shaped time-conversion curves, the authors surprisingly claimed that these reactions obeyed first order kinetics. This unfortunately disqualifies any further kinetic and mechanistic argument. The assignment of visible peaks recorded during the polymerisation was also erroneous and paid no heed to the work of Prosser and Young in this domain[667]. Further investigation of this system was reported in a second paper but the solvent was changed for nitrobenzene. Apart from the same misinterpretation of the origin of the visible spectra, it was proved that the polymerisation was effectively cationic. Pre-electrolysis of the supporting salt in the absence of monomer produced in fact an acidic initiator. The third communication dealing with this system does not add any relevant information concerning the fundamentals of the polymerisation process.

D. Anethole

Cerrai's group has studied the electroinitiated cationic polymerisation of this monomer in a series of elegant papers. When perchlorate salts were used as supporting electrolytes in methylene chloride it was shown that perchloric acid was formed at the anode[799]. The authors however could not exclude the concurrent oxidation of the monomer at the same electrode as an alternative route for initiation. In a subsequent paper[800] the group restudied this system by cyclic voltametry with a potential range in which the solvent-electrolyte couple alone is not substantially affected. In this case, anodic initiation was shown to proceed exclusively through the direct oxidation of the monomer giving eventually dicationic propagating species:

$$M - e^- \xrightarrow{E_1} \dot{M}^+ \xrightarrow{+M} \dot{M}M^+ \xrightarrow{E_2} {}^+MM^+ \ .$$

The kinetics of polymerisation were followed both during electrolysis and after the passage of current. No termination reactions were detected and a treatment of the data obtained produced a k_p value of 3 M^{-1} sec^{-1} at 25°C. This very low value for a propagation rate constant referring to such a basic monomer is probably due once again to the presence of the supporting electrolyte. Cerrai and collaborators have also studied the electroinitiated polymerisation of anethole in the presence of tertabutylammonium tri-iodide[801]. As in the case of their study of vinyl ethers in the same conditions, it was found that the initiator of this polymerisation was molecular iodine produced at the anode by oxidation of the anion.

E. Conclusions

The brief survey of the most relevant work carried out in the last decade which has been presented above clearly illustrates that the field of electroinitiated cationic polymerisation has contributed rather sparingly to the knowledge of the general problems enumerated in the introduction to this review. In fact, the major route of initiation in these processes consists in the anodic synthesis of a typical cationic initiator such as perchloric acid, boron trifluoride and iodine. The electrolytic reaction serves therefore the sole purpose of preparing a catalyst which in more classical studies is simply added to the monomer solution. If one considers moreover that the background electrolyte present in excess in these reactions has, as we have seen, a detrimental effect on the rate of polymerisation, and complicates considerably the overall mechanistic pattern, it becomes even more apparent that these studies have failed to deepen our understanding of the fundamentals of cationic polymerisation. In those investigations where direct anodic oxidation of the monomer is believed to take place, the situation is certainly more interesting but more detailed work needs to be done to prove the formation of cation radicals and to follow the reactions of these intermediates. For the moment the initiation schemes proposed are not adequately supported by solid experimental evidence. As for the kinetic analysis, the reported values of the propagation rate constants indicate that large errors, perhaps of a systematic nature, are present in the calculations or in the basic assumptions, since these values are definitely too low, sometimes by several factors of ten.

While the technique is therefore too complex to be exploited in a fundamental approach, this does not imply that its usefulness has been insignificant. Its practical applications have in fact given excellent results in such technological fields as polymer coating of metal surfaces etc. This type of research was recently discussed at the Pisa meeting[780].

A very recent paper by Cerrai and coworkers[802] came to our attention after most of this review had been written, but its importance calls for inclusion in this chapter. Indeed the authors question the very nature of the initiating process in the classical electrolytic polymerisation. They reached this conclusion after a very thorough study of the electrochemical polymerisation of cyclohexylvinyl ether in ethylene chloride with tetrabutylammonium tetrafluoroborate and perchlorate, having shown that initiation could not be attributed to the anodic oxidation of the electrolyte anion, of the solvent, or of the monomer. The acid formed at the anode compartment was therefore throught to originate from the electrolysis of residual moisture in the system. This conclusion was supported by the fact that under the most rigorous experimental conditions the rates of polymerisation were considerably lower than when the runs had been performed under nitrogen. Obviously, these findings cast serious doubts about the interpretation of the initiation mechanism in most of the previous work in which the formation of the initiator was not studied in detail, and open the doors to future investigations aimed at confirming of refuting this novel interpretation.

F. Electrodialysis

D'yachkovskii and collaborators[803-806] have developed this technique and applied it to the initiation of cationic polymerisation. A three-compartment electrolytic cell is set up in which the partitions between the chambers are cellophane membranes permeable to ionic species. The anodic compartment contains an initiator solution, the middle one a monomer solution and the cathodic one a supporting electrolyte. When an electric field is applied between the two electrodes the carbenium ions formed at the anode migrate to the middle chamber and promote the cationic polymerisation of the monomer contained therein. It has been argued by the authors that the failure to obtain polymerisation when Brønsted or Lewis acids were placed in the anodic chamber and styrene in the middle one, in contrast to the successful polymerisation of this monomer when a source of carbocations such as trityl chloride was placed at the anode, are a proof against the existence of pseudocationic polymerisation. This argument lacks validity because while it shows that electrolytically generated carbenium ions are initiators for the polymerisation of styrene, a fact which is by no means a novelty, it certainly fails to disprove that addition of perchloric acid to styrene in appropriate solvents brings about a polymerisation propagated by non-ionic species.

The technique of electrodialysis appears to us as a very complicated and unrewarding way of studying cationic polymerisations and we feel that the propagation rate constants and other supposedly fundamental studies reported by D'yachkovskii's group[807] are not entirely reliable because of the large number of assumptions required for the treatment of the experimental data.

G. Electric Field Effects on Cationic Polymerisation

Approximately ten years ago a controversy arose between Ise's school at Kyoto and Giusti's at Pisa concerning the role of a strong electric field applied to a cationic polymerisation system. This debate has been adequately summarised by Giusti[778]. Suffice it to say here that the work carried out at Pisa has proved beyond doubt that the enhancement of the polymerisation rate sometimes observed when the field is applied (continuosly or intermittently, but not reversed), is simply due to electrolytic phenomena which generate active species at the anode in addition of those already formed by the catalytic system. These new active species were thought to originate from the electrolysis of the counterions produced in the normal polymerisation. A recent paper by Akbulut et al.[808] has confirmed the validity of these conclusions.

VIII. Photoinitiation

A special instance of photoinitiation was already discussed in Sect. VI-B, within the context of "bare" cation formation. In those systems in fact, high-energy photons were used to produce the ejection of an electron from the monomer molecule. In this chapter we will briefly review other photochemical techniques involving the formation of electronically excited intermediates, which in turn generate suitable species for the initiation of cationic polymerisation. Crivello[809] has recently reviewed this topic and other authors have published more specialised monographs on some specific aspects of cationic photoinitiation. We will therefore reduce our coverage to the basic premises on which the various methods are founded and to a few comments on the more recent contributions and mechanistic interpretations.

A. Donor-Acceptor Photolysis

The principle behind this type of initiation is the formation of a charge-transfer complex between the monomer (electron donor) and a suitable electron acceptor, and its subsequent photolysis to give the monomer radical cation as primary product. The photodecomposition of the excited complex can also generate free radicals and the double nature of the ensuing polymerisation process, i.e. cationic and radical, has been proved in several systems. The mechanistic aspects of these interactions have been extensively studied with N-vinylcarbazole, styrene and α-methylstyrene. Shirota and Mikawa[810] recently reviewed these polymerisations with particular emphasis on the behaviour of styrene, while several monographs and essays by Irie, Hayashi and collaborators deal in particular with the other two[741, 811–813]. Ledwith[144, 814, 815] has also surveyed this field from the point of view of the fundamental processes involved in such systems, and Davidson[816] has exhaustively treated the basic topic of the photochemical properties of charge-transfer complexes. Given this adundant literature, we direct the interested reader to the above general references. Some of these studies employed new sophisticated techniques, which have recently become available to photochemists (fast-pulse excitation with multiple-frequency laser radiation, coupled with time-resolved spectroscopic detection systems), whose application to cationic polymerisation is a qualitative step forward towards the detailed investigation of the nature of the active species (and other intermediates), their concentration, and the kinetics and mechanism of initiation. Recent work by Yamamoto and collaborators[817] has contributed already to a better understanding of the reactivity of the cation radicals of styrene and some of its derivatives, and undoubtedly future research along these lines will provide much fundamental information.

While the working principle of the above studies involves the use of a strong electron acceptor (e.g., tetracyanobenzene or pyromellitic dianhydride) to generate the charge-transfer complex with the olefinic monomer, Marek and coworkers found that certain Lewis acids can give rise to photosensitive complexes with isobutene and butadiene, leading to a new type of cationic photopolymerisation. Since this work has not been reviewed in the various monographs previously mentioned but has given rise to some controversies, we will discuss it in some detail here. Marek and Toman[818] first reported in 1972 that quiescent solutions of isobutene in n-heptane at $-78°C$ containing vanadium or titanium tetrachloride could be activated for polymerisation by illumination with ultraviolet, visible or near-infrared light. Similar results were obtained with undiluted isobutene. These novel and interesting observations prompted the Czech group to carry out a detailed investigation of various aspects related to the mechanism of the photoinitiation process and a series of papers followed the original discovery[448, 449, 819–825]. The most relevant findings of these studies are summarised below (note that the authors' attention was specifically focussed on visible-light excitation with vanadium tetrachloride):

— Illumination of the non-polymerising solutions results in a slow initiation. The rate of monomer consumption grows then steadily and reaches a maximum value after an acceleration period (and *not* an induction period, as termed by the authors) the duration of which varies from a few minutes to about thirty minutes depending on the catalyst and monomer concentrations. High monomer-to-catalyst or catalyst-to-monomer concentration ratios give low acceleration periods.

— The interruption of illumination at limited conversions brings about a progressive deceleration of monomer consumption until polymerisation ceases completely. A new irradiation revives the process, again with an acceleration phase.

— Keeping all other parameters constant, the photoinitiation is favoured by a decrease in temperature (VCl_4, between -20 and $-78°C$). At temperatures above about $-20°C$ the polymerisation in the dark becomes progressively more important and at about $10°C$ light ceases to have any influence on the process.

— At constant illumination and $-20°C$, the maximum rate of polymerisation takes the kinetic form $-d[M]/dt = k[VCl_4]^{0.4}[M]$. At the same temperature, that rate is directly proportional to the irradiation intensity.

— Complexation between isobutene and VCl_4 (claimed to give a $1:1$ adduct) gives rise to a broad absorption region between 500 and 700 nm. The strength of this association is modest as testified by the value of the equilibrium constant (increasing as temperature decreases) and of the extinction coefficient of the complex ($100-200 \ M^{-1}$ cm^{-1}). Spectroscopic evidence for similar complexes with $TiCl_4$ (about 350 nm, tailing into the visible) and $SnCl_4$ (much weaker, around 270 nm) has been also reported, but unfortunately the details of that study have not been published yet.

— Addition of oxygen to the isobutene-Lewis acid mixture *bleaches the absorption due to complexation* and inhibits the photoinitiated polymerisation.

— While no ESR signals can be detected during these photopolymerisation, in frozen matrices containing the reaction components irradiation gives rise to signals which could be assigned to the isobutene radical cation and the corresponding Lewis-acid radical anion.

– TiCl$_3$ and VCl$_3$ are found after the photopolymerisations by TiCl$_4$ and VCl$_4$ respectively.

The mechanism of initiation proposed by Marek and coworkers for these systems envisages the excitation of the Lewis acid-isobutene charge-transfer complex followed by electron transfer giving the monomer radical cation. The latter is supposed to dimerise rapidly to yield the dimeric dication which is therefore the active species for propagation. The main arguments supporting this scheme were: the inhibiting effect of oxygen (which is not observed in standard cocatalysed cationic polymerisations) and the detection of ESR active entities in model experiments performed with glassy matrices. We feel that the occurrence of an electron-transfer step is unlikely and that counter arguments can be readily found, which suggest that the actual behaviour of these polymerisations runs counter to the formation of radical cations following electronic excitation of the complex. Williams[826] already pointed out at Akron that "it is dangerous to rely on computer simulation of ESR spectra without paying attention to the magnitude of coupling constants and relating them to spin distribution" and wondered if the assignement of the spectra obtained in the matrices to the isobutene radical cation was in fact in conformity "with the theoretical spin distribution". This query did not receive a satisfactory answer at that meeting. More importantly, Williams questioned the validity of extrapolating evidence obtained in a solid to interpret a mechanism which takes place in solution. This is a very important point, which again was not contested by Marek. In the various papers published by his group, the lack of signals due to cation radicals during polymerisation was attributed to the very low concentration of these species arising from their extremely short lifetime before dimerisation. It can be concluded that no very direct evidence of the existence of such primary products has in fact been offered for the systems in which photoinitiation takes place. The indirect argument about the inhibiting effect of oxygen seems at first sight valid. However, a close inspection of the authors' reports[819, 820] reveals that, as we underlined above, the complexes themselves are bleached by the presence of oxygen, i.e., oxygen destroys the association of the Lewis acid with isobutene, probably by giving stronger interactions with the former. In view of this scavenging effect, it is not surprising that the photopolymerisation does not take place in the presence of oxygen, but the real reason is not the quenching of radical cations by this compound. What happens is simply that the concentration of the absorbing complex is drastically reduced when oxygen is added and therefore there is no appreciable light absorption (and excitation) when these solutions are illuminated. In other words, photoinitiation cannot take place in the absence of the entities to be photolysed. Thus, the poisoning effect of oxygen operates before the primary act of excitation, since it affects the very formation of the charge-transfer complex, and any argument concerning its deactivating role upon cation radicals through formation of peroxy radicals in therefore void. The above discussion does not exclude the formation of radical cations following excitation, but only shows the weakness of the evidence offered by the authors in support of such mechanism. There is however one important observation which seems to us inconsistent with electron transfer initiation. The acceleration periods noticed in these photopolymerisations are not compatible with the formation of active species following two extremely fast processes, viz. photochemical ionisation and radical coupling. This sequence of events would be expected to produce a maximum *initial* rate of polymerisation, particularly since termination is operative as indicated by the cessation of activity soon after the interruption of the irradiation. Photochemical interactions lead to rapidly attained steady-state conditions and there is no reason why it should take several minutes or more to arrive at maximum efficiency. The argument of possible impurities which have to be scavenged before the active species can operate freely is not valid, since, as we have seen in the description of the results, a second irradiation of the solution after a period of darkness gives rise to a second acceleration period in a system which should have been prepurified in the first partial photopolymerisation. It can therefore be concluded that these accelerations are an intrinsic feature of the behaviour of the systems in question. We wish to propose the following mechanism as an alternative to that involving the formation of cation radicals produced by electron transfer in the excited donor-acceptor complex. Our proposal is a qualitative one, which makes use of all available experimental evidence. The photolysis of the complex produces a chemical reaction between the olefin and the Lewis acid eventually leading to the reduction of the transition element in the latter. The formation of $TiCl_3$ and VCl_3 testifies to this reduction. The details of this interaction arising from the excited complex are unknown, but a cocatalyst could be generated by it (e.g. chlorine) and this would allow polymerisation to start, first,slowly, then, as the concentration of this cocatalytic substance increases, more rapidly. The energetics of this process are obviously modest, since at temperatures above about $-10°C$ the corresponding thermal reaction begins to be important. The cocatalyst is consumed in the polymerisation as indicated by the termination in the absence of irradiation and the new acceleration period noticed during a second irradiation. In conclusion, we are proposing that these systems are typically cationic, and light intervenes to give a suitable cocatalyst through a photochemical pathway involving the reduction of the Lewis acid.

We would like to underline a rather surprising observation reported in the above publication. Marek and collaborators managed to obtain systems containing $TiCl_4$ and isobutene at low temperature giving no polymerisation (0% conversion). This feat is difficult to explain in view of the traditional difficulty encountered in Plesch's, Sigwalt's and Kennedy's laboratories to minimise the so-called prepolymerisation in such systems, let alone achieving no polymerisation at all. This is all the more puzzling considering that the techniques used by Marek's group were not particularly elaborate (injection of reagents with syringes, no high vacuum). We have no explanation for this important discrepancy.

A recent report by Kennedy and Diem[827, 899] describes the promoting effect of a glow discharge (Telsa coil) or ultraviolet irradiation upon the low temperature polymerisation of isobutene by titanium tetrachloride. The authors ascribed the yield increases observed to the formation of cocatalytic amounts of chlorine or hydrogen chloride arising from the decomposition of the catalyst into $TiCl_3$ and a chlorine atom. The interpretation is probably valid, but we point out that HCl is not a cocatalyst for the polymerisation of isobutene by $TiCl_4$ in hexane, one of the solvents used in that work[377]. We prefer to think that a combination of factors, difficult to control or separate, should be held responsible for these phenomena: apart from the formation of cocatalytic substances arising from chlorine atoms, it is in fact known that both the glow discharge and ultraviolet light can effectively "degas" the vessel walls introducing new impurities (moisture) into the reaction mixture. These phenomena complicate considerably the interpretation of the results and we feel that no fundamental conclusions can be reached from such a study. A critique of the photochemical arguments expounded in the second paper[899] to attempt a rationalisation of the polymerisation by condensation has been submitted[901].

B. Photolysis of Arylonium Salts

Crivello and Lam[828–834] have recently reported several studies on the photoinitiation of cationic polymerisation promoted by in-situ generation of strong Brønsted acids arising from the photolysis of various arylonium salts. Following an initial investigation on the activity of diaryliodonium salts with counterions such as BF_4^-, SbF_6^-, etc., these authors have extended their interest towards other halonium, sulphonium and selenonium salts with the same range of anions. The general mechanism underlying all these photolyses is based on a decomposition pathway retaining the original MtX_{n+1}^- ion which is presumed to associate with a proton formed in the rearrangement of the salt to give the corresponding Brønsted acid. This mechanism has been justified by photolysis product analysis which seems to exclude the formation of the Lewis acid MtX_n from the anion. Thus, for example:

$$(Ar)_2 I^+ BF_4^- \longrightarrow Ar - \overset{+}{I^\cdot} + Ar^\cdot + BF_4^-$$

$$2\, Ar^\cdot \longrightarrow Ar - Ar$$

$$Ar - \overset{+}{I^\cdot} + SH \longrightarrow Ar - \overset{+}{I} - H + S^\cdot \ (2\, S^\cdot \longrightarrow S_2)$$

$$Ar - \overset{+}{I} - H + BF_4^- \longrightarrow Ar - I + HBF_4 \ .$$

where SH is the solvent and HBF_4, the effective catalyst for the cationic polymerisation of alkenyl and heterocyclic monomers. This type of process applies equally well to the other families of onium salts studied. Since the salts are not initiators in the dark, monomer, solvent (if any) and salt can be mixed and stored indefinitely, until the polymerisation is desired. Irradiation of the mixture with the adequate frequency produces immediate initiation. The excitation of the salt can also be photosensitised, thus reducing the energy required for the photolysis[835]. The essentially practical interest of these systems is all too obvious to need comment. From the point of view of more fundamental implications, it is difficult to assess what could be gained from such investigations in terms of a better knowledge of initiation mechanism. We must also note that the general principle behind this technique is not new, since Schlesinger[836] had reported in 1974 that aryldiazonium salts behaved in essentially the same way.

C. Other Techniques

Woodhouse et al.[837] uncovered a simple and promising new way of initiating cationic polymerisation by a photochemical process. They achieved the bulk polymerisation of tetrahydrofuran and other cyclic ethers by irradiating their stable solutions of $AgBF_4$, $AgPF_6$, $AgSbF_6$ and $AgSO_3CF_3$ with 254 nm light at 25°C. Copper (I and II) salts were also successfully used with longer wavelengths. No thermal polymerisation was observed with these systems in the absence of irradiation. In all polymerisations the finely divided metal precipitated during the process. Although the monomers used in this work fall outside the scope of the present review, we quote this interesting discovery because of its obvious potentials for future fundamental research. The simplicity of the systems should in fact allow in-depth studies of the initiation mechanism. The results of experiments with olefins have just begun[838].

Ledwith[839, 840] has recently speculated about the possibility of initiating cationic polymerisation using common free radical sources coupled with oxidising cations. This idea has already been put into practice in his laboratory and the successful polymerisation of vinyl ethers and tetrahydrofuran has been achieved with two systems[841, 842]. In the first study 2,2-dimethoxy-2-phenyl acetophenone was photolysed at 366 nm in n-butylvinyl ether in the presence of di-p-tolyl iodonium hexafluorophosphate as oxidising salt. The free radicals produced in the photolysis were transformed into cationic active species for the polymerisation of the vinyl ether by the electron transfer to the iodonium ion. In the second report, various radical sources were photolysed in the presence of the monomer and silver hexafluorophosphate, the latter acting as one-electron oxidant.

Diem et al.[843] investigated the photolysis of adamantyl iodide[844] as a possible source of carbocationic initiation for the polymerisation of isobutene. Although no polymer was obtained with this simple system (or with t-butyl iodide as photolyte), addition of iodine scavengers such as zinc, zinc iodide or both together gave some polymerisation, indicating that the carbocations produced in the photolysis of alkyl iodides possess a modest initiating power if generated in the presence of isobutene.

IX. Initiation from a Charge-Transfer Complex

In the preceding chapter we briefly examined the photolytic excitation of charge-transfer complexes involving an alkenyl monomer as electron donor as a means of initiation of cationic polymerisation. In certain instances, namely with highly nucleophilic monomers possessing relatively low ionisation potentials, suitable charge-transfer complexes can be prepared which undergo one electron transfer processes "spontaneously", i.e. by thermal means. The reactions following the formation of the monomer radical cation probably involve radical coupling and propagation by a dicationic species. A great deal of work has been devoted to this type of initiation, particularly with N-vinylcarbazole and the vinyl ethers. This field has been very adequately reviewed both from the point of view of the basic interactions of donors with acceptors and the properties of the resulting charge-transfer complexes[64, 65, 814], and in the more specific domain of cationic polymerisations where the monomer plays the role of the donor molecule[810, 845, 846]. These monographs are exhaustive and cover the literature up to very recently.

An interesting study which is somewhat outside the field just defined, but which nevertheless involved the use of electron acceptors as cocatalysts, was communicated at Akron in 1976 by Cesca and collaborators[847]. These authors reported that the polymerisation of isobutene (and its copolymerisation with isoprene) in chloroform at $-40°C$ could be effectively achieved by using the catalytic combination Et_2AlCl (or Me_2AlCl)-electron acceptor, the most effective acceptors being chloranil and 2,4,5,7-tetranitrofluoren-9-one. A detailed investigation was made of the interaction between the organometallics and chloranil. No chlorine exchange was detected and on the basis of product analysis and characterisation, and ESR spectra of the reacting solutions, it was concluded that the semiquinonic structure

was the most probable result of the radical decomposition of the charge-transfer complex between the two reactants. It was moreover argued that since the ESR signal due to the semiquinonic radical decreased appreciably when isobutene was added to the catalytic mixture, an electron transfer could be envisaged between the donor (monomer) and the acceptor (radical). This reaction would thus generate an isobutene radical cation:

which, after coupling to give the dimeric dication, would initiate polymerisation. The absence of chlorine and aromatic end-groups in the polymers and in the dimers of 2,4,4 -trimethypentene-1 and 1,1-diphenylethylene was taken as evidence for the "indirect" participation of the electron acceptors in the initiation mechanism. Finally, the fact that some 1,1,4,4-tetraphenylbutadiene was isolated among the products of the interaction of these catalyst combinations with 1,1-diphenylethylene, suggested that indeed the dimeric dication could have been present as active species (it does seem that only such precursor could give the tail-to-tail diunsaturated dimer by double spontaneous transfer). The latter observation was the only (indirect) evidence in favour of an initiation process involving the formation of the monomer cation radical. It seems obvious to us that more work would be required to confirm such hypothesis, given its unlikelihood on energetic grounds, particularly in the case of isobutene. These systems are particularly interesting in view of their novelty and efficiency in terms of yield and especially for the high DP's obtained at relatively "high" temperatures, and future investigation, announced by the authors[847], should provide enlightening information.

X. Initiation from a Polymer

The use of polymeric initiators or coinitiators to induce the polymerisation of a second monomer by a cationic mechanism is a particularly attractive possibility, since it would permit the synthesis of block and graft copolymers. The search for adequate systems in this context has been intensive, but only very recently has it met with some success, and this is far from being as spectacular as the achievements obtained in the same area with anionic systems. The main difficulties to be surmounted have been discussed in the general introduction to this review (see Chap. I), and have to do with the ubiquity of transfer and termination reactions in cationic polymerisation. Nevertheless, the advances of the last few years seem encouraging and one would expect that the near future will provide considerable progress, both quantitative and qualitative.

This chapter deals with three different approaches to the preparation of block and graft copolymers, their common feature being the initiation of the second polymerisation by an active or activated first polymer.

A. Cationic Block Copolymerisation

The first (partly) successful attempts to prepare block copolymers by double cationic initiation involved the preparation of the first block, its isolation and transformation into a macroinitiator and the subsequent blocking with the second monomer. Thus, Jolivet and Peyrot[848] reported in 1973 the synthesis of a poly(isobutene-b-styrene) based on the preparation of a terminally benzylated polyisobutene, the chloromethylation of the aromatic end groups and the polymerisation of styrene onto these $-CH_2Cl$ moieties catalysed by diethylaluminium chloride. The yield of block copolymer was limited due to transfer reactions in both the first and the second polymerisation, i.e. appreciable amounts of homopolymers were also obtained. A similar procedure was used by Kennedy and Melby[614] a few years later to prepare the same type of copolymer.

More recently, Kennedy and collaborators[632,849,850] used boron chloride with different cocatalysts to prepare homopolymers with chlorine end groups and induced the blocking of the second monomer with diethylaluminium chloride. About 50% efficiency was claimed for poly(isobutene-b-styrene), the rest of the product being a mixture of the two homopolymers[850], and a higher unspecified value in the case of poly(isobutene-b-α-methylstyrene)[632]. The mechanism of initiation from the macromolecule of the first monomer is not altogether clear, in view of the observations concerning the real role of $Et_2AlCl-RCl$ systems in the generation of active species which we developed in Sect. IV-C-5, and which will be briefly applied to grafting in Sect. C of this chapter.

The latest application of the macroinitiator technique to double cationic polymerisation comes from Seung and Young[851] who prepared and isolated terminally iodinated polystyrene and then used it to initiate the polymerisation of 2-methyl-2-oxazoline which is sensitive to alkyl iodide. Of course only those polystyrene molecules which contained iodine end groups were effective in promoting block copolymerisation.

In situ cationic block copolymerisation has been claimed by Rooney et al.[678] for the system N-vinylcarbazole-vinyl ethers when using stable carbocationic initiators. Our doubts about the validity of that claim have already been expounded in Sect. V-A-4-e). In a joint effort between Dublin and Ghent a block copolymer was prepared by adding 1-t-butyl aziridine to polystyrene perchlorate[852]. This work made an ingenious use of two important factors: (i) the polymerisation of styrene by perchloric acid at very low temperatures suffers negligible transfer and most of the chains formed possess a terminal perchlorate group, and (ii) certain N-substituted aziridines polymerise cationically following a nearly living mechanism[26]. Thus, the "dormant" perchlorate end groups were used as initiators for the very nucleophilic arizidines and sequential polymerisation was quite seccessful.

The latest report on in situ cationic block copolymerisation, and indeed the first one dealing with two alkenyl monomers, comes from Higashimura's laboratory[363]. In this study the best conditions were sought for a near-living polymerisation of p-methoxystyrene by iodine in carbon tetrachloride and methylene chloride. A progressive increase in DP as the polymerisation proceeded and after a second monomer addition, indicated that transfer reactions had been minimised. If isobutylvinyl ether was introduced at the end of the first polymerisation of p-methoxystyrene, the DP of the product also increased and selective solvent extractions showed that a block copolymer had been obtained. The authors assumed that in the first polymerisation all the iodine had been consumed according to an initiation mechanism involving the selfionisation of the halogen and the attack of I^+ on the monomer, I_3^- being the counterion. This is however doubtful, since the polymerised solutions were reported to be wine-red in colour, which seems to indicate free or complexed iodine in these systems. Also, the specific mode of initiation is purely speculative both kinetically and mechanistically, and in view of the fundamental work of Giusti on the role of iodine in the cationic polymerisation of olefins (see Sect. III-E-14-a) those suggestions are probably incorrect and the polymerisation should instead be considered as pseudocationic, particularly in carbon tetrachloride.

Di Maina et al.[900] recently described a direct synthesis of styrene-isobutene block copolymers based on the use of $AlCl_3$ in CH_2Cl_2. This brief report underlines that the polymerisation of styrene, initiated by traces of isobutene in the presence of $AlCl_3$, proceeds without appreciable termination and transfer reactions. A more detailed account of these interesting experiments was announced by the authors and should provide an explanation of some still intriguing features, particularly concerning the role of traces of isobutene in promoting the onset of styrene polymerisation.

B. Anionic-Cationic Block Copolymerisation

Within the context of stable carbenium salts initiation, we already examined a very interesting and successful study on the block copolymerisation of α-methylstyrene with cyclopentadiene performed by Vairon and Villesange (see Sect. V-A-4-b). The preparation of the product required three basic operations: (i) the living anionic polymerisation of α-methylstyrene to give "monodisperse" macromolecules, (ii) transformation of their end groups into stable carbocationic moieties, and (iii) initiation of the polymerisation of cyclopentadiene from these active ends under conditions of minimal transfer and termination reactions. Thus, the macroinitiators in the second polymerisation were generated by a controlled anionic polymerisation and allowed the synthesis of a triblock near-isomolecular copolymer.

A similar principle was followed by Burgess et al.[853−855] when they discussed the general possibility of preparing block copolymers by a mixed anionic-cationic mechanism. The technique they adopted consisted in preparing living polystyrene or polybutadiene by standard techniques, killing them with bromine or xylylene bromide to give "monodisperse" bromine-terminated products,

$$\sim\sim\sim M_1^- \, Li^+ \; + \; Br_2 \; \longrightarrow \; LiBr \; + \; \sim\sim\sim M_1 Br$$

$$\sim\sim\sim M_1^- \, Li^+ \; + \; BrCH_2\text{---}(C_6H_4)\text{---}CH_2Br \; \longrightarrow \; LiBr \; + \; \sim\sim\sim M_1\text{---}CH_2\text{---}(C_6H_4)\text{---}CH_2Br \; ,$$

or via a Grignard intermediate. Thereafter, a solution of the second monomer, to be polymerised cationically (tetrahydrofuran was the only one used in this work), in the presence of silver perchlorate was mixed with a solution of the brominated polymer and the following metathesis generated the macroinitiators for the block copolymerisation:

$$\sim\sim\sim M_1 Br \; + \; AgClO_4 \; \longrightarrow \; AgBr \; + \; \sim\sim\sim M_1^+ \, ClO_4^-$$

(or the corresponding esters)

$$\sim\sim\sim CH_2\text{---}(C_6H_4)\text{---}CH_2Br \; + \; AgClO_4 \; \longrightarrow \; AgBr \; + \; \sim\sim\sim CH_2\text{---}(C_6H_4)\text{---}CH_2^+ \, ClO_4^-$$

The above ideas open promising avenues for the preparation of block copolymers. A major problem of course remains, namely finding the appropriate conditions for the effective living polymerisation of the second monomer onto the cationic macroinitiators. The only olefin approaching this behaviour which has been use to the present is cyclopentadiene. Among cyclic n-donor monomers the choice is wider, but these considerations are outside the scope of the present review.

C. Cationic Grafting

The original idea of cationic grafting produced by initiation from a polymer containing cocatalytic moieties was proposed by Plesch in 1958[856]. Basically, the principle of this branching reaction relies on activation of a C–X bond of the polymer by a Lewis acid in a process thought to proceed by halide ion abstraction:

$$-\overset{\displaystyle |}{\underset{\displaystyle |}{C}}-X \; + \; MtX_n \; \longrightarrow \; -\overset{\displaystyle |}{\underset{\displaystyle |}{C}}{}^+ \; MtX_{n+1}^- \; ,$$

the active sites thus produced on the polymer chain being the starting points for the growth of branches in the presence of a suitable monomer which can be polymerised cationically. Plesch[856] was the first to show that his idea could be put into practice when he grafted styrene onto a vinyl chloride-vinylidene chloride copolymer using aluminium chloride as catalyst. Other studies followed, but the grafting efficiency was always low, owing to important transfer reactions which yielded a considerable amount of homopolymer of the added monomer.

Kennedy and coworkers improved considerably the degree of grafting by using diethylaluminium chloride at low temperature and claimed that this achievement was due to the essential absence of transfer reactions in the branching polymerisation. The work of this prolific group has recently been the object of an entire issue of Applied Polymer Symposium[857] and the interested reader will find in those papers a general introduction to cationic grafting and the specific contributions and applications achieved at Akron.

In view of this abundant information and monographs, we will restrict our observation to one fundamental aspect of cationic grafting, namely the problem of whether all branches originate from initiation on the original polymer (grafting *from*), or if a nonnegligible fraction of them arises from other reactions such as the attack of a growing homopolymer onto a double bond of the original polymeric substrate or chain transfer from the same growing polymer onto a dead chain of the substrate (grafting *onto*). These reactions have been discussed and examined most appropriately in Sigwalt's laboratory in the last few years[613,858] and a large body of evidence has been gathered to show that the high grafting efficiencies obtained in the Kennedy-type systems are not only due to the single process of grafting *from* (accompanied by clean, transfer-less polymerisation), but most probably by a complex combination of various mechanisms, where grafting *onto* reactions must play an important role. In particular the French workers successfully grafted polyindene onto unsaturated polymers which did not contain any halogenated moiety[613]. It must therefore be concluded that the behaviour of these systems is not yet fully understood and that initial oversimplifications about the grafting mechanism are currently being revised.

Another type of cationic grafting involves the initiation from a halogenated polymer submitted to ionising radiation. Chapiro[859] has summarised the little work done in this field and reported some work of his own on the grafting of styrene and isobutene onto polyvinylchloride. The initiation mechanism invoked for such processes was as follows:

$$\text{H}-\overset{\displaystyle\{}{\underset{\displaystyle\{}{\text{C}}}-\text{Cl} \quad \text{\small MWW} \longrightarrow \quad \text{H}-\overset{\displaystyle\{}{\underset{\displaystyle\{}{\text{C}}}{}^{+} \quad + \quad \text{Cl}^{-} \quad \xrightarrow[\;+\;\text{M}\;]{\text{grafting}}$$

We have omitted from the above brief survey those studies dealing with the cationic grafting of a polymer onto another since these Friedel-Crafts alkylation reactions, although proceeding via carbenium ion intermediates, do not *initiate* any polymerisation and fall therefore outside the realm of the present review. Kennedy has recently summarised this type of work[860].

XI. Miscellaneous Initiators

This final chapter is devoted to a brief survey of systems which, while bearing typical cationic connotations, cannot be classified in any of the categories of initiation already discussed. Since little is known about the mechanism through which active species are formed with most of these peculiar catalytic combinations, the present approach will be more descriptive than critical.

A. Metal Perchlorates

In 1947[863], Lilley and Forster used magnesium perchlorate to dry styrene and noticed that this salt induced a slow heterogeneous polymerisation. A few years later, Eley and Richards[864] briefly studied the polymerisation of octylvinyl ether by silver perchlorate and concluded that initiation was probably due to both the unhydrous and the monohydrate salt. At the same time, Salomon[865] reported that this salt could induce the polymerisation of styrene, vinyl acetate, methyl methacrylate and other monomers and that these processes were not inhibited by typical free-radical quenchers. Walling et al.[866] copolymerised styrene and methyl methacrylate with magnesium perchlorate and concluded from the even composition of the product that the mechanism of this process must have been radicalic. Burton and Praill[867] showed quite conclusively in 1953 that silver perchlorate is not a "Lewis acid" since under carefully controlled conditions it does not catalyse acetylation of aromatics or the polymerisation of styrene at room temperature. Losev and Zakharova[868] investigated the polymerisation of styrene under the influence of potasium, ammonium, barium and magnesium perchlorate and found the latter to be the most active, but they did not propose any specific mechanism for these processes. Hodgdon[869] showed that anhydrous lithium perchlorate is an effective initiator for the heterogeneous polymerisation of several vinyl monomers. They opted for a cationic mechanism because an attempted copolymerisation of styrene and methyl methacrylate produced only polystyrene, and amines retarded or inhibited all processes. However, acrylamide was found to polymerise in the presence of $LiClO_4$, a fact which seems to be incompatible with a cationic mechanism. Hermans and Smets[870] reported a kinetic study of the polymerisation of styrene and methyl methacrylate by silver perchlorate and concluded that a coordinated mechanism, promoted by the silver ion, was responsible for propagation following initiation by a salt-monomer complex. They rejected the hypothesis of a cationic polymerisation in view of the equimolar composition of the copolymer obtained when a 1 : 1 mixture of the two monomers was treated with AgClO$_4$. Contrary to these conclusions are those of Solomon and collaborators[871, 872] who studied the polymerisation of N-vinylcarbazole by various metal perchlorates. Beletskaya and Ryabtsev's[873] results on the polymerisation of styrene by mercuric perchlorate and phenyl and pentafluorophenyl mercuric perchlorate are also consistent with a cationic process. Indeed, the mechanism proposed by them is probably the soundest of all those encountered in this specific area. The last published study describing the specific catalytic action of metal perchlorates for the polymerisation of vinyl monomers is Maroy's thesis[874], which was devoted to a wider investigation, but dealt abundantly with this topic. The most interesting aspect of this work is certainly the fact that isobutene, a purely "cationic" monomer, was found to polymerise, particularly by magnesium perchlorate, in heterogeneous conditions. These results have recently been confirmed in our laboratory[875] and there seems to be no doubt about the cationic nature of the initiation, although the details of its mechanism remain to be understood.

The panorama sketched above clearly shows that a great deal of disagreement exist as to the real origin of the mechanism giving rise to vinyl polymerisation when metal perchlorates are used as catalysts both in the sodid state and in solution. Probably there is no general mechanism for these processes and the specific experimental conditions play a fundamental role in determining the occurrence or not of polymerisation and its character. We refer in particular to such factors as the pretreatment of the salt, its residual moisture content, the temperature, the possible specific interactions between the salt, the monomer and the solvent, and of course the presence of such "impurities" as traces of perchloric acid or active oxygen. Each system is thus rather unique and difficult to relate to others, however similar they might seem. We are presently engaged in studying these peculiarities in our laboratory and hope to have some preliminary conclusions soon.

B. Metal Chlorides

Turton et al.[876] conducted an excellent investigation on the initiating capacity of magnesium chloride in the solid phase vis-a-vis isobutene and other vinylic monomers. The cationic nature of these polymerisations was clearly proved and electrophilic sites present on the surface of the fine grains of the anhydrous salt were thought to be responsible for the heterogeneous initiation. Water was found to be a "cocatalyst" if added is small doses, but became a poison in higher amounts. Thorough milling of the catalyst improved its performance very considerably. We must note that Iwasaki et al.[877] had already reported that vinyl ethers can be polymerised by magnesium halides.

C. Grignard Reagents

Biswas and John[878,879] recently found that N-vinylcarbazole can be polymerised by phenylmagnesium bromide and *n*-butylmagnesium bromide in benzene solution, the former initiator giving higher rates and DP's. These processes were retarded by water and basic additives and it was concluded that the polymerisation must be cationic:

$$2\,RMrX \rightleftharpoons RMg^+\ RMgX_2^-\quad,$$

$$RMg^+ + M \longrightarrow RMgM^+\quad.$$

The authors reported that about 90% of the oligomers obtained contained a magnesium atom, a fact which strongly favours their proposed initiation mechanism.

D. Phosphorus and Arsenic Compounds

Biswas and Mishra[880] reported that arsenic trichloride can initiate the polymerisation of α-methylstyrene in benzene and methylene chloride at $25°C$. No exhaustive study was made of this system. The same authors[881] also reported that PCl_3, PBr_3 and $POCl_3$ are effective initiators for the polymerisation of α-methylstyrene, particularly in nitrobenzene. The latter catalyst was selected for a more detailed investigation: water was considered as detrimental to initiation, which was postulated to occur "directly" as

$$CH_2{=}CPh(CH_3) + POCl_3 \longrightarrow Cl_2OP\text{-}CH_2\text{-}\overset{+}{C}Ph(CH_3)\,Cl^-\quad.$$

In the same vein, Taninaka and Minoura[882] found that N-vinylcarbazole polymerises in benzene in the presence of catalytic amounts of various phosphorous derivatives, such as PCl_3, PBr_3, $POCl_3$, etc. These authors have also studied the polymerisation of styrene by PCl_3[883]. They found that the initial rate of monomer consumption was proportional to the second power of the monomer concentration and the first power of the initiator concentration. This rate was the higher, the higher the dielectric constant of the medium and virtually zero in benzene. One phosphorus atom was found statistically in each polymer chain even after 10 reprecipitations. The authors concluded that initiation must follow selfionisation of PCl_3, although this process must give essentially ion pairs, since conductivity measurements failed to show appreciable signals. They thus wrote:

$$2\,PCl_3 \rightleftharpoons PCl_2^+ PCl_4^-$$

$$PCl_2^+ PCl_4^- + CH_2{=}CHPh \longrightarrow Cl_2P\text{-}CH_2\text{-}\overset{+}{C}HPh\ PCl_4^-\quad.$$

Taninaka et al.[884] recently published a similar study on the polymerisation of styrene by $POCl_3$. Although S-shaped conversion-time curves were obtained, the maximum rate was found to depend upon the first power of the initiator concentration and the second power of the monomer concentration. An increase in dielectric constant of the solvent produced a marked increase in the maximum rate of polymerisation. They also observed that both in nitrobenzene and methylene chloride water was a "cocatalyst", but did not propose a mechanism for this interaction. No phosphorus analysis was apparently performed on these polymers. The cationic nature of the process was deduced from copolymerisation experiments with methyl methacrylate. We think that $POCl_3$ is not the real initiator here, but instead the hydrolytic process:

$$POCl_3 + H_2O \longrightarrow POCl_2OH + HCl$$

generates the true initiating species (Brønsted acids). Indeed even in the experiments performed without added water, the authors affirmed that the concentration of residual moisture could be as high as 1×10^{-3} M. Probably the facile hydrolysis of $POCl_3$, the three chlorine atoms of which have different reactivity, produced the initiating species.

E. Other Initiators

Isobutyl vinyl ether has recently been polymerised by iron sulphate[885], following the calcination of the initiator at 700°C. The same treatment at 750°C gives an inactive salt. Obviously, the key point about the activity of this compound has to do with some particular crystalline structure which is modified between these two temperatures.

Biswas and collaborators have investigated the polymerisation of N-vinylcarbazole by various oxides, and vanadates[886–888]. All these heterogeneous reactions were claimed to be cationic in nature, and tentative coordination-type mechanisms proposed.

Bittles et al.[903] carried out a detailed kinetic and mechanistic study of the polymerisation of styrene by commercial catalytic-cracking acid clays (Filtrol). This heterogeneous process gave low DP's and displayed all the typical features of a cationic (or pseudocationic) polymerisation.

Biswas and Maity[904] recently reviewed the polymerisation of alkenyl monomers by molecular sieves. The cationic nature of these systems seems well established, but the mechanism of initiation and the possible role of residual water as cocatalyst need further investigation.

Several more qualitative reports mentioning the use of curious initiators are to be found scattered in the literature, but often the supposedly cationic mechanism involved in these polymerisation is not clearly established.

Acknowledgements. We wish to thank many colleagues for providing us with unpublished manuscripts and valuable information. We are particularly grateful to Professors P. H. Plesch and P. Sigwalt for stimulating discussions on some of the topics covered in this monograph. Mrs. Joan Gandini's constant readiness to help with linguistic advice and with typing greatly smoothed our progress. To her goes our deep gratitude.

Notes Added in Proof

Since the manuscript was submitted many relevant papers appeared in the literature or are about to be published. Also, an International Symposium on Cationic and Other Ionic Polymerisations was held in Kyoto[905]. We include the following notes, which also deal with a few earlier investigations which had escaped our search. The headings below refer to the specific sections of the main text to which each note should be added.

I

Reviews concerning ring-opening polymerisations of heterocyclic monomers were published by Billingham[906], Penczek[907] and Goethals and Schacht[908].

II-C

Griffith and Pugh[909] compiled useful data relating to solvent polarity and proposed new means of predicting polarity scale values.

II-G-2

Lyerla et al.[910] have applied pulsed ^{13}C NMR spectroscopy and magic-angle spinning techniques to obtain the first spectra of carbenium ions in the solid state. Low-temperature spectra of immobilised reactive entities allow the characterisation of otherwise elusive species.

II-G-3

Penczek[911] discussed at Kyoto the interesting problem of carbenium ion − onium ion equilibria and summarised data and concepts elaborated recently in his laboratory. The aspect most relevant here has to do with the cationic polymerisation of π − and n-donor monomers in which the active carbenium ions can equilibrate with the corresponding onium ions both intramolecularly (isomerisation of the active species) and intermolecularly (reaction with the n-donor atom of a polymer chain). Owing to the marked differences in stability and reactivity of the two types of cations, these interactions would of course bring about important kinetic complications in systems involving such monomers as vinyl ethers.

III-E-6-a)

Obrecht and Plesch[912] thoroughly reexamined the system styrene-trifluoroacetic acid-methylene chloride at 40°C. They found that residual moisture played a determining role on the kinetic behaviour of the polymerisations. Both rigorously dried and "wet" systems and the effect of added 1-phenylethyl trifluoroacetate were studied. The authors concluded that true cationic and pseudocationic polymerisation coexist in this system, the former giving high DP's and being sensitive to water and other bases, the latter proceeding even in the presence of small amounts of nucleophiles and giving oligomers containing trifluoroacetate end groups.

III-E-6-c)

Bolza and Treloar[913] investigated the kinetics and mechanism of the polymerisation of *iso*butyl vinyl ether by trifluoroacetic acid in ethylene chloride between −2.5 and 35°C. The evidence led the authors to support a pseudocationic mechanism. Some of the chemical aspects discussed need clarification, e.g. why acid expulsion from a growing ester should be considered a termination step, rather than a transfer reaction. As for the intimate nature of the chain carriers, the authors suggested that the ester might need to be activated by a molecule of acid (a suggestion adequately supported by their results), but later considered this activation unnecessary to explain the absence of polymerisation at low acid concentrations. Perhaps it would be more reasonable to assume that the ester alone is not a chain carrier. Finally, they ignored the possibility of homoconjugation between the trifluoroacetate anion and its parent acid both in the selfionisation of the acid and in their experiments involving the addition of tetraethylammonium trifluoroacetate in relatively high concentrations. Since homoconjugation is very pronounced[263–268] even in the specific conditions used in this study, we cannot explain that Bolza and Treloar found a negligible effect of the added salt on the experimental rate constant.

III-E-7

Yamazaki and collaborators[914] studied the homo- and co-dimerisation of styrenes by diphenylphosphate. These reactions were shown to proceed through a pseudocationic mechanism by the isolation of the active ester intermediate. With *p*-methoxystyrene oligomers were obtained, i.e. the ester was in this case more active towards propagation (very nucleophilic monomer) and less prone to spontaneous transfer.

III-E-12-a)

Kunitake and Takarabe[915,916] published further results on the flash polymerisation of styrene by triflic acid followed by a uv stop-flow technique. The rate constants of each elementary step were obtained[915] by following simultaneously the concentration changes of both ionic active species and monomer during polymerisations conducted in ethylene chloride at −1 to 30°C. At 20°C, $k_i \sim 20\,\text{M}^{-1}\text{s}^{-1}$ and $k_p \sim 1 \times 10^5\,\text{M}^{-1}\text{s}^{-1}$ (the latter being the composite value for free ions and ion pairs). The activation energies for both steps were around 7 kcal mole^{-1}. The lifetime of the ionic chain carriers was calculated to be of the order of several milliseconds. The authors underlined that in all experiments less than 10% of the added acid was consumed in the initiation reaction by the time all the styrene had been polymerised. Unfortunately, no attempt was made to detect the presence of *free* acid during these reactions. We believe that at least part of the acid not accounted for by the formation of carbenium ions was consumed to give the 1-phenylethyl triflate and to homoconjugate with the triflate anion formed in the protonation reaction. Under the conditions of this work, particularly the very high acid concentrations, the predominance of ionic chain carriers is a normal feature. The authors are justified in considering the mechanism a straightforward one. Indeed, any ester contribution in their system would have been entirely masked by the very high activity of the ionic species. However, their comments on previous studies on the same system, but under different experimental conditions, neglected that the more complex behaviour (Chmelir's work) and the much lower k_p's (Higashimura's work) arose from the concomitant contribution of ester and ionic chain carriers, the former predominating in media of low dielectric constant.

In their second paper[916] Kunitake and Takarabe tackled the delicate problem of determining the propagation rate constants of free ions and ion pairs, by studying the influence of added tetrabutylammonium triflate on the kinetics of polymerisation and estimating the values of the equilibrium constants relating free and paired species. The salt concentrations used were at least one order of magnitude lower than those of the acid, so that homoconjugation problems can be ignored here. It was found that k_p^+ was only 6 to 24 times higher than $k_p^{\pm}$ between 0 and 30°C, and $E_p^+ = 14$ kcal mole^{-1} and $E_p^{\pm} = 5$ kcal mole^{-1}. The relatively small difference in activity between free ions and pairs confirms previous results obtained by Lorimer and Pepper[202] and Sigwalt's group[671]. It seems that in the cationic polymerisation of alkenyl monomers the propagating capability of ionic species is high, but the difference in reactivity between free ions and ion pairs is modest, in contrast to most anionic polymerisations. An explanation must await a better definition of the intimate nature of these ionic species in terms of degree of solvent stabilisation, etc. The fairly high activation energy for free ion propagation obtained by Kunitake and Takarabe is surprising in view of the results obtained in radiation-induced polymerisations where this parameter for bare cations is always much lower, and often close to zero. This comparison is valid, since the values of the propagation rate constants are very similar in the two types of experiments performed on the same monomer[916].

Sawamoto et al.[917] studied the spectroscopic stop-flow behaviour of the polymerisation of styrene and some of its derivatives by triflic acid in ethylene chloride at 30°C. They obtained overall values of the propagation rate constant of about $1 \times 10^5 M^{-1}s^{-1}$ for styrene, p-chlorostyrene and p-methylstyrene. As previously in the same laboratory with p-methoxystyrene, "invisible species" were held responsible for some or most of the propagation, particularly in media of low dielectric constant, i.e. pseudocationic polymerisation occurs in all these systems and predominates when the polarity of the solvent does not favour ion formation.

Both investigations[915-917], as several previous ones, assumed that the carbenium ions derived from styrene and its methyl and chloro derivatives possess an extinction coefficient of $10^4 M^{-1}cm^{-1}$. It would certainly make these studies more complete if those coefficients were determined for each carbocation in the media used for the polymerisation studies.

Chmelir et al.[918,919] continued their work on the system styrene – triflic acid – CH_2Cl_2. In a first communication[918] they reach the now familiar conclusion that "nonionic growth processes" accompany and sometime dominate over ionic ones. This recognition of pseudocationic polymerisation reflects a general trend, particularly manifest at the Kyoto meeting. In the second communication[919] the authors studied the conductivity of the system $EtPh-CF_3SO_3H-CH_2Cl_2$ and found that the presence of ethylbenzene enhances the amount of ionic species present in the acid solution. They attributed this to the protonation of the aromatic component but did not check spectroscopically the presence of the corresponding carbocations. Their conclusions rejecting the validity of Higashimura's interpretation of the effect of the dielectric constant on the behaviour of the polymerisation of styrene[313] is unfounded, since the latter work was conducted at *constant styrene concentration* and with added CCl_4, a compound which cannot be protonated. The second paper by Chmelir et al.[919] merely shows that large amounts of ethylbenzene induce a modest increase in conductivity which could be due to nucleophilic impurities present in that additive.

Hasegawa and Higashimura[920] showed that perfluorinated "resinsulphonic acid", Nafion-H, is an effective catalyst for the oligomerisation of styrene.

III-E-12-b) and c)

Sigwalt and Sauvet[921] used ^1H and ^{19}F NMR spectroscopy to show that the triflate anion homoconjugates with one and two molecules of parent acid. In particular, they detected this in the protonation of 1,1-diphenylethylene and 3-phenylindene. This study confirms the necessity of considering homoconjugation as an integral part of the phenomenology of cationic polymerisation.

III-E-12-d)

Sawamoto et al.[922] applied a fast conductimetric stop-flow technique (about 1 ms time resolution) to the study of the initiation reaction of *p*-methoxystyrene by various initiators including friflic acid. Within 20 ms following the mixing of the reactants in ethylene chloride at 30°C the conductivity increased to an asymptotic value, then it slowly decreased. The initial increase was attributed to the formation of ionic chain carriers and since it apparently obeyed a semilogarithmic law (excess of monomer), the pseudofirst order constants obtained were equated to the product $k_i[M]_0$. The values of k_i thus obtained agreed closely with those previously determined by a spectroscopic technique[205,891]. However, the kinetic analysis on the experimental conductivity-time traces (such as the one published) does not allow a clear-cut determination of the order in view of the fairly poor signal-to-noise ratio. Moreover, several assumptions were implicitly made to arrive at the k_i values: (i) the ratio of free ions to ion pairs remains constant during the initiation reaction (as the concentration of ionic species increases this ratio can vary appreciably); (ii) the acid is only consumed in the ionogenic protonation reaction (if however esterification takes place concurrently, the acid concentration will decrease following a law different from that postulated); (iii) the conductivity changes due to the acid self dissociation was ignored.

The authors did not analyze the exponential decay of the conductivity following the fast rise. Those traces were much less marred by noise and could provide useful information on the termination processes.

III-E-14-a)

Cerrai and coworkers[923] confirmed the presence of ionic chain carriers in the polymerisation of anethole by iodine. They detected a maximum at 385 nm in the uv spectrum of polymerising solutions, similar to the behaviour of p-methoxystyrene with the same initiator[89].

Higashimura et al.[924] have studied the polymerisation of N-vinylcarbazole by iodine in chlorinated hydrocarbons. They found that the polymer DP's increased during the reactions and increased further after a second addition of monomer. This "living" behaviour was particularly evident at low temperature, since the DP's obtained were consistent with an initiation involving one iodine molecule *per* macromolecule formed, i.e. no transfer and no termination. This system looks very promising.

IV-B-2

Cerrai and Tricoli[925] examined the polymerisation of anethole by $BF \cdot Et_2O$ in ethylene chloride under the influence of an electric field. They were able to characterise spectroscopically both the carbocation derived from the monomer at 385 nm (sensitive to the presence of the field as already reported[923]) and a catalyst-monomer (or polymer) complex at 333 nm, not affected by the applied field.

IV-B-4-c)

Grattan and Plesch[926] published an important paper on the polymerisation of isobutene by $AlCl_3$ and $AlBr_3$ using conductimetric and radiotracer techniques. Their results clearly showed that direct initiation in these systems is caused by the AlX_2^+ cation formed in the self-ionisation of the Lewis acid in alkyl halides. They also referred to a recent interesting study by Brueggeller and Mayer[927] concerning the formation and NMR characterisation of the t-butyl cation from t-butyl bromide and excess aluminium bromide in brominated hydrocarbons and of dimethylhalonium ions from CH_3X and aluminium bromide. Grattan and Plesch discussed in their paper the alternative initiation routes and concluded that their findings rule out such pathways as the Hunter-Yohé mechanism. While agreeing with these authors in the context of aluminium-based Lewis acids, we feel that their similar reasoning concerning $TiCl_4$ is weak. In fact, the evidence given in Sect. IV-B-4-b) strongly suggests that with this Lewis acid direct acid-base initiation to give the corresponding zwitterion can take place.

IV-C-2-a—c)

Marek and collaborators[928] studied the apparent cocatalytic effect of HF in the polymerisation of isobutene initiated by various chlorine-based Lewis acids. It was shown that HF readily exchanged halogenes with the catalysts to give more active mixed-halogen Lewis acids. In the same paper HCl was shown not to be a cocatalyst with $TiCl_4$ (heptane, $-20°C$), just like HCl and HF with TiF_4 in the same conditions. These results, together with other similar work by the same group on mixed metal halides[407,498,499] suggest that direct initiation following self-ionisation is important in these systems and favoured by the halide-exchange reactions.

IV-C-3-e)

Sangalov et al.[929] examined the interaction of water with chloroalkyl aluminium compounds in the presence and absence of isobutene. A 1 : 1 complex between $EtAlCl_2$ and water was observed and claimed to possess no Brønsted acidity and only slight Lewis acidity. The transformation of this complex into an "alumoxane aquacomplex" was inferred from a kinetic analysis of its polymerisation of isobutene. The conclusion that the effect of water is different when isobutene is present in the reaction mixture is interesting. More research is needed to settle these odd chemical problems.

IV-C-4-c)

Guenzet and Camps[930] reexamined in detail their results on the esterification of olefins by carboxylic acids in the presence of stannic chloride. In the case of AcOH a 4:1 complex with the Lewis acid was held responsible for the formation of (mainly) acetate and chloride, the latter product leaving the species $SnCl_3CH_3COO \cdot (CH_3COOH)_3$. The collapse of the ion pair to give a neutral species with the anion derived from the Lewis acid must be viewed as a typical termination reaction in the context of cationic polymerisation. Guenzet and Camps also showed by kinetic isotope studies that the protonation is the rate-determining step in these processes and wrote:

$$SnCl_4(CH_3COOH)_4 + \overset{\backslash}{\underset{/}{C}} = \overset{/}{\underset{\backslash}{C}} \overset{slow}{\longrightarrow} H - \overset{|}{\underset{|}{C}} - \overset{|}{\underset{|}{C}}{}^+ \; SnCl_4 \cdot (AcOH)_3 \cdot AcO^-$$

without being more explicit about the mechanism of this interaction. We think that the concerted attack of the two molecules of protonated AcOH onto the olefin could give an Ad_E3-type intermediate, precursor to the carbenium ion. The rapid collapse of the latter due to its reaction with the acetate or chloride moiety of the complex anion $SnCl_4(CH_3COO)_2^{--}$ would then give the observed products (see p. 162).

IV-C-5-b)

The cationic oligomerisation of butene-1 was studied in Cesca's laboratory[931] with different catalyst-cocatalyst combinations or mixtures of Lewis acids. The mixture $EtAlCl_2$ + $SnCl_4$ proved more effective (97% yield in n-heptane at 20°C) than $AlCl_3$ + $TiCl_4$ (60%, same conditions). The combination $EtAlCl_2$–t-butyl chloride also gave high yields, while Et_2AlCl–i-propyl chloride was a poor initiator. Aluminium chloride "alone" was only moderately active and neither $EtAlCl_2$ nor Et_2AlCl initiated the process in the absence of a copromoter. The authors rationalised these results in terms of chloride ion exchange between the organoaluminium chlorides and the alkyl chloride to give $AlCl_3$ which, generated in situ, is more active than when added as a reagent. These conclusions reflect closely our own interpretation of the phenomenology of polymerisations involving alkylaluminium halides and alkyl or aralkyl halides expounded throughout Sect. IV-C-5.

Further support for these mechanistic views comes from a paper which had escaped our attention and which Prof. Cesca kindly provided us with in an Italian translation. In 1971 Vilim et al.[932] studied the oligomerisation of propene at 50°C in n-heptane catalysed by organoaluminium chlorides in conjunction with t-butyl chloride. They showed that maximum rates were obtained in the following conditions: $[t\text{-BuCl}]:[Et_3Al] = 4$, $[t\text{-BuCl}]:[Et_2AlCl] = 3$ and $[t\text{-BuCl}]:[EtAlCl_2] = 2$. These results were interpreted according to the following general mechanism:

$$Et_x AlCl_{3-x} + x \, t\text{-BuCl} \rightarrow AlCl_3 + x \, t\text{-BuEt}$$

$$AlCl_3 + t\text{-BuCl} \rightleftharpoons t\text{-Bu}^+ \, AlCl_4^-,$$

i.e. chloride ion exchange to produce nascent (and soluble) aluminium chloride followed by ionisation of the cocatalyst. While the first step was clearly proved in this paper, the formation of active species could have followed another pathway, viz. direct initiation after the self-ionisation of $AlCl_3$. However, given the high temperature used, the low polarity of the medium and the unspecified moisture level, no definite conclusion can be reached on this latter point.

Prof. Cesca also reminded us of an investigation carried out in his laboratory[933] concerning the chloride exchange reaction in the systems $Et_{3-x}AlCl_x (x = 0,1)/t\text{-BuCl}$ and the use of the resulting products for the oligomerisation of ethylene. Quantitative yields of precipitated aluminium chloride were obtained in methylene chloride at $-78°C$ using stoicheiometric amounts of $t\text{-BuCl}$, but in non-polar solvents this reaction was incomplete. It was also noticed that $AlCl_3$ and $t\text{-BuCl}$ formed a $1:1$ crystalline complex. The authors stated that an equimolar mixture of Et_2AlCl and $t\text{-BuCl}$ did not produce any $AlCl_3$. We think that although no precipitation of this product was observed, some might have been formed and kept in solution thanks to the presence of ethylene formed in the reaction mixture from hydridation processes. Indeed, $AlCl_3$ can be solubilised by complexation with olefins[445]. The main product of the $1:1$ interaction of Et_2AlCl with $t\text{-BuCl}$ was $C_6H_{13}Cl$ even in the absence of *added* ethylene; the same product was formed when $AlCl_3 \cdot t\text{-BuCl}$ was made to react with ethylene. This seems a strong indication that some $AlCl_3$ was indeed generated in the $1:1$ interaction between Et_2AlCl and $t\text{-BuCl}$.

Reibel et al.[934] investigated the polymerisation of isobutene promoted by the combination Et_3Al–p-methylbenzyl chloride (molar ratio $20:1$) in methylene chloride and in mixtures CH_2Cl_2–n-pentane. Unexpectedly low conversions were obtained ($<20\%$) despite the relatively high Et_3Al concentration (0.2 M). We think the generation of $EtAlCl_2$ and/or $AlCl_3$ must have been very slow so that the build-up of these catalytic substances was insufficient to give high yields. Accordingly a decrease in the polarity of the medium also reduced the poly(isobutene) yields, due to an even slower rate of chloride-ion exchange. Obviously this interpretation does not exclude that some of the products obtained when $2,4,4$-trimethylpentene-1 was used, instead of isobutene, could have arisen from the p-methylbenzylic cation formed through heterolytic ionisation of the cocatalyst by Et_3Al.

Sivaram[936] reviewed the role of organoaluminium compounds in the field of cationic polymerisation. Apart from the section devoted to the use of alkyl and aralkyl halides as cocatalysts, already amply discussed here, we comment on other active combinations mentioned by the author. In the case of Et_2AlCl–RSO_2Cl Sivaram claimed that its activity for the polymerisation of isobutene was due to the RSO_2^+ ion formed through the classical Kennedy initiation scheme. When organic acid halides were used as cocatalysts Sivaram speculated that their efficiency was due to the formation of acylium ions capable of initiating the polymerisation of isobutene. Our views on the above interpretations are in line with the general mechanism proposed throughout this section, namely halide exchange to give $EtAlCl_2$ and $AlCl_3$ which are the real catalysts. Sivaram's proposal of isobutene initiation by acylium ions is questionable, since it is known that iso-butene is not polymerised by CH_3CO^+, but gives instead a triethylpyrilium salt by ac-ylation[936,937]. Sivaram et al.[938] reported in the same vein that styrene and α-methyl-styrene are polymerised by the combination $Et_2AlCl-POCl_3$ through the formation of the initiating species $POCl_2^+$ and quoted phosphorus incorporation in the polymers as supporting evidence for that mechanism. Here too, halide exchange seems to us a more plausible interpretation of the observations. Phosphorus incorporation is better rationa-lised by electrophilic substitution of $POCl_3$ on the aromatic rings of the polymers. In conclusion, Sivaram's claims[935,938] about the formation and initiating capability of such ions as $POCl_2^+$, RSO_2^+, RCO^+ must await further evidence and in particular the structural verification of the initial addition compounds, preferably conducted on mod-el olefins.

IV-C-5-e)

Recent developments in Kennedy's laboratory on the use of boron chloride in conjunc-tion with alkyl and aralkyl halides for the synthesis of polymers with predetermined structure were discussed by that author at two recent meetings[939,940]. Kennedy also kindly provided us with manuscripts of forthcoming papers which are discussed below.

Kennedy and Smith[941] synthesised and characterised an α, ω-di(t-chloro)polyiso-butene using BCl_3 with p-dicumyl chloride in methylene chloride. One fragment of the bifunctional cocatalyst was found in each polymer molecule as measured by double-detection (uv and refractive index) GPC analysis. This result proved that initiation had taken place according to the following mechanism at each C-Cl bond:

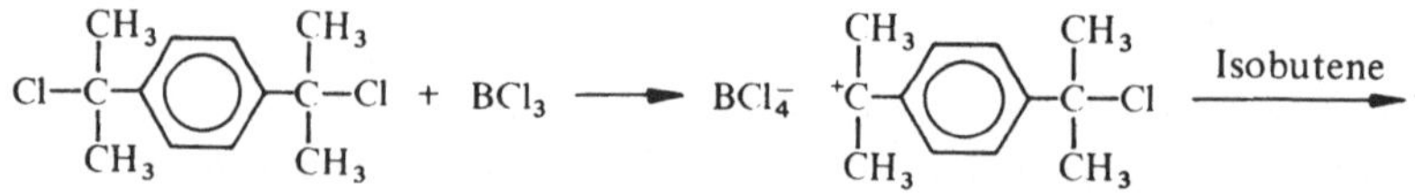

On the other hand, ^1H NMR spectroscopy focussed at 1.65 ppm (protons belonging to geminal methyl groups next to the terminal tertiary chlorines) provided evidence for the presence of two Cl atoms per macromolecule, i.e. for termination according to the reac-tion below, applied to both active ends:

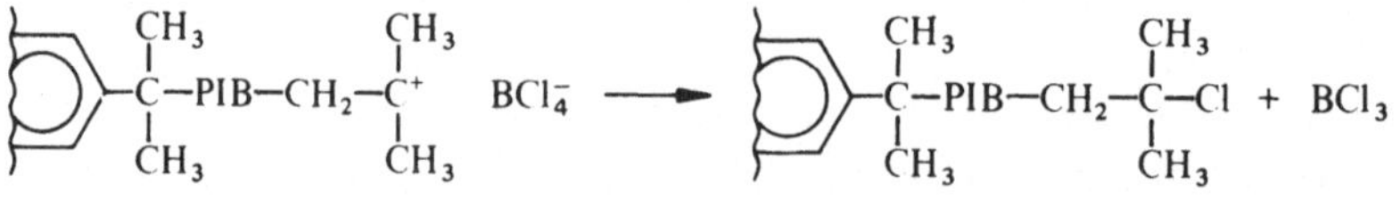

This work is probably the best demonstration of the clean and straightforward behav-iour of boron chloride in both initiation and termination reactions. Indeed, the phenyl content per polymer molecule was unity $\pm 10\%$. Also, an interesting test confirmed the validity of the mechanism proposed: a plot of $2[M]_0/\overline{DP}_n$ vs. the cocatalyst concentra-tion [I] gave a straight line as predicted by the equation

$$2[M]_0/\overline{DP}_n = k_{trI}[I]/k_p + k_t/k_p$$

elaborated from Kennedy's mechanism. The authors did not look into the possibility of termination by indanylic cyclisation, viz.:

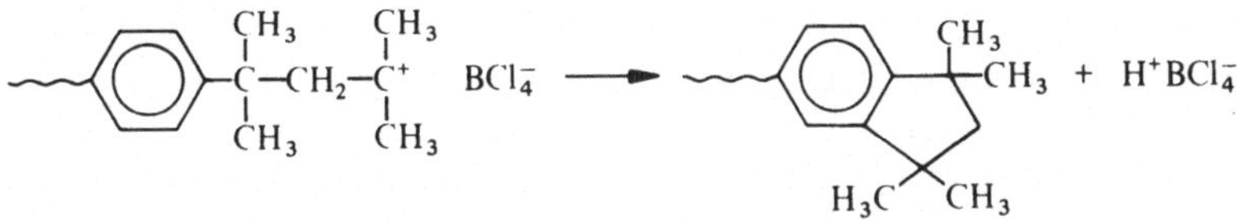

but probably this reaction is slower in these systems than in those involving aromatic monomers with conventional cationic initiators.

The evidence presented in this work is much more convincing than that offered in the context of chloroalkyl aluminium initiators. The chemistry of BCl_3-RCl combinations represents a major success of the Akron group.

V-A-2-a)

Velichkova et al.[942] studied the photolyses of various trityl salts ($\sim 10^{-3}$ M in CH_2Cl_2) by visible light and identified a number of products including triphenylmethane, fluorenone and phenylfluorenones. With anions containing chlorine, e.g. $SbCl_6^-$, triphenylmethyl chloride was a major product.

V-A-3-b)

Heublein and collaborators[943] claimed in a recent paper that ethylene chloride solutions of many trityl salts derived from Lewis acids were unstable due to the reaction

$$Ph_3C^+MtX_{n+1}^- \xrightarrow{k_z} Ph_3CX + MtX_n$$

and measured k_z spectroscopically and conductimetrically. The origin of that decay reaction is not clear, since a large number of the salts investigated has been found to be stable (i.e. fully ionised) in solution if kept away from moisture and light, preferably in vacuo (see Sect. V-A). We suspect that impurities or visible light might have caused the above interaction. The authors reported that k_z decreased with increasing salt concentration, which seems to corroborate that an impurity causes the decay. Water was found to have little effect on the first-order rate constant k_z up to concentrations of about 10^{-4} M against a salt concentration about three times smaller. This is surprising and in contradiction with previous observations pointing to the poisoning effect of moisture as a nucleophile for the trityl ion. Indeed, a technique has been developped to determine residual water in "dry" systems by using these salts (see Sect. V-A-1-a)).

V-B

Pask and Plesch[944] recently confirmed that the trityl ion is incapable of initiating the polymerisation of isobutene (counterion SbF_6^-). They found however that $(p\text{-ClPh})_3C^+$ SbF_6^- and $Ph_2HC^+SbF_6^-$ are active towards that monomer as predicted by thermodynamical considerations.

V-D-1

Hasegawa and Higashimura[945] reported an interesting application of their recent studies on the selective dimerisation of styrenes. By choosing the conditions which give an optimum yield of linear dimer, they polymerised divinylbenzene with acetyl perchlorate to a linear polymer with the structure

resulting from succesive protonation, dimerisation and proton expulsion cycles.

V-D-2

Boekhoff[946] polymerised styrene with $CH_3CO^+SbF_6^-$ and $(CH_3CO)_3O^+SbF_6^-$ and claimed without adequate proof that in solvent acetic anhydride oligomers were formed possessing a telechelic structure, viz. a CH_3CO group at one end and a CH_3COO one at the other.

Plesch et al.[947] have studied the kinetics of acenaphthylene polymerisation by $EtCO^+SbF_6^-$ and $EtCO^+PF_6^-$ and calculated a propagation rate constant of 23 $M^{-1}s^{-1}$ in nitrobenzene at 25°C with both initiators, which suggests that free ions are the main chain carriers.

VI-A

Tabata and collaborators[948,949] reported more information on the spectra and reactivity of the cations and cation-radicals formed as intermediate in the radiation-induced polymerisation of N-vinylcarbazole.

Deffieux et al.[950] studied the radiation-induced polymerisation of p-methoxystyrene in bulk and in methylene chloride. The values of $k_p = 3 \times 10^6$ $M^{-1}s^{-1}$ and $E_p \sim 0$ were obtained for the bulk process.

VIII-A

Yamamoto et al.[951] pursued their studies on the photoinduced dimerisation and polymerisation of olefins in the presence of electron acceptors. A mechanism was discussed involving the excitation of the EDA complex to give the olefin cation-radical which can then undergo various reaction pathways with itself and with the monomer.

Marek and Toman[952] published a new brief study of the system VCl_4-isobutene illuminated by visible monochromatic light (578 and 589 nm) not absorbed by the Lewis acid on its own. Polymerisations occurred in bulk only when the system was illuminated. The authors concluded that initiation must have taken place following the exitation of the monomer-catalyst charge-transfer complex to give the isobutene radical-cation.

VIII-C

Ledwith[953] expounded at the Mainz meeting his new ideas and results concerning the photolysis of free-radical sources in the presence of oxidising agents to generate initiators for cationic polymerisation.

Kinstle and Tufts[954)] irradiated 1-iodocyclohexene with a medium-pressure mercury arc in the presence of cyclohexene oxide. Low yields of polymer were obtained and their formation attributed to the cyclohexenyl cation generated in the photolysis of the parent iodide.

Smets[955)] reviewed the topic of photoinitiated cationic polymerisation at the Kyoto meeting and described some novel interesting work from his own laboratory. Diazonium salts were used to photoinitiate the polymerisation of N-vinylcarbazole or to cross-link linear copolymers of methyl methacrylate and glycidyl methacrylate through the epoxy groups of the second comonomer.

X-A

Giusti[956)] reported at Kyoto further work on the synthesis of block copolymers of styrene and isobutene. The general technique consists in preparing a solution of one of these monomers in a mixture of methylene chloride and n-heptane at $-50°C$ and then adding the catalyst combination $Et_2AlCl–Cl_2$. Complete conversion is reached within about one minute and the second monomer is then added. A high yield of block copolymer is obtained. The success of these experiments seems to be due to a careful adjustment of the experimental parameters through mathematical modelling. The near "living" character of these processes is thus optimised. Three-block copolymers were also prepared following the addition sequence styrene-isobutene-styrene. Full details of this work are eagerly awaited.

Various block copolymerisations have been described by Kennedy and collaborators[939–941)] following the successsful preparation of telechelic isobutene polymers. Thus, Kennedy and Smith[941)] used an α, ω-di(t-chloro)polyisobutene as starting material for the preparation of triblock copolymer with two poly(α-methylstyrene) blocks attached to the central poly(isobutene) chain. Initiation from the difunctional polymer was achieved with Et_2AlCl. Following selective extractions on the product, no detectable amounts of homopoly(isobutene) was obtained, and less than 7% homopoly(α-methylstyrene) was isolated. Molecular weight determinations seemed to confirm the minimal occurrence of homopolymerisation processes. It would be important to show that blocking in the presence of added poly(α-methylstyrene) would not change the features of the process, i.e. the added polymer should be recovered as such after extraction. We believe that this test is essential to prove that initiation from the telechelic poly(isobutene) proceeded via the cationogenic reaction induced by Et_2AlCl. Otherwise the blocking reaction might have occurred after the homopolymerisation of α-methylstyrene through a polymer-polymer interaction.

Hallensleben and Möller[957)] prepared block copolymers of styrene and polyesters by mixing the latter with monomeric styrene in methylene chloride and adding $BF_3 \cdot Et_2O$. The authors claimed that growing polystyrene chains transferred to the polyesters to give dialkoxycarbenium ions capable of initiating the polymerisation of styrene to give block copolymers.

X-C

Ambrose and Newell[958)] studied the synthesis of graft copolymers prepared from chlorinated poly(butadiene) and α-methylstyrene in toluene. They reported that Me_3Al was the best initiator for this process, giving a high grafting efficiency even at high α-methyl-styrene conversions. With Et_2AlCl the grafting efficiency did not vary appreciably with conversion and remained at about 65%, but the reaction was very fast and difficult to control. Et_3Al was found to be a poor initiator. Unfortunately the authors did not report on the purification of the various catalysts used, and it seems that in fact they used them as received. In the case of Et_2AlCl it can be argued therefore that low concentrations of $EtAlCl_2$ impurities could have induced the polymerisation of α-methyl-styrene even in a low-polarity medium. The study of the effect of water on the system involving Me_3Al is interesting. Control experiments showed that α-methylstyrene did not polymerise with Me_3Al alone or with added amounts of water. These results confirm that water is not a cocatalyst for trialkyl aluminium compounds in non-polar solvents. In the presence of chlorinated butadiene the grafting rate of α-methylstyrene by Me_3Al was strongly enhanced by water additions up to $[Me_3Al]:[H_2O] \sim 10$. The above evidence strongly suggests that in the *absence* of water no grafting would have occurred, i.e. no ionisation of the C–Cl bonds on the chlorinated poly(butadiene) in toluene. The effect of water can be rationalised in terms of an enhanced acidity of the interaction products between water and Me_3Al. These more acidic species would promote the chloride ion exchange between polymer and catalyst giving e.g. $MeAlCl_2$ which can initiate directly the polymerisation of α-methylstyrene. The growing chains of this monomer would then graft *onto* polybutadiene. This alternative interpretation is in line with our ideas on initiation with organoaluminium halides in the presence of organic halides (see Sect. IV-C-5). The fact that Et_3Al was found to initiate the grafting with very low efficiency simply means that the exchange reactions giving chlorinated ethylaluminium derivatives were too slow in toluene, even in the presence of traces of moisture.

Denes et al.[959] recently reported the grafting of surface-chlorinated polypropylene fibers with styrene in the presence of Et_2AlCl. The extent of grafting was very low, but apparently the polystyrene chains covered the polypropylene fibers quite effectively.

XI-A

As mentioned in the main text, we have been carrying out a detailed investigation of the initiating capabilities of a variety of metal salts, in particular perchlorates and trifluoromethanesulphonates (triflates). The first part of this study was devoted to heterogeneous systems[960,961], i.e. systems in which the initiator is insoluble in the reaction medium. Isobutene was chosen as test monomer since it only polymerises via a cationic mechanism, but several other monomers and model olefins were also used. Reactions were conducted, after a thorough baking of the salts *in vacuo*, by adding the undiluted monomer or its solutions in methylene chloride or hexane to the salt grains under vigorous stirring. The following salts showed efficiencies which followed the order:

$$Co(ClO_4)_2 \geqslant Al(ClO_4)_3 \sim Ni(ClO_4)_2 \geqslant Al(CF_3SO_3)_3 > Ga(CF_3SO_3)_3 > Mg(ClO_4)_2.$$

Polymerisations with the most active salts were extremely fast. Moisture was found to be detrimental to these processes and traces of alcohols and amines inhibited them completely. Tests with hindered pyridines proved that the active sites on the salt surface possessed a Lewis character, most probably due to exposed metal cations. On the basis of all the experimental evidence gathered we proposed an initiation mechanism based on the following scheme:

where Mt^+ is a metal cation and B^- a perchlorate or triflate anion, both exposed on the salt surface. This direct coordinated initiation mechanism probably possesses wider implications.

Our investigation on the activity of metal salts as initiators of cationic polymerisation has recently continued with homogeneous systems[962]. Solvents such as sulphur dioxide, the nitroalkanes and acetonitrile dissolved sufficient amounts of the salts to allow a study of these processes in the absence of solid catalyst. Also, some monomers (dioxolane, THF, etc.) were found to be good solvents for the salts. Reactions were carried out under high vacuum conditions and with carefully purified and dried reagents. Only the salts which had exhibited the highest activities in heterogeneous systems were studied. They all acted as very efficient initiators in solution and it was shown that these reactions were not due to acidic impurities in the catalysts. Aluminium triflate ($AlTf_3$) was used in spectroscopic tests to measure its capacity of generating carbenium ions with olefins. In reactions with an excess of 3-phenylindene, an olefin incapable of polymerising because of steric hindrance, the concentration of carbenium ions formed reached about 50% of the initial concentration of $AlTf_3$. These results and other considerations suggest the following initiation mechanism:

$$2\ AlTf_3 \rightleftharpoons Tf_2Al^+ + AlTf_4^-$$

$$Tf_2Al^+ + \ \overset{\diagdown}{\underset{\diagup}{C}} = \overset{\diagup}{\underset{\diagdown}{C}} \ \rightarrow Tf_2Al-\overset{|}{\underset{|}{C}}-\overset{|}{\underset{|}{C}}{}^+$$

i.e. a direct initiation similar to that encountered with the aluminium halides (see Sect. IV-B-4-c)).

We also synthesised the salt $AlTf_2SbF_6$ and found its activity even more pronounced than that of the most efficient perchlorates. Almost explosive polymerisations were observed with most monomers tried in nitromethane. This salt does not need to "self-ionise" in solution since it is ionic in the solid state and can be considered the equivalent of such carbocationic salts as $Ph_3C^+SbF_6^-$ and tropylium hexafluorophosphate. We expect therefore that its initiation interaction with olefins follows a simple 1:1 addition.

Further work is in progress with these novel initiators, which should provide valuable fundamental information.

XI-E

Biswas and Maity[963] polymerised N-vinylcarbazole with molecular sieves and concluded that Brønsted sites on the catalyst were responsible for initiation when the sieves had been preheated at 200°C. With thermal treatments exceeding that temperature the rare-earth cations present in the sieves were considered as possible electrophilic initiators.

Kučera and collaborators have prepared stable silenium salts and used them as initiators for the polymerisation of styrene[964] and heterocyclic monomers[965].

Conclusions

Some of the very recent work reviewed confirms that many of the goals enumerated in the Introduction are now within reach. The best examples of qualitative advance are:

— The general consensus on the validity of pseudocationic polymerisation, attained through much experimental evidence extending the original findings to a variety of systems.
— The remarkable control that can be exercised on certain processes. Predetermined DP's, narrow DP distributions, and the synthesis of telechelic polymers and block copolymers are now possible in some instances, and living cationic systems are being perfected. These achievements close the gap between anionic and cationic polymerisation.
— The determination of reliable values of the initiation and propagation rate constants, including the specific contribution of free ions and ion pairs to chain growth.
— The understanding of direct-initation mechanisms. In this context, the use of metal salts of strong acids has opened a new area of fundamental studies. Indeed, we believe that the chemical insertion of the olefin between the metal cation and the anion following its adsorption on the salt surface closely resembles recent mechanistic proposals put forward by Zambelli and collaborators[966,967] to rationalise initiation in stereospecific polymerisation (propene with vanadium-based catalysts). We are inclined to conclude that the cationic nature of stereospecific polymerisations of the Ziegler-Natta type is supported by our findings[960−962], viz. electrophilic attack by a metal cation as the first phase of the monomer insertion process. This type of initiation gives a metal-to-carbon bond which is a basic moiety. The fact that CO_2 has been shown to react with Ziegler-Natta active centers is therefore not surprising and does not prove the anionic nature of those processes, since CO_2 would obviously insert into the Mt−C bond of our catalysts too. In heterogeneous processes characterised by rate-determining diffusion and complexation of the monomer on the active sites[968], monomer structure and reactivity play a much diminished role compared to the readiness to polymerise. Thus, ethylene will be adsorbed much more readily than other olefins and polymerise despite its reluctance to do so in cationic homogeneous systems.

XII. References

1. Plesch, P. H. (ed.): Cationic polymerisation and related complexes. Cambridge: Heffer 1953
2. Plesch, P. H. (ed.): The chemistry of cationic polymerisation. Oxford: Pergamon Press 1963
3. Kennedy, J. P.: Cationic polymerization of olefins: a critical inventory. New York: John Wiley 1975
4. Pepper, D. C.: Cationic polymerization. In: Friedel-Crafts and related reactions, Vol. 2 (Olah, G. A., ed.). New York: Interscience 1964, Ch. 30
5. Kennedy, J. P., Langer Jr., A. W.: Fortschr. Hochpolym.-Forsch. *3*, 508 (1964)
6. Eastham, A. M.: Cationic polymerization. In: Encyclopedia of polymer science and technology, Vol. 3. New York: Interscience 1965, p. 35
7. Colclough, R. O.: Chem. Ind. (London) *1966*, 1660
8. Plesch, P. H., Gandini, A.: New views in cationic polymerisation. In: SCI Monograph n° 20. London: Soc. Chem. Ind. 1966, p. 107
9. Plesch, P. H.: Pure Appl. Chem. *12*, 117 (1966)
10. Plesch, P. H.: Cationic polymerization. In: Progress in high polymers, Vol. 2 (Robb, J. C., Peaker, F. W., eds.). London: Iliffe Books 1968, p. 137
11. Higashimura, T.: Rate constants of elementary reactions in cationic polymerisation. In: Structure and mechanism in vinyl polymerization (Tsuruta, T., O'Driscoll, K. F., eds.). New York: Marcel Dekker 1969, p. 313
12. Zlàmal, Z.: Mechanisms of cationic polymerization. In: Vinyl polymerization, Part 2 (Ham, G. E., ed.). New York: Marcel Dekker 1969, p. 231
13. Tsukamoto, A., Vogl, O.: Cationic polymerisation. In: Progress in polymer science, Vol. 3 (Jenkins, A. D., ed.). New York: Pergamon Press 1971, p. 199
14. Plesch, P. H.: Adv. Polym. Sci. *8*, 137 (1971)
15. Plesch, P. H.: Brit. Polym. J. *5*, 1 (1973)
16. Ledwith, A., Sherrington, D. C.: Cationic polymerisation. In: Reactivity, mechanism and structure in polymer chemistry. (Jenkins, A. D., Ledwith, A., eds.). London: John Wiley 1974, p. 244
17. Pepper, D. C.: J. Polym. Sci., Polym. Symposia *50*, 51 (1975)
18. Kubisa, P., Penczek, S.: Cationic polymerisation. In: Encyclopedia of polymer science and technology, Suppl. Vol. 2. New York: Interscience 1977, p. 161
19. Dunn, D. J.: The cationic polymerisation of vinyl monomers. In: Developments in polymerisation (Haward, R. N., ed.). Barking (U.K.): Applied Science 1979
20. Various Authors: J. Macromol. Sci., Chem. *A6*, 919 (1972)
21. Prepr. Intern. Symp. Cationic Polymerization. Rouen (France): Sept. 1973
22. Internat. Symp. Cationic Polymerization. Rouen (France): 1973. Makromol. Chem. *175* (4) (1974)
23. Fourth Internat. Symp. Cationic Polymerization. Akron (USA): 1976. J. Polym. Sci., Polym. Symposia *56* (1976)
24. Saegusa, T., Kobayashi, S.: Ref. [23], p. 241
25. Penczek, S., Matyjaszewski, K.: Ref. [23], p. 255
26. Goethals, E. J.: Ref. [23], p. 271
27. Yamashita, Y.: Ref. [23], p. 447
28. Plesch, P. H.: Ricerca Scient. (Milan), Sup. Simp. Macromol. *25*, 140 (1955)

29. Plesch, P. H.: Europ. Symp. Electric Phenomena in Polymer Sci. Pisa (Italy): March 1978. Main Lecture L 1

30. Cundall, R. B.: Ref. [2], Ch. 15, p. 549

31. Kennedy, J. P., Marechal, E.: Vinyl carbocationic polymerization. New York: Wiley, in press

32. Marechal, E.: J. Macromol. Sci.-Chem. *A7*, 433 (1973)

33. Heublein, G., Wondraczek, R.: Ref. [23], p. 359

34. Schildknecht, C. E.: Ind. Eng. Chem. *50*, 107 (1958)

35. Tsuruta, T.: Structure and reactivity of vinyl monomers. In: Structure and mechanism in vinyl polymerization. (Tsuruta, T., O'Driscoll, D. F., eds.). New York: Marcel Dekker 1969, p. 27

36. Engelbrecht, A., Rode, B. M.: Monatsh. Chem. *103*, 1315 (1972)

37. Fujinaga, T., Sakamoto, I.: J. Electroanal. Chem. *85*, 185 (1977)

38. Rode, B. M., Engelbrecht, A., Schantl, J.: Z. Phys. Chem. *253*, 17 (1973)

39. Bos, M., Dahmen, E. A. M. F.: Anal. Chim. Acta *63*, 185 (1973)

40. Coutagne, D.: Doctorate Thesis, Grenoble Univers. 1973, p. 190

41. Bolza, F., Treloar, F. E.: J. Chem. Eng. Data *17*, 197 (1972)

42. Kolthoff, I. M., Bruckenstein, S., Chantooni, M. K.: J. Am. Chem. Soc. *83*, 3927 (1961)

43. Jasinski, T., El-Harakany, A. A., Halaka, F. G., Sadek, H.: Croat. Chem. Acta *51*, 1 (1978)

44. Gandini, A., Plesch, P. H.: J. Chem. Soc. *B 1966*, 7

45. Gandini, A., Plesch, P. H.: Europ. Polym. J. *4*, 55 (1968)

46. Cardona, N., Schulz, G. V.: Makromol. Chem. *177*, 2797 (1976)

47. Sigwalt, P.: ibid. *175*, 1017 (1974)

48. Jensen, W. B.: Chem. Revs. *78*, 1 (1978)

49. Klopman, G.: J. Am. Chem. Soc. *90*, 223 (1968)

50. Gutman, V.: Coord. Chem. Rev. *18*, 225 (1976)

51. Satchell, D. P. N., Satchell, R. S.: Quart. Rev. *25*, 171 (1971)

52. Drago, R. S.: Structure and Bonding *15*, 73 (1973)

53. Marks, A. P., Drago, R. S.: Inorg. Chem. *15*, 1800 (1978)

54. Pearson, R. G.: Surv. Progr. Chem. *5*, 1 (1969) and refs. therein

55. Cook, D.: Can. J. Chem. *41*, 522 (1963)

56. Kuhn, S. J., McIntyre, J. S.: ibid. *43*, 375 (1965)

57. Tomita, A. et al.: J. Inorg. Nucl. Chem. *29*, 105 (1967)

58. Kuran, W., Pasynkiewicz, S., Florjanczyk, Z.: Makromol. Chem. *154*, 71 (1972)

59. Furukawa, J., Kobayashi, E., Nagata, S., Moritani, T.: J. Polym. Sci., Polym. Chem. Ed. *12*, 1799 (1974)

60. Satchell, D. P. N., Satchell, R. S.: Chem. Revs. *69*, 251 (1969)

61. Olah, G. A., Germain, A., White, A. M.: Acylium ions (acyl cations). In: Carbonium ions, Vol. 5 (Olah, G. A., von R. Schleyer, P., eds.). New York: Wiley Interscience 1976, Ch. 35

62. Olah, G. A.: Halonium ions. New York: Wiley-Interscience 1975

63. Olah, G. A.: Ref. [62], pp. 30–33

64. Foster, R. (ed.): Molecular complexes. London: Elek Science 1973 (Vol. 1) and 1974 (Vol. 2)

65. Foster, R. (ed.): Molecular association, Vol. 1. London: Academic Press 1975

66. Mulliken, R. S.: J. Chim. Phys. *61*, 20 (1964)

67. Panayotov, I. M., Heublein, G.: J. Macromol. Sci.-Chem. *A11*, 2065 (1977)

68. Heublein, G., Schubert, G., Hallpap, P.: J. Prakt. Chem. *320*, 291 (1978)

69. Weissberger, A., Proskauer, E. S.: Organic solvents. New York: Interscience 1955, p. 270

70. Matsuda, M.: Progr. Polym. Sci. Japan *2*, 49 (1971)

71. Burow, D. F.: Liquid sulfur dioxide. In: The chemistry of non-aqueous solvents (Lagowski, J. J., ed.). New York: Academic Press 1970, p. 138

72. Cheradame, H., Mazza, M., Hung, N. A., Sigwalt, P.: Europ. Polym. J. *9*, 375 (1973)

73. Natsuume, T. et al.: Chem. Comm. *1969*, 189

74. Matsuume, T. et al.: ibid. *1969*, 289

75. Hamid, M. A., Nowakowska, M., Plesch, P. H.: Makromol. Chem. *132*, 1 (1970)

76. Olah, G. A., Pittman Jr., C. U.: Adv. Phys. Org. Chem. *4*, 305 (1966)

77. Olah, G. A.: Angew. Chem., Int. Ed. *12*, 173 (1973) and refs. therein

78. Ahlberg, P., Engdahl, C.: Chem. Scripta *11*, 95 (1977)
79. Olah, G. A., von R. Schleyer, P. (eds.): Carbonium ions. New York: Wiley-Interscience 1968 (Vol. I), 1970 (Vol. II), 1972 (Vol. III), 1973 (Vol. IV), 1975 (Vol. V)
80. Olah, G. A.: J. Am. Chem. Soc. *94*, 808 (1972)
81. Dorfman, L. M., Sujdak, R. J., Bockrath, B.: Acc. Chem. Res. *9*, 352 (1976)
82. De Sorgo, M., Pepper, D. C., Szwarc, M.: Chem. Comm. *1973*, 419
83. Bertoli, V., Plesch, P. H.: J. Chem. Soc. B *1968*, 1500
84. Olah, G. A., Pittman, C. U., Symons, M. C. R.: Ref. [79], Vol. I, p. 173
85. Jones, R. L., Dorfman, L. M.: J. Am. Chem. Soc. *96*, 5715 (1974)
86. Bywater, S., Worsfold, D. J.: Can. J. Chem. *44*, 1671 (1966)
87. Kunitake, T., Takarabe, K.: Polym. J. *10*, 105 (1978)
88. Olah, G. A. et al.: J. Am. Chem. Soc. *88*, 1488 (1966)
89. Sawamoto, M., Higashimura, T.: Macromolecules *11*, 328 (1978)
90. Higashimura, T.: Polym. Prepr. *20*, 161 (1979)
91. Fleischfresser, B. E. et al.: J. Am. Chem. Soc. *90*, 2172 (1968)
92. Bywater, S., Worsfold, D. J.: Can. J. Chem. *55*, 85 (1977)
93. Cheradame, H., Hung, N. A., Sigwalt, P.: Ref. [23], p. 335
94. Hung, N. A. et al.: Tetrahedron *34*, 335 (1978)
95. Young, R. N.: J. Chem. Soc. *B 1969*, 896
96. De Mola, A. H. et al.: J. Polym. Sci. Polym. Chem. Ed. *16*, 761 (1978)
97. Bertoli, V., Plesch, P. H.: Spectrochim. Acta *25A*, 447 (1969)
98. Franklin, J. L.: Ref. [79], Vol. I, p. 77
99. Burstall, M. L., Treloar, F. E.: Ref. [2], p. 15
100. Lossing, F. P., Semeluk, G. P.: Can. J. Chem. *48*, 955 (1970)
101. Lossing, F. P., Maccoll, A.: ibid. *54*, 990 (1976)
102. Blint, R. J., McMahon, T. B., Beauchamp, J. L.: J. Am. Chem. Soc. *96*, 1269 (1974)
103. Staley, R. H., Wieting, R. D., Beauchamp, J. L.: ibid. *99*, 5964 (1977)
104. Hehre, W. J. et al.: ibid. *96*, 7162 (1974)
105. Arnett, E. M., Larsen, J. W.: Ref. [79], Vol. I, p. 441
106. Arnett, E. M., Petro, C.: J. Am. Chem. Soc.: *100*, 5402 (1978)
107. Arnett, E. M., Petro, C.: ibid. *100*, 5408 (1978)
108. Hatch, F., Munson, B.: J. Phys. Chem. *82*, 2362 (1978) and refs. therein
109. Attinà, M. et al.: Chem. Comm. *1978*, 938 and refs. therein
110. Sujdak, R. J., Jones, R. L., Dorfman, L. M.: J. Am. Chem. Soc. *98*, 4875 (1976)
111. DePalma, V. M., Wang Y., Dorfman, L. M.: ibid. *100*, 5416 (1978)
112. Dorfman, L. M., DePalma, V. M.: Pure Appl. Chem. *51*, 123 (1979)
113. Williams, F.: Principles of radiation-induced polymerisation. In: Fundamental processes in radiation chemistry (Ausloos, P., ed.). New York: Interscience 1978, p. 515
114. Wang, Y., Dorfman, L. M.: Macromolecules *13*, 63 (1980)
115. Kubisa, P., Penczek, S.: Makromol. Chem. *144*, 169 (1971)
116. Berman, E. L. et al.: Vysokomol. Soed. *B 20*, 546 (1978)
117. Goethals, E.: private communication
118. Pepper, D. C.: Ref. [22] p. 1077
119. Sawamoto, M., Higashimura, T.: Macromolecules *11*, 501 (1978)
120. Burton, H., Munday, D. A., Praill, P. F. G.: J. Chem. Soc. *1956*, 3933, and previous papers in that series
121. Praill, P. F. G.: ibid. *1957*, 3162
122. Radell, J., Connolly, J. W., Paymond, A. J.: J. Am. Chem. Soc. *83*, 3958 (1961)
123. Pocker, Y., Kevill, D. N.: ibid. *87*, 5060 (1965)
124. Hoffman, D. N.: J. Org. Chem. *36*, 1716 (1971)
125. Baum, K., Beard, C. D.: J. Am. Chem. Soc. *96*, 3233 (1974)
126. Baum, K.: J. Org. Chem. *41*, 1663 (1976)
127. Gandini, A., Plesch, P. H.: J. Chem. Soc. *1965*, 4826
128. Kagiya, T. et al.: J. Polym. Sci. *A-1 7*, 917 (1969)
129. Gramstad, T., Haszeldine, R. N.: J. Chem. Soc. *1957*, 4069

130. Olah, G. A., Nishimura, J.: J. Am. Chem. Soc. *96*, 2214 (1974)
131. Beard, C. D., Baum, K.: J. Org. Chem. *39*, 3875 (1974)
132. Howells, R. D., McCown, J. D.: Chem. Rev. *77*, 69 (1977)
133. Lemieux, R. U., Kondo, T.: Carbohydr. Res. *35*, C4 (1974)
134. Lee, C. C., Unger, D.: Can. J. Chem. *51*, 1494 (1973)
135. Souverain, D.: Thesis, Paris Univers. VI, 1978
136. Su, T. M., Sliwinski, W. F., von R. Schleyer, P.: J. Am. Chem. Soc. *91*, 5386 (1969), and refs. therein
137. Matyjaszewski, K., Penczek, S.: J. Polym. Sci. Polym. Chem. Ed. *12*, 1905 (1974)
138. Kobayashi, S., Danda, H., Saegusa, T.: Macromolecules *7*, 415 (1974)
139. Szwarc, M.: Carbanions, living polymers and electron transfer processes. New York: Interscience 1968
140. Hemery, P., Boileau, S., Sigwalt, P.: J. Polym. Sci. *C 52*, 189 (1975)
141. Boileau, S.: personal communication
142. Szwarc, M. in: Ions and ion pairs in organic reactions. Vol 2. (Szwarc, M., ed.). New York: Wiley-Interscience 1974, p. 376
143. Bard, A. J., Ledwith, A., Shine, H. J.: Adv. Phys. Org. Chem. *13*, 155 (1976)
144. Ledwith, A. et al.: Ref. [23] p. 483
145. Gandini, A. et al.: Chem. Ind. (London) *1965*, 1225
146. Biddulph, R. H., Plesch, P. H.: ibid. *1959*, 1482
147. Cheradame, H., Vairon, J. P., Sigwalt, P.: Europ. Polym. J. *4*, 13 (1968)
148. Sauvet, G., Vairon, J. P., Sigwalt, P.: J. Polym. Sci. Symp. *52*, 173 (1975)
149. Plesch, P. H.: Ref. [23], p. 373 and refs. quoted therein
150. Kabir-ud-din, Plesch, P. H.: J. Electroanal. Chem. *93*, 29 (1978)
151. Saegusa, T., Matsumoto, S.: J. Macromol. Sci.-Chem. *A4*, 873 (1970) and refs. therein
152. Brown, C. P., Mathieson, A. R.: J. Chem. Soc. *1957*, 3620
153. Slomkowski, S., Penczek, S.: Ref. [21], p. C 35
154. Rutherford, P. P.: Chem. Ind. (London) *1962*, 1614
155. de la Mare, P. B. D., Bolton, R.: Electrophilic additions to unsaturated systems. Amsterdam: Elsevier 1966
156. Nowlan, V. J., Tidwell, T. T.: Acc. Chem. Res. *10*, 252 (1977)
157. Brown, H. C., Okamoto, Y.: J. Am. Chem. Soc. *80*, 4979 (1958)
158. Schubert, W. M., Keeffe, J. R.: ibid. *94*, 559 (1972)
159. Knittel, P., Tidwell, T. T.: ibid. *99*, 3408 (1977)
160. Chwang, W. K., Nowlan, V. J., Tidwell, T. T.: ibid. *99*, 7233 (1977)
161. Fahey, R. C.: Topics Stereochem. *3*, 237 (1968)
162. Bolton, R. in: Comprehensive chemical kinetics, Vol. 9 (Bamford, C. H., Tipper, C. F. H., eds.). Amsterdam: Elsevier 1973, p. 1
163. Richardson, W. H., O'Neal, H. E. in: Comprehensive chemical kinetics, Vol. 5 (Bamford, C. H., Tipper, C. F. H., eds.). Amsterdam: Elsevier 1972, p. 381
164. Swinbourne, E. S.: Ref. [163], p. 149
165. Pocker, Y., Stevens, K. D., Champoux, J. J.: J. Am. Chem. Soc. *91*, 4199 (1969)
166. Pocker, Y., Stevens, K. D.: ibid. *91*, 4205 (1969)
167. Fahey, R. C., McPherson, C. A.: ibid. *91*, 3865 (1969)
168. Fahey, R. C., Monahan, M. W., McPherson, C. A.: ibid. *92*, 2810 (1970)
169. Fahey, R. C., Monahan, M. W.: ibid. *92*, 2816 (1970)
170. Fahey, R. C., McPherson, C. A.: ibid. *93*, 2445 (1971)
171. Fahey, R. C., McPherson, C. A., Smith, R. A.: ibid. *96*, 4534 (1974)
172. Pasto, D. J., Meyer, G. R., Lepeska, B.: ibid. *96*, 1858 (1974)
173. Brown, H. C., Brady, J. D.: ibid. *74*, 3570 (1952)
174. Cook, D., Lupien, Y., Schneider, W. G.: Can. J. Chem. *34*, 957 (1956)
175. Shilov, E. A., Mironova, D. F.: Proc. Acad. Sci. USSR Phys. Chem. Sec. *115*, 531 (1957)
176. Fahey, R. C., Lee, D. J.: J. Am. Chem. Soc. *89*, 2780 (1967)
177. Fahey, R. C., Payne, M. T., Lee, D. J.: J. Org. Chem. *39*, 1124 (1974)
178. Mollard, M. et al.: Bull. Soc. Chim. France *1966*, 83

179. Mollard, M. et al.: ibid. *1966*, 1186
180. Corriu, R., Guenzet, J.: Tetrahedron *26*, 671 (1970)
181. Corriu, R., Guenzet, J., Reye, C.: Bull. Soc. Chim. France *1970*, 1099
182. Guenzet, J., Toumi, M., Toumi, A.: Tetrahedron *30*, 159 (1974)
183. Guenzet, J., Toumi, A., Toumi, A.: Bull. Soc. Chim. France *1976*, 577
184. Roberts, R. M. G.: J. Chem. Soc. Perkin II *1976*, 1183
185. Gandini, A., Prieto, S.: to be published
186. Peterson, P. E., Allen, G.: J. Am. Chem. Soc. *85*, 3608 (1963)
187. Peterson, P. E. et al.: ibid. *87*, 5163 (1965)
188. Mason, T. J., Norman, R. O. C.: J. Chem. Soc. Perkin II *1973*, 1840
189. Brown, H. C., Liu, K. T.: J. Am. Chem. Soc. *97*, 2469 (1975)
190. Roberts, R. M. G.: J. Chem. Soc. Perkin II *1976*, 1374
191. Latrémouille, G. A., Eastham, A. M.: Can. J. Chem. *45*, 11 (1967)
192. Christian, S. D., Stevens, T. L.: J. Phys. Chem. *76*, 2039 (1972)
193. Fraenkel, G., Bartlett, P. D.: J. Am. Chem. Soc. *81*, 5582 (1959)
194. Sumrell, G. et al.: Can. J. Chem. *42*, 2710 (1964)
195. Ayres, R. L., Michejda, C. J., Rack, E. P.: J. Am. Chem. Soc. *93*, 1389 (1971)
196. Chao, C. S. et al.: Hua Hsueh *1965*, 95. Chem. Abstr. *65*, 3691 (1966)
197. Giusti, P., Puce, G., Andruzzi, F.: Makromol. Chem. *98*, 170 (1966)
198. Ledwith, A., Sherrington, D. C.: Polymer *12*, 344 (1971)
199. Johnson, A. F., Young, R. N.: Ref. [23], p. 211
200. Bernasconi, C. F., Boyle Jr., W. J.: J. Am. Chem. Soc. *96*, 6070 (1974)
201. Bulgrin, V. C., Lookhart, G. L.: ibid. *96*, 6077 (1974)
202. Lorimer, J. P., Pepper, D. C.: Proc. Roy. Soc. *A 351*, 551 (1976)
203. Kunitake, T., Takarabe, K.: Ref. [23], p. 33
204. Higashimura, T., Sawamoto, M.: Polym. Bull. *1*, 11 (1978)
205. Sawamoto, M., Higashimura, T.: Polym. Prepr. *20*, 727 (1979)
206. Benoit, R. L., Buisson, C.: Electrochim. Acta *18*, 105 (1973) and refs. therein
207. Talarmin, J., L'Her, M., Courtot-Coupez, J.: J. Chem. Res. (S) *1977*, 28; (M) *1977*, 327
208. Tokura, N., Shiina, K., Tearashima, T.: Bull. Chem. Soc. Japan *35*, 1986 (1962)
209. Tuck, D. G.: Progr. Inorg. Chem. *9*, 161 (1968)
210. Martin, J. S., Fujiwara, F. Y.: Can. J. Chem. *49*, 3071 (1971)
211. Fujiwara, F. W., Martin, J. S.: J. Chem. Phys. *56*, 4091 (1972)
212. Fujiwara, F. W., Martin, J. S.: J. Am. Chem. Soc. *96*, 7625 and 7632 (1974)
213. Mayo, F. R., Katz, J. J.: ibid. *69*, 1339 (1947)
214. Mayo, F. R., Savoy, M. G.: ibid. *69*, 1348 (1947)
215. Maass, O., Wright, C. H.: ibid. *46*, 2664 (1924)
216. Brooks, V. T.: ibid. *56*, 1998 (1934)
217. Sangalov, Y. A., et al.: Polym. Sci. USSR. *20*, 215 (1978)
218. Pepper, D. C., Somerfield, A. E.: Ref. [1], p. 75. Chem. Ind. (London) *1954*, 42
219. Tsuda, Y.: Makromol. Chem *36*, 102 (1960)
220. Asami, R., Tokura, N.: J. Polym. Sci. *42*, 553 (1960)
221. Tokura, N., Kawahara, T.: Bull. Chem. Soc. Japan *35*, 1902 (1962)
222. Giusti, P., Andruzzi, F.: J. Polym. Sci. *C 16*, 3797 (1968)
223. Pocker, Y. et al.: J. Am. Chem. Soc. *86*, 5011 and 5012 (1964)
224. Dewar, M. J. S., Fahey, R. C.: ibid. *85*, 2248 (1963)
225. Dewar, M. J. S., Fahey, R. C., ibid. *85*, 2245 (1963)
226. Giusti, P., Andruzzi, F.: Chim. Ind. (Milan) *48*, 1079 (1966)
227. Stobbe, H., Faerber, E.: Chem. Ber. *57B*, 1838 (1924); Dolinski, J., Dziewonski, K.: ibid. *48* 1917 (1915)
228. Andruzzi, F., Cerrai, P., Giusti, P.: Chim. Ind. (Milan) *52*, 466 (1970)
229. Schostakowski, M. F., et al.: Dokl. Acad. Nauk SSSR *173*, 593 (1967)
230. Pielichowski, J.: J. Polym. Sci. Symp. *42*, 451 (1973)
231. Meyers, R. A., Christman, E. M.: J. Polym. Sci. A-1 *6*, 945 (1968)
232. Biswas, M., Chakravarty, D.: J. Polym. Sci. Polym. Chem. Ed. *11*, 7 (1973)

233. Reeves, L. W., Schneider, W. G.: Trans. Faraday Soc. *54*, 314 (1958)
234. Speakman, J. C., Mills, H. H.: J. Chem. Soc. *1961*, 1164
235. Biswas, M., Kar, I.: Indian J. Chem. *5*, 119 (1967)
236. Scott, H., Konen, T. P., Labes, M. M.: Polym. Lett. *B 2*, 689 (1964)
237. Gandini, A., Prieto, S.: J. Polym. Sci. Polym. Lett. Ed. *15*, 337 (1977)
238. Brown, C. P., Mathieson, A. R.: J. Phys. Chem. *58*, 1057 (1954)
239. Reeves, L. W.: Can. J. Chem. *39*, 1711 (1961)
240. Brown, C. P., Mathieson, A. R.: J. Chem. Soc. *1958*, 3445
241. Gandini, A., Prieto, S.: 19th Canadian high polymer forum, 1977. To be published
242. Thyrion, F., Decroocq, D.: Comptes Rend. Acad. Sci. *260*, 2797 (1965)
243. Brown, C. P., Mathieson, A. R.: J. Chem. Soc. *1957*, 3608
244. Brown, C. P., Mathieson, A. R.: ibid. *1957*, 3612
245. Brown, C. P., Mathieson, A. R.: ibid. *1957*, 3625
246. Brown, C. P., Mathieson, A. R.: ibid. *1957*, 3631
247. Brown, C. P., Mathieson, A. R.: ibid. *1958*, 3507
248. Mathieson, A. R.: Ref. [2], p. 268
249. Blakely, C. F., et al.: J. Chem. Soc. *1961*, 1939
250. Blakely, C. F., Wassermann, A.: ibid. *1961*, 1946
251. French, P. V., Roubinek, L., Wassermann, A.: ibid. *1961*, 1953
252. Murphy, J., Roubinek, L., Wassermann, A.: ibid *1961*, 1964
253. Cooper, W.: Ref. [2], p. 358
254. Evans, A. G., Jones, N., Thomas, J. H.: J. Chem. Soc. *1955*, 1824
255. Evans, A. G., et al.: ibid. *1956*, 2757
256. Evans, A. G., Jones, P. M. S., Thomas, J. H.: ibid. *1957*, 104
257. Evans, A. G., Jones, P. M. S., Thomas, J. H.: ibid. *1958*, 4563
258. Leftin, H. P., Hall, W. K.: J. Phys. Chem. *66*, 1457 (1962)
259. Sauvet, G.: Doctorate thesis, Univers. Paris VI, 1974
260. Sakota, N., Nakamura, H., Nishihara, K.: Makromol. Chem. *129*, 56 (1969)
261. Murty, T. S. S. R., Pitzer, K. S.: J. Phys. Chem. *73*, 1426 (1969)
262. Kirszenbaum, M., Corset, J., Josien, M. L.: J. Phys. Chem. *75*, 1327 (1971) and following paper by Murty, T. S. S. R.
263. Klemperer, W., Pimentel, G. C.: J. Chem. Phys. *22*, 1399 (1954)
264. Fujiwara, H., Yoshida, N., Ikenoue, T.: Bull. Chem. Soc. Japan *48*, 1970 (1975)
265. Kreevoy, M. M., Liang, T., Chang, K. C.: J. Am. Chem. Soc. *99*, 5207 (1977)
266. Arnett, E. M., Chawla, B.: ibid. *100*, 217 (1978)
267. Samoilenko, A. A., et al.: Dokl. Akad. Nauk. SSSR *239*, 388 (1978)
268. Gandini, A., Martinez, R., Cheradame, H.: to be published
269. Armour, M., et al.: J. Polym. Sci. *A-1 5*, 1527 (1967)
270. Kresta, J., Livingston, H. K.: J. Polym. Sci. *B 8*, 795 (1970)
271. Throssell, J. J. et al.: J. Am. Chem. Soc. *78*, 1122 (1956)
272. Nikolayev, A. F., Belogorodskaya, K. V., Babushkina, I. M.: Zh. Prikl. Khim. *41*, 1815 (1968)
273. Nikolayev, A. F. et al.: Vysokomol. Soed. *B 12*, 24 (1970)
274. Sawamoto, M. et al.: Makromol. Chem. *178*, 389 (1977)
275. Gandini, A., Plesch, P. H.: unpublished
276. Bourne, E. J. et al.: J. Chem. Soc. *1958*, 3268
277. Belogorodskaya, K. V. et al.: Polym. Sci. USSR *18*, 11 (1976)
278. Nikolayev, A. F. et al.: J. Appl. Chem. USSR *48*, 1477 (1975)
279. Belogorodskaya, K. V. et al.: Polym. Sci. USSR *19*, 1499 (1977)
280. Alvarez, R., Gandini, A., Martinez, R.: J. Polym. Sci. Polym. Lett. Ed. *13*, 385 (1975)
281. Gandini, A., Martinez, R.: Ref. [23], p. 79
282. Martinez, R.: Doctorate thesis, Univers. Havana 1975
283. Gandini, A., Martinez, R.: to be published
284. Fontana, C. M.: Ref. [2] p. 213
285. Faithful, B. D. et al.: J. Chem. Soc. *A 1966*, 1185
286. Tazuke, S., Tjoa, T. B., Okamura, S.: J. Polym. Sci. *A-1 5*, 1911 (1967)

287. Van Looy, H., Hammett, L. P.: J. Am. Chem. Soc. *81*, 3872 (1959)
288. Burton, R. E., Pepper, D. C.: Proc. Roy. Soc. *A 263*, 58 (1961)
289. Hayes, M. G., Pepper, D. C.: ibid. *A 263*, 63 (1961)
290. Albert, A., Pepper, D. C.: ibid *A 263*, 75 (1961)
291. Jenkinson, D. H., Pepper, D. C.: ibid. *A. 263*, 82 (1961)
292. Mathieson, A. R.: Ref. [2], p. 265
293. Peniche, C., Gandini, A.: Rev. CENIC (Cuba): *4*, 59 (1973); C. A. *84*, 5472 (1976)
294. Ikeda, K., Higashimura, T., Okamura, S.: Kobunshi Kagaku *26*, 364 (1969). Chem. Abs. *71*, 39431 (1969)
295. Yamamoto, S., Tatsumi, M., Shomoto, K.: Nippon Kagaku Kaishi *1976*, 1132
296. Kohjiya, S., Terada, A., Yamashita, S.: Chem. Lett. *1972*, 671
297. Masuda, T., Sawamoto, M., Higashimura, T.: Makromol. Chem. *177*, 2981 (1976)
298. Kolthoff, I. M., Chantooni, Jr., M. K.: J. Am. Chem. Soc. *87*, 4428 (1965)
299. Pepper, D. C.: Proc. Int. Symp. Macromol. Chem., Wiesbaden, paper III, A 9 (1959)
300. Pepper, D. C.: personal communication
301. Goypiron, A., de Villepin, J., Novak, A.: J. Chim. Phys. *76*, 267 (1979)
302. Josson, C., Deporcq-Stratmains, M., Vast, P.: Bull. Soc. Chim. France *1977*, 820
303. Buck, H. M.: Rec. Trav. Chim. Pays-Bas *89*, 794 (1970)
304. Kolthoff, I. M., Chantooni, Jr., M. K.: J. Am. Chem. Soc. *95*, 8539 (1973)
305. Chemlir, M., Cardona, N., Schulz, G. V.: Makromol. Chem. *178*, 169 (1977)
306. Le Borgne, A.: Thesis, Univers. Paris VI 1974
307. Mathias, E.: Ph. D. thesis, Keele Univers. 1970
308. Chmelir, M.: Ref. [21] paper C-7
309. Cardona, N., Chmelir, M.: IUPAC Symp. Macromolecules, Madrid 1974. Comm. I. 3-24
310. Chmelir, M.: ibid. Comm. I 3-25
311. Chmelir, M.: Makromol. Chem. *176*, 2099 (1975)
312. Chmelir, M.: Ref. [23], p. 317
313. Sawamoto, M., Masuda, T., Higashimura, T.: Makromol. Chem. *177*, 2995 (1976)
314. Sawamoto, M., Masuda, T., Higashimura, T.: ibid. *178*, 1497 (1977)
315. Sawamoto, M. et al.: J. Polym. Sci. Polym. Lett. Ed. *13*, 279 (1975)
316. Higashimura, T., Nishii, H.: ibid. *15*, 1179 (1977)
317. Takeda, T., Sawamoto, M., Higashimura, T.: Polym. J. *9*, 377 (1977)
318. Cardona-Suetterlin, N.: Polym. Bull. *1*, 149 (1978)
319. Cardona-Suetterlin, N.: ibid. *1*, 307 (1979)
320. Chmelir, M., Schulz, G. V.: ibid. *1*, 355 (1979)
321. Cardona-Suetterlin, N.: ibid. *1*, 361 (1979)
322. Farnum, D. G.: J. Am. Chem. Soc. *89*, 2970 (1967)
323. Hung, N. A.: Doctorate thesis, Univers. Paris VI 1976
324. Kolthoff, I. M.: Anal. Chem. *46*, 1992 (1974)
325. Pavia, A. C.: Rev. Chim. Min. *7*, 471 (1970) and refs. therein
326. Rosolovskii, V. Y.: The chemistry of anhydrous perchloric acid (in Russian). Moscow: Nauka 1966
327. Coutagne, D. M.: J. Am. Chem. Soc. *93*, 1518 (1971)
328. Grigorovich, Z. I., Malov, Y. I., Rosolovskii, V. Y.: Bull. Acad. Sci. USSR Chem. *1973*, 257
329. Karelin, A. I., Grigorovich, Z. I., Rosolovskii, V. Y.: ibid. *1974*, 1157
330. Gandini, A., Plesch, P. H.: J. Chem. Soc. *1965*, 6019
331. Mathias, E., Plesch, P. H.: Chem. Ind. (London) *1971*, 1043
332. Tauber, S. J., Eastham, A. M.: J. Am. Chem. Soc. *82*, 4888 (1960)
333. Plesch, P. H.: Ref. [23], p. 322. Plesch, P. H.: private communication
334. Pepper, D. C., Reilly, P. J.: J. Polym. Sci. *58*, 639 (1962)
335. Pepper, D. C., Reilly, P. J.: Proc. Roy. Soc. *A 291*, 41 (1966)
336. Gandini, A., Plesch, P. H.: Polym. Lett. *3*, 127 (1965)
337. Hamman, S. D. et al.: J. Macromol. Sci.-Chem. *A6*, 771 (1972)
338. Darcy, L. E., Millrine, W. P., Pepper, D. C.: Chem. Comm. *1968*, 1441
339. MacCarthy, B., Millrine, W. P., Pepper, D. C.: ibid. *1968*, 1442

340. Lorimer, J. P., Pepper, D. C.: Ref. [21], paper C23
341. Pepper, D. C., Ref. [23], p. 39
342. Dunn, D. J., Mathias, E., Plesch, P. H.: Europ. Polym. J. *12*, 1 (1976)
343. Kucera, M.: Ref. [21], paper C21
344. Kucera, M.: J. Macromol. Sci.-Chem. *A 7*, 1611 (1973)
345. Dunn, D. J.: M. Sc. Dissertation, Univers. Bradford 1970
346. Brown, G. R., Pepper, D. C.: Polymer *6*, 497 (1965)
347. Solomon. O. F., Ciuta, I. Z., Cobianu, N.: Polym. Lett. *2*, 311 (1964)
348. Eley, D. D., Isack, F. L., Rochester, C. H.: J. Chem. Soc. *A 1968*, 1651
349. Okamura, S., Kanoh, N., Higashimura, T.: Makromol. Chem. *47*, 19 (1961)
350. Kanoh, N., Higashimura, T., Okamura, S.: ibid. *56*, 65 (1962)
351. Giusti, P.: Chim. Ind. (Milan) *48*, 435 (1966)
352. Giusti, P., Andruzzi, F.: ibid. *48*, 442 (1966)
353. Giusti, P., Andruzzi, F.: J. Polym. Sci. *C 16*, 3797 (1968)
354. Cerrai, P., Andruzzi, F., Giusti, P.: Makromol. Chem. *117*, 128 (1968)
355. Andruzzi, F. et al.: Chim. Ind. (Milan) *52*, 22 (1970)
356. Andruzzi, F. et al.: ibid. *52*, 146 (1970)
357. Cerrai, P. et al.: Europ. Polym. J. *10*, 1141 (1974)
358. Higashimura, T., Kishiro, O., Takeda, T.: J. Polym. Sci. Polym. Chem. Ed. *14*, 1089 (1976)
359. Yamamoto, K., Higashimura, T.: ibid. *14*, 2621 (1976)
360. Higashimura, T., Yamamoto, K.: ibid. *15*, 301 (1977)
361. Higashimura, T. et al.: ibid. *16*, 503 (1978)
362. Higashimura, T., Kishiro, O.: Polym. J. *9*, 87 (1977)
363. Higashimura, T., Mitsuhashi, M., Sawamoto, M.: Macromolecules *12*, 178 (1979)
364. Stoicescu, C., Dimonie, M.: Rev. Roum. Chim. *13*, 109 (1968)
365. Gandini, A.: Adv. Polym. Sci. *25*, 47 (1977)
366. Gandini, A., Peniche, C., Galego, N.: unpublished
367. Magagnini, P. L.: private communication
368. Evans, A. G., Jones, P. M. S., Thomas, J. H.: J. Chem. Soc. *1957*, 2095
369. Eley, D. D.: Ref. [2], p. 385 and refs. quoted therein
370. Eley, D. D., Isack, F. L., Rochester, C. H.: J. Chem. Soc. *A 1968*, 872 and refs. therein
371. Okamura, S., Kanoh, N., Higashimura, T.: Makromol. Chem. *47*, 35 (1961)
372. Parnell, R. D., Johnson, A. F.: IUPAC Symp. Kinetics and Mechanism of Polyreactions, 1969. Paper 2/18
373. Janjua, K. M., Johnson, A. F.: Ref. [21], paper C15
374. Sawamoto, M., Masuda, T., Higashimura, T.: J. Polym. Sci. Polym. Chem. Ed. *16*, 2675 (1978)
375. Higashimura, T.: Ref. [23], p. 71 and refs. therein
376. Higashimura, T., Hiza, M., Hasegawa, H.: Macromolecules *12*, 217 (1979)
377. Plesch, P. H., Polanyi, M., Skinner, H. A.: J. Chem. Soc. *1947*, 257
378. Biddulph, R. H., Plesch, P. H., Rutherford, P. P.: Proc. Int. Symp. Macromol. Chem., Wiesbaden, paper III, A 10 (1959)
379. Plesch, P. H., Rutherford, P. P.: J. Chem. Soc. *1964*, 2566
380. Biddulph, R. H., Plesch, P. H., Rutherford, P. P.: ibid. *1965*, 275
381. Williams, G., Bardsley, H.: ibid. *1952*, 1707
382. Kennedy, J. P.: J. Polym. Sci. A-1 *6*, 3139 (1968)
383. Wichterle, O., Marek, M., Trehoval, I.: Proc. Int. Symp. Macromol. Chem., Wiesbaden, paper III, A 13 (1959)
384. Wichterle, O., Marek, M., Trehoval, I.: J. Polym. Sci. *53*, 281 (1961)
385. Eley, D. D., Johnson, A. F.: J. Chem. Soc. *1964*, 2238
386. Colclough, R. O., Dainton, F. S.: Trans. Faraday Soc. *54*, 886 (1958)
387. Colclough, R. O., Dainton, F. S.: ibid. *54*, 894 (1958)
388. Colclough, R. O., Dainton, F. S.: ibid. *54*, 898 (1958)
389. Colclough, R. O., Dainton, F. S.: ibid. *54*, 901 (1958)
390. Chmelir, M., Marek, M., Wichterle, O.: J. Polym. Sci. *C 16*, 833 (1967)
391. Evans, A. G., Meadows, G. W.: Trans. Faraday Soc. *46*, 327 (1950)

392. Ghanem, N. A., Marek, M.: Europ. Polym. J. *8*, 999 (1972)
393. Plesch, P. H.: J. Chem. Soc. *1950*, 543
394. Imanishi, Y., Higashimura, T., Okamura, S.: Chem. High Polym. Japan *18*, 333 (1961)
395. Cheradame, H., Sigwalt, P.: Bull. Soc. Chim. France *1970*, 843
396. Cheradame, H.: Doctorate thesis, Univers. Paris 1966
397. Chau, P. M., Cheradame, H., Sigwalt, P.: unpublished
398. Cheradame, H. et al.: Bull. Soc. Chim. France *1970*, 849
399. Moszynska, B.: Compt. Rend. Acad. Sci. *256*, 1261 (1963)
400. Masuda, Y.: J. Phys. Soc. Japan *11*, 670 (1956)
401. Hunter, W. H., Johé, R. V.: J. Am. Chem. Soc. *55*, 1248 (1933)
402. Gantmakher, A. R., Medvedev, S. S.: Dokl. Akad. Nauk SSSR *106*, 1031 (1956)
403. Longworth, W. R., Plesch, P. H., Rutherford, P. P.: Proc. Chem. Soc. *1960*, 68
404. Kennedy, J. P.: J. Macromol. Sci.-Chem. *A 6*, 329 (1972)
405. Plesch, P. H.: Macromol. Chem. (Suppl. Pure Appl. Chem.) *8*, 305 (1973). IUPAC Symp. Macromolecules, Helsinki 1972
406. Plesch, P. H.: Ref. [1], p. 85
407. Marek, M., Chmelir, M.: J. Polym. Sci. *C 23*, 223 (1968)
408. Maslinska-Solich, J., Chmelir, M., Marek, M.: Coll. Czech. Chem. Comm. *34*, 2611 (1969)
409. Di Maina, M. et al.: Makromol. Chem. *178*, 2223 (1977)
410. Longworth, W. R., Panton, C. J., Plesch, P. H.: J. Chem. Soc. *1965*, 5579
411. Jordan, D. C., Mathieson, A. R.: ibid. *1952*, 2354. Ref.[1], p. 90
412. Bourne-Branchu, R., Hung, N. A., Cheradame, H.: Bull. Soc. Chim. France *1976*, 1349
413. Villesange, M. et al.: J. Macromol. Sci.-Chem. *A 11*, 391 (1977)
414. Hung, N. A., Cheradame, H., Sigwalt, P.: Europ. Polym. J. *9*, 385 (1973)
415. Polton, A., Sigwalt, P.: Bull. Soc. Chim. France *1970*, 131
416. Sauvet, G., Vairon, J. P., Sigwalt, P.: J. Polym. Sci. Polym. Chem. Ed. *16*, 3047 (1978)
417. Masure, M., Sauvet, G., Sigwalt, P.: ibid. *16*, 3065 (1978)
418. Bywater, S., Worsfold, D. J.: Can. J. Chem. *56*, 2093 (1978)
419. Lewis, I. C., Singer, L. S.: J. Chem. Phys. *43*, 2712 (1965)
420. Bracke, W. et al.: J. Am. Chem. Soc. *91*, 203 (1969)
421. Holmes, J., Pettit, R.: J. Org. Chem. *28*, 1695 (1963)
422. Dewar, M. J. S.: Bull. Soc. Chim. France *18*, C71 (1951)
423. Corriu, R., Coste, C.: ibid. *1969*, 3272
424. Habersberg, W. I., Hoytenk, G. J.: J. Chem. Soc. *1959*, 3055
425. Heublein, G., Helbig, M.: Zeitschr. Chem. *9*, 427 (1969)
426. Gerbier, J.: Compt. Rend. Acad. Sci. *B 262*, 685 (1966)
427. Gerbier, J.: ibid. *B 264*, 444 (1967)
428. Corriu, R., Coste, C., Sournia, A.: Bull. Soc. Chim. France *1970*, 2998
429. Brown, H. C., Wallace, W. J.: J. Am. Chem. Soc. *75*, 6265 (1953)
430. Choi, S. U., Brown, H. C.: ibid. *88*, 903 (1966)
431. Perkampus, H. H., Weiss, W.: Ber. Bunsen Ges. *75*, 446 (1971)
432. Fairbrother, F., Nixon, J. F.: J. Chem. Soc. *1958*, 3224
433. Elliot, B., Evans, A. G., Owen, E. D.: ibid. *1962*, 689
434. Goates, J. R. et al.: J. Phys. Chem. *68*, 2617 (1964)
435. Plesch, P. H., Brackman, D. S.: J. Chem. Soc. *1953*, 1289
436. Longworth, W. R., Plesch, P. H., Rutherford, P. P.: Int. Conf. Coord. Chem.: Chem. Soc. Spec. Publ. *13*, 115 (1959)
437. Plesch, P. H.: Ref.[20], p. 980
438. Dijgraaf, C.: J. Phys. Chem. *69*, 666 (1965)
439. Hammond, P. R.: Chem. Comm. *1968*, 987
440. Krauss, H. L., Nickl, J.: Z. Naturforsch. *206*, 630 (1965)
441. Goldstein, I. P. et al.: Dokl. Akad. Nauk SSSR *138*, 839 (1961)
442. Nakane, R., Watanabe, T., Kurihara, O., Oyama, T.: Bull. Chem. Soc. Japan *36*, 1376 (1963)
443. Elegant, L., Cassan, J., Azzaro, M.: Bull. Soc. Chim. France *1968*, 2675
444. Kennedy, J. P., Milliman, G. E.: Adv. Chem. Ser. *91*, 287 (1969)

445. Kennedy, J. P., Thomas, R. M.: J. Polym. Sci. *46*, 233 (1960)
446. Egger, K. W., Cocks, A. T.: J. Am. Chem. Soc. *94*, 1810 (1972)
447. Bogomolova, T. B., Gantmakher, A. R., Lyudvig, Y. B.: Polym. Sci. USSR *14*, 2587 (1972)
448. Toman, L., Marek, M., Jokl, J.: J. Polym. Sci. Polym. Chem. Ed. *12*, 1897 (1974)
449. Marek, M.: Ref. [23], p. 149
450. Baddeley, J., Eley, D. D., King, P. J.: Ref. [1], p. 16
451. Brown, H. C., Pearsall, H. W.: J. Am. Chem. Soc. *74*, 191 (1952)
452. Brown, H. C., Wallace, W. J.: ibid. *75*, 6268 (1953)
453. Grodzicki, A., Potier, A.: J. Inorg. Nucl. Chem. *35*, 61 (1973)
454. Peach, M. E., Tracy, V. L., Waddington, T. C.: J. Chem. Soc. *A 1969*, 366
455. Manteghetti, A.: Thesis, Univers. Montpellier 1973
456. Germain, A., Pascal, J. L., Potier, J.: Can. J. Chem. *55*, 3096 (1977)
457. Gillespie, R. J., Moss, K. C.: J. Chem. Soc. *1966*, 1170
458. Davies, R. K., Moss, K. C.: ibid. *A 1970*, 1054
459. Bacon, J., Dean, P. A. W., Gillespie, R. J.: Can. J. Chem. *49*, 1276 (1971)
460. Commeyras, A., Olah, G. A.: J. Am. Chem. Soc. *91*, 2979 (1969)
461. Brownstein, S., Paasivirta, J.: Can. J. Chem. *43*, 1645 (1965)
462. Harris, J. J.: Inorg. Chem. *5*, 1627 (1966)
463. Hartman, J. S., Stilbs, P.: Chem. Comm. *1975*, 566
464. Ignatov, M. I., Ilin, I. G., Buslaev, Y. A.: Dokl. Akad. Nauk SSSR *245*, 604 (1979)
465. Brownstein, S.: J. Inorg. Nucl. Chem. *35*, 3567 (1973)
466. Brownstein, S.: ibid. *35*, 3575 (1973)
467. Brownstein, S., Latrémouille, G.: Can. J. Chem. *52*, 2236 (1974)
468. Velichkova, R. S., Panayotov, I. M.: Makromol. Chem. *138*, 171 (1970)
469. Grattan, D. W., Plesch, P. H.: J. Chem. Soc. Dalton *1977*, 1734
470. Kennedy, J. P., Squires, R. G.: J. Macromol. Sci.-Chem. *A 1*, 805 (1967)
471. Beard, J. H., Plesch, P. H., Rutherford, P. P.: J. Chem. Soc. *1964*, 2566
472. Priola, A., Cesca, S., Ferraris, G.: Makromol. Chem. *160*, 41 (1972)
473. Matyjaszewski, K., Diem, T., Penczek, S.: Makromol. Chem. *180*, 1817 (1979) and refs. therein
474. Vairon, J. P., Sigwalt, P.: Bull. Soc. Chim. France *1971*, 559
475. Chau, P. M.: Thesis, Univers. Paris VI, 1973
476. Imanishi, Y., Imamura, H., Higashimura, T.: Kobunshi Kagaku *27*, 251 (1970)
477. Plesch, P. H.: J. Chem. Soc. *1953*, 1653
478. Mathieson, A. R.: Ref. [2], p. 245
479. Mathieson, A. R.: Ref. [2], p. 246
480. Cheradame, H., Hung, N. A., Sigwalt, P.: Compt. Rend. Acad. Sci. *268 C*, 476 (1969)
481. Bourne-Branchu, R., Cheradame, H., Sigwalt, P.: ibid. *268 C*, 1292 (1969)
482. Sauvet, G., Vairon, J. P., Sigwalt, P.: Bull. Soc. Chim. France *1970*, 4031
483. Griffith, J. E.: J. Chem. Phys. *9*, 642 (1941)
484. Hung, N. A., Cheradame, H., Sigwalt, P.: unpublished
485. Jacobsen, C. E., Dumarevska, B. B.: Mem. Inst. Chem. Ukrain. Acad. Sci. *7*, 15 (1946)
486. Kinsella, E., Coward, J.: Spectrochim. Acta *A 24*, 2139 (1968)
487. Mole, T., Jeffery, E. A.: Organoaluminium compounds. Amsterdam: Elsevier 1972, pp. 18 and 95
488. Hay, J. N., Hooper, P. G., Robb, J. C.: Trans. Faraday Soc. *65*, 1365 (1969)
489. Longworth, W. R., Plesch, P. H.: J. Chem. Soc. *1959*, 1887
490. Grattan, D. W., Plesch, P. H.: J. Electroanal. Chem. *103*, 81 (1979)
491. Plesch, P. H.: Ref. [22], p. 1065
492. Eley, D. D., Monk, D. F., Rochester, C. H.: J. Chem. Soc. Faraday I *1976*, 1584
493. Magagnini, P. L. et al.: Makromol. Chem. *178*, 2235 (1977)
494. Allison, J. A., Ridge, D. P.: J. Am. Chem. Soc. *99*, 35 (1977)
495. Grattan, D. W., Plesch, P. H.: to be published
496. Grattan, D. W.: Ph. D. thesis, Keele Univers. 1973
497. Schmidt, P., Chmelir, M., Marek, M., Schneider, B.: Coll. Czec. Chem. Comm. *33*, 1604 (1968)

498. Chmelir, M., Marek, M.: J. Polym. Sci. *C 22*, 177 (1968)
499. Lopour, V. P., Marek, M.: Makromol. Chem. *134*, 23 (1970)
500. Sauvet, G.: private communication
501. Olah, G. A.: Ref. [22], p. 501
502. For a review on these topics see Kennedy, J. P.: Ref. [3], chapter 1
503. Sigwalt, P.: XXIIIrd IUPAC congress, Boston 1971. London: Butterworth 1971, Vol. 4, p. 495
504. Sigwalt, P., Lapeyre, W., Cheradame, H.: Ref. [21], paper C33
505. Lieser, K. H., Guetlich, P.: Ber. Bunsenges. *67*, 445 (1963)
506. Bauer, D., Foucault, A.: J. Electroanal. Chem. *39*, 377 (1972)
507. Dorofeyeva, N. G., Kudra, O. K.: Ukrain. Khim. Zhur. *27*, 306 (1961); C. A. *56*, 5455 (1962)
508. Ahmed, W., Gerrard, W., Maladkar, V. K.: J. Appl. Chem. *20*, 109 (1970)
509. Barnes, A. J., Halcam, H. E., Scrimshaw, G. F.: Trans. Faraday Soc. *65*, 3159 (1969)
510. Kevdina, I. B., Shantarovich, V. P., Shivedchikov, A. P.: Khim. Vys. Energ. *10*, 375 (1976)
511. Brown, H. C., Melchiore, J. J.: J. Am. Chem. Soc. *87*, 5269 (1965)
512. Pocker, Y.: J. Chem. Soc. *1960*, 1972
513. Pocker, Y.: ibid. *1960*, 1292
514. Pocker, Y.: ibid. *1959*, 1179
515. Pocker, Y.: Trans. Faraday Soc. *55*, 1266 (1959)
516. Heald, K., Williams, G.: Ref. [1] p. 78 and refs. therein
517. Williams, G., Bardsley, H.: J. Chem. Soc. *1952*, 1707
518. Heald, K., Williams, G.: ibid. *1954*, 357
519. Heald, K., Williams, G.: ibid. *1954*, 362
520. Evans, A. G., Lewis, J.: ibid. *1959*, 1946
521. Evans, A. G., Price, D.: ibid. *1959*, 2982
522. Fontana, C. M., Kidder, G. A.: J. Am. Chem. Soc. *70*, 3745 (1948)
523. Kennedy, J. P., Squires, R. G.: J. Macromol. Sci.-Chem. *A 1*, 995 (1967)
524. Kennedy, J. P., Sivaram, S.: J. Org. Chem. *38*, 2262 (1973)
525. Evans, A. G., James, E. A., Owen, E. D.: J. Chem. Soc. *1961*, 3532
526. Sal'nikov, V. V., Druyan, E. A., Makarova, F. N.: Vysokomol. Soed. *3*, 1730 (1961)
527. Pepper, D. C.: Ref. [1], p. 70
528. Evans, A. G., Lewis, J.: J. Chem. Soc. *1957*, 2975
529. Overberger, C. G., Ehrig, R. J., Marcus, R. A.: J. Am. Chem. Soc. *80*, 2456 (1958)
530. Metz, D. J.: J. Polym. Sci. *50*, 497 (1961)
531. Higashimura, T., Moribe, I.: J. Polym. Sci. Polym. Lett. Ed. *12*, 391 (1974)
532. Lyuvig, E. B., Gantmakher, H. R., Medvedev, S. S.: Dokl. Akad. Nauk SSSR *156*, 1163 (1964)
533. Mathieson, A. R.: Ref. [2], p. 235
534. Williams, G.: J. Chem. Soc. *1940*, 775
535. Pfeiffer, P.: Ber. *38*, 2466 (1905)
536. Kucera, M. Svàbik, J., Majerova, K.: Coll. Czech. Chem. Comm. *37*, 2708 (1972)
537. Bogomolova, T. B., Gantmakher, A. R.: Dokl. Akad. Nauk SSSR *217*, 369 (1974)
538. Bogomolova, T. B., Gantmakher, A. R.: ibid. *230*, 117 (1976)
539. Bogomolova, T. B., Gantmakher, A. R.: Polym. Sci. USSR *20*, 1480 (1979)
540. Pepper, D. C.: Trans. Faraday Soc. *45*, 397 and 404 (1949)
541. Okamura, S., Higashimura, T.: J. Polym. Sci. *21*, 289 (1956)
542. Higashimura, T., Okamura, S.: Chem. High Polym. Japan *13*, 338 (1958)
543. Dainton, F. S., Tomlinson, R. H., Batke, T. L.: Ref. [1], p. 80
544. Norrish, R. G. W., Russell, K. E.: Nature *160*, 543 (1947)
545. Norrish, R. G. W., Russell, K. E.: Trans. Farraday Soc. *48*, 91 (1952)
546. Russell, K. E.: Ref. [1], p. 114
547. Plesch, P. H.: Ref. [2], p. 174
548. Laflair, R. T.: M.Sc. thesis, Queens Univers. Kingston Canada 1962
549. Biddulph, R. H., Plesch, P. H.: J. Chem. Soc. *1960*, 3913
550. Plesch, P. H.: Ref. [2], p. 183

551. Longworth, W. R., Plesch, P. H.: Proc. Int. Symp. Macromol. Chem., Wiesbaden, paper III, A 11 (1959)
552. Cheradame, H., Sigwalt, P.: Comp. Rend. Acad. Sci. *259*, 4273 (1964)
553. Cheradame, H., Sigwalt, P.: ibid. *260*, 159 (1965)
554. Mondal, M. A. S., Young, R. N.: Europ. Polym. J. *7*, 523 (1971)
555. Clayton, J. M., Eastham, A. M.: J. Am. Chem. Soc. *79*, 5368 (1957)
556. Kennedy, J. P., Huang, S. Y., Feinberg, S. C.: J. Polym. Sci. Polym. Chem. Ed. *15*, 2801 (1977)
557. Kennedy, J. P. Huang, S. Y., Feinberg, S. C.: ibid. *15*, 2869 (1977)
558. Plesch, P. H.: Ref. [2], pp. 158 and 167
559. Pino, P., Ciardelli, F., Lorenzi, G. P.: Makromol. Chem. *70*, 182 (1964)
560. Higashimura, T., Hirokawa, Y.: J. Polym. Sci. Polym. Chem. Ed. *15*, 1137 (1977)
561. Ueshima, T. et al.: Makromol. Chem. *98*, 58 (1966)
562. Masuda, T., Hasegawa, K,, Higashimura, T.: Macromolecules *7*, 728 (1974)
563. Bhattacharya, T. K., Mukherjee, A.: Indian J. Chem. *14 A*, 941 (1976)
564. Evans, A. G., Polanyi, M.: J. Chem. Soc. *1947*, 252
565. Evans, A. G., Meadows, G. W.: J. Polym. Sci. *4*, 359 (1949)
566. Dainton, F. S., Sutherland, G. B. B. M.: ibid. *4*, 37 (1949)
567. Okamura, S., Higashimura, T., Ogawa, Y.: Chem. High Polym. Japan *16*, 239 (1959)
568. Derouault, J., Dziembowska, T., Forel, M. T.: J. Mol. Struct. *47*, 59 (1978)
569. Clayton, J. M., Eastham, A. M.: Can. J. Chem. *39*, 138 (1961)
570. Allcock, H. R., Eastham, A. M.: ibid. *41*, 932 (1963)
571. Zlamal, Z., Kazda, A.: J. Polym. Sci. A-1 *1*, 3199 (1963)
572. Plesch, P. H.: Ref. [2], pp. 171 and 197
573. Bauer, R. F., Laflair, R. T., Russell, K. E.: Can. J. Chem. *48*, 1251 (1970)
574. Bauer, R. F., Russell, K. E.: J. Polym. Sci. A-1 *9*, 1451 (1971)
575. Russell, K. E., Vail, L. G. M. C.: Ref. [23], p. 183
576. Russell, K. E., Vail, L. G. M. C.: Can. J. Chem. *57*, 2355 (1979)
577. Wunderly, H. L., Sowa, F. J.: J. Am. Chem. Soc. *59*, 1010 (1937)
578. Zagorodnil, S. V.: Trud. Voronezh. Grosud. Univ. *10*, 241 (1938)
579. Guenzet, J., Camps, M.: Tetrahedron *30*, 649 (1974) and refs. therein
580. Raoult, M.: Ann. Chem. *2*, 66 (1884)
581. Sumaroskova, I. T., Litviak, I. G.: Izv. Sekt. Drug. Blagorod. Metal. *27*, 127 (1952)
582. Satchell, D. N. P., Wardell, J. L.: Trans. Faraday Soc. *61*, 1132 (1965)
583. Osipov, O. A., Somafala, G. S., Glusko, E. I.: Zh. Obshch. Khim. *27*, 1428 (1957)
584. Guenzet, J., Camps, M.: Tetrahedron Lett. *1972*, 2647
585. Guenzet, J., Camps, M.: Bull. Soc. Chim. France *1973*, 3167
586. Plesch, P. H.: J. Chem. Soc. *1950*, 543
587. Clayton, J. M., Eastham, A. M.: J. Chem. Soc. *1963*, 1636
588. Brownstein, S.: Can. J. Chem. *56*, 343 (1978)
589. Prosser, H. J., Young, R. N.: Europ. Polym. J. *11*, 403 (1975)
590. Mach, K., Drahoradova, E.: Coll. Czech. Chem. Comm. *40*, 326 (1975)
591. Matyska, B. et al.: ibid. *44*, 1262 (1979)
592. Imoto, M., Aoki, S.: Makromol. Chem. *63*, 141 (1963)
593. Zlamal, Z., Kazda, A.: J. Polym. Sci. A-1 *4*, 1783 (1966)
594. Giusti, P. et al.: Makromol. Chem. *136*, 97 (1970)
595. Kennedy, J. P.: J. Polym. Sci. *38*, 263 (1959)
596. Olah, G. A., Quinn, H. W., Kuhn, S. J.: J. Am. Chem. Soc. *82*, 426 (1960)
597. Kennedy, J. P. in: Polymer Chemistry of synthetic elastomers. (Kennedy, J. P., Tornqvist, E. G. M., eds.). New York: Interscience 1968, Part I, Ch. 5
598. Van Dyke, R. E.: J. Am. Chem. Soc. *73*, 2018 (1951)
599. Brown, H. C., Wallace, W. J.: ibid. *75*, 6279 (1953)
600. Walker, D. G.: J. Phys. Chem. *64*, 939 (1960)
601. Ziegler, K., Gellert, H. G.: Brennstoff.-Chem. *33*, 193 (1952)
602. Cesca, S., Giusti, P., Magagnini, P. L., Priola, A.: Makromol. Chem. *176*, 2319 (1975)
603. Sinn, H., Winter, H., Tirpitz, W.: Makromol. Chem. *48*, 59 (1961)

604. Zeiss, H.: Organometalic chemistry. New York: Reinhold 1960, p. 238
605. Nesmeyanov, A. N., Kocheshkov, K. A.: Methods of elemento-organic chemistry. Amsterdam: North Holland 1967, Vol. I, p. 46
606. Priola, A., Cesca, S., Ferraris, G., Di Maina, M.: Makromol, Chem. *176*, 2289 (1975)
607. Kennedy, J. P., Trivedi, P. D.: Adv. Polym. Sci. *28*, 83 (1978) and refs. therein
608. Kennedy, J. P., Trivedi, P. D.: ibid. *28*, 113 (1978) and refs. therein
609. Kennedy, J. P., Gillham, J. K.: ibid. *10*, 1 (1972)
610. Kennedy, J. P., Rengachary, S.: ibid. *14*, 1 (1974)
611. Cesca, S. et al.: Ref. [23], p. 162
612. Kennedy, J. P., Shinkawa, A., Williams, F.: J. Polym. Sci. A-1 *9*, 1551 (1971)
613. Sigwalt, P., Polton, A., Miskovic, M.: Ref. [23], p. 13
614. Kennedy, J. P., Melby, E. G.: J. Polym. Sci. Polym. Chem. Ed. *13*, 29 (1975)
615. Polton, A.: private communication
616. Giusti, P. et al.: Makromol. Chem. *176*, 2303 (1975)
617. Cesca, S. et al.: ibid. *176*, 2339 (1975)
618. Priola, A. et al.: ibid. 2271 (1975)
619. Reibel, L. C., Kennedy, J. P., Chung, Y. L.: J. Org. Chem. *42*, 690 (1977)
620. Evans, R. M., Satchell, R. S.: J. Chem. Soc. *B 1970*, 298
621. Evans, R. M., Satchell, R. S.: ibid. *B 1970*, 300
622. Fairbrother, F.: ibid. *1941*, 293
623. Fairbrother, F.: ibid. *1945*, 503
624. Howald, R. A., Willard, J. E.: J. Am. Chem. Soc. *77*, 2046 (1955)
625. Howald, R. A., Willard, J. E.: ibid. *78*, 6217 (1956)
626. Brezhneva, N. E., Roginskii, S. Z., Shilinskii, A. I.: J. Phys. Chem. USSR *10*, 368 (1937)
627. Kistiakowsky, G. B., Van Wazer, J. R.: J. Am. Chem. Soc. *65*, 1829 (1943)
628. Kennedy, J. P., Desai, N. V., Sivaram, S.: ibid. *95*, 6386 (1975)
629. Cesca, S., Priola, A., Ferraris, G.: Makromol. Chem. *156*, 325 (1972)
630. Pecka, J., Lopour, P., Marek, M.: ibid. *176*, 1725 (1975)
631. Okuhara, K.: J. Org. Chem. *43*, 2745 (1978)
632. Kennedy, J. P., Huang, S. Y., Smith, R. A.: Polym. Bull. *1*, 371 (1979)
633. Longworth, W. R.: Mason, C. P.: J. Chem. Soc. *A 1966*, 1164
634. Bowyer, P. M., Ledwith, A., Sherrington, D. C.: ibid. *B 1971*, 1511
635. Gogolczyk, W., Slomkowski, S., Penczek, S.: J. Chem. Soc. Perkin II 1977, 1729
636. Vairon, J. P., Villesange: personal communication
637. Price, E., Lichtin, N. N.: Tetrahedron Lett. *1960* (18), 10
638. Kessler, H., Walter, A.: Angew. Chem., Intern. Ed. *12*, 773 (1973)
639. Baaz, M., Gutmann, V., Masaguer, J. R.: Monatsh. Chem. *92*, 582 and 590 (1961)
640. Bos, S. S., Treloar, F. E.: Austr. J. Chem. *31*, 2445 (1978)
641. Ledwith, A., Sherrington, D. C.: Adv. Polym. Sci. *19*, 1 (1975)
642. Brauman, J. I., Archie, Jr., W. C.: J. Am. Chem. Soc. *92*, 5981 (1970)
643. Slomkowski, S., Kubisa, P., Penczek, S.: Symp. macromolecules, Helsinki 1972, paper I-135
644. Slomkowski, S., Penczek, S.: J. Chem. Soc. Perkin II *1974*, 1718
645. Kabir-ud-Din, Plesch, P. H.: ibid. *1978*, 937
646. Harmon, K. M.: Cycloheptatrienylium (tropenylium) ions. In Ref. [79] Vol. IV, p. 1579
647. Dauben, Jr., H.: J. Org. Chem. *25*, 1442 (1960)
648. van Tamelen, E. E., Cole, Jr., T. M.: J. Am. Chem. Soc. *93*, 6158 (1971)
649. van Tamelen, E. E. et al.: ibid. *90*, 1372 (1968)
650. Slomkowski, S., Penczek, S.: Chem. Comm. *1970*, 1347
651. Bonthrone, W., Reid, D. H.: J. Chem. Soc. *1959*, 2773
652. Olah, G. A., Svoboda, J. J.: J. Am. Chem. Soc. *95*, 3794 (1973)
653. Heck, R. H., Magee, P. S., Winstein, S.: Tetrahedron Lett. *1964*, 2033
654. Kursanov, D. N., Vol'pin, M. E., Akhrem, I. S.: Dokl. Akad. Nauk SSSR *120*, 531 (1958)
655. Vol'pin, M. E., Akhrem, I. S., Kursanov, D. N.: J. Gen. Chem. USSR *30*, 170 (1960)
656. Bawn, C. E. H. et al.: Polymer *12*, 119 (1971)
657. Ledwith, A., Lockett, E., Sherrington, D. C.: ibid. *16*, 31 (1975)

658. Chung, Y. J. et al.: ibid. *16*, 527 (1975)
659. Subira, F., Sauvet, G., Vairon, J. P., Sigwalt, P.: Ref. [23], p. 221
660. Stannett, V. et al.: Ref. [23], p. 233
661. Sauvet, G., Vairon, J. P., Sigvalt, P.: Compt. Rend. Acad. Sci. *C 265*, 1090 (1967)
662. Sauvet, G., Vairon, J. P., Sigwalt, P.: J. Polym. Sci. A-1 *7*, 983 (1969)
663. Sauvet, G., Vairon, J. P. Sigwalt, P.: Europ. Polym. J. *10*, 501 (1974)
664. Gaylord, N. G., Svestka, M.: J. Macromol. Sci.-Chem. *A 3*, 897 (1969)
665. Kunitake, T., Ochiai, T., Ohara, O.: J. Polym. Sci. Polym. Chem. Ed. *13*, 2581 (1975)
666. Eckard, A. D., Ledwith, A., Sherrington, D. C.: Polymer *12*, 444 (1971)
667. Prosser, H. J., Young, R. N.: Europ. Polym. J. *8*, 879 (1972)
668. Subira, F., Polton, A., Sigwalt, P.: Ref. [21], paper C 36
669. Burns, F. W. et al.: Symp. macromolecules, Helsinki, 1972, paper I-30
670. Cotrel, R. et al.: Ref. [21], paper C 8
671. Cotrel, R. et al.: Macromolecules *9*, 931 (1976)
672. Goka, A. M., Sherrington, D. C.: Polymer *16*, 819 (1975)
673. Rooney, J. M.: Makromol. Chem. *179*, 2419 (1978)
674. Bawn, C. E. H., Fitzsimmons, C., Ledwith, A.: Proc. Chem. Soc. *1964*, 391
675. Bowyer, P. M., Ledwith, A., Sherrington, D. C.: Polymer *12*, 509 (1971)
676. Rooney, J. M.: Ref. [23], p. 47
677. Rooney, J. M.: Makromol. Chem. *179*, 165 (1978)
678. Rooney, J. M., Squire, D. R., Stannett, V. T.: J. Polym. Sci. Polym. Chem. Ed. *14*, 1877 (1976)
679. Turchi, G., Matera, F., Magagnini, P. L.: Makromol. Chem. *170*, 75 (1973)
680. Aso, C., Kunitake, T., Matsuguma, Y., Imaizumi, Y.: J. Polym. Sci. A-1 *6*, 3049 (1968)
681. Higashimura, T., Fukushima, T., Okamura, S.: J. Macromol. Sci.-Chem. *A 1*, 683 (1967)
682. Johnson, A. F., Pearce, D. A.: Ref. [23], p. 57
683. Yamamoto, K., Hugashimura, T.: Polymer *16*, 815 (1975)
684. Bonner, T. G., Clayton, J. M., Williams, G.: J. Chem. Soc. *1958*, 1705
685. Sambhi, M. S., Treloar, F. E.: Polym. Lett. *3*, 445 (1965)
686. Gandini, A., Galego, N.: unpublished results with trityl perchlorate
687. Sambhi, M. S.: Macromolecules *3*, 351 (1970)
688. Plesch, P. H.: Ref. [23], p. 68
689. Plesch, P. H., Pask, S.: personal communication
690. Sherrington, D. C.: Ref. [23], p. 323
691. Littlejohn, M. A. et al.: J. Macromol. Sci.-Chem. *A 11*, 1603 and 1613 (1977)
692. Heublein, G.: Intern. symp. macromolecular chem., Tashkent 1978, Vol. 2, p. 111
693. Sangalov, Y. A. et al.: Polym. Sci. USSR *20*, 1508 (1979)
694. Otsu, T., Ko, M., Sato, T.: J. Polym. Sci. A-1 *8*, 789 (1970)
695. Ko, M., Nakanishi, T., Sato, T., Otsu, T.: Mem. Fac. Eng. Osaka Univ. *14*, 153 (1973)
696. Hernàndez, C., Gandini, A.: Rev. CNIC (Cuba) *4*, 49 (1973); C. A. *84*, 5471 (1976)
697. Cook, D.: Can. J. Chem. *37*, 48 (1959)
698. Cook, D.: ibid. *40*, 480 (1962)
699. Olah, G. A. et al.: J. Am. Chem. Soc. *84*, 2733 (1962)
700. Olah, G. A. et al: ibid. *85*, 1328 (1963)
701. Gates, P. N., Steele, D.: J. Mol. Struct. *1*, 349 (1967–68)
702. Goetz, J. G., Leroy, M. J. F.: Z. Anorg. Allg. Chem. *424*, 59 (1976)
703. Goetz, G. J., Leroy, M. J. F.: ibid. *427*, 281 (1976)
704. Goetz, G. J., Leroy, M. J. F.: Indian J. Chem. *16*, 394 (1978)
705. Goetz-Grandmont, G. J., Leroy, M. J. F.: J. Chem. Res., in press
706. Nuyken, O., Plesch, P. H.: Chem. Ind. (London) *1973*, 379
707. Germain, A., Commeyras, A., Casadevall, A.: Bull. Soc. Chim. France *1973*, 2537 and preceeding papers in that series
708. Lyubinskaya, O. V. et al.: Izv. Akad. Nauk SSSR. Ser. Khim. 1978, 397
709. Burton, H. Praill, P. F. G.: J. Chem. Soc. *1950*, 2034
710. Jander, G., Surawski, H.: Z. Electrochem. *65*, 384, 469 (1961)
711. Avedikian, A. M., Commeyras, A.: Bull Soc. Chim. France *1970*, 1258

712. Longworth, W. R., Plesch, P. H.: Proc. Chem. Soc. *1958*, 117
713. Masuda, T., Higashimura, T.: Polym. Lett. *9*, 783 (1971)
714. Masuda, T., Higashimura, T.: J. Polym. Sci. A-1 *9*, 1563 (1971)
715. Masuda, T., Higashimura, T.: J. Macromol. Sci.-Chem. *A 5*, 550 (1971)
716. Higashimura, T., Kishiro, O.: J. Polym. Sci. Polym. Chem. Ed. *12*, 967 (1974)
717. Higashimura, T. et al.: ibid. *13*, 1393 (1975)
718. Takeda, T., Sawamoto, M., Higashimura, T.: Polymer J. *9*, 377 (1977)
719. Higashimura, T., Nishii, H.: J. Polym. Sci. Polym. Chem. Ed. *15*, 329 (1977)
720. Kohjiya, S., Yamashita, S.: Chem. Lett. *1973*, 1007
721. Hasegawa, K., Asami, R., Higashimura, T.: Macromolecules *10*, 585 and 592 (1977)
722. Hasegawa, K., Asami, R.: J. Polym. Sci. Polym. Chem. Ed. *16*, 1449 (1978)
723. Biswas, M., Kabir, G. M. A.: Polymer, *19*, 595 (1978)
724. Franta, E., Reibel, L., Lehmann, J., Penczek, S.: Ref. [23], p. 139
725. Afshar-Taromi, F. et al.: Makromol. Chem. *179*, 849 (1978)
726. Sioda, R. E.: J. Phys. Chem. *72*, 2322 (1968)
727. Manning, G., Parker, V. D., Adams, R. N.: J. Am. Chem. Soc. *91*, 4584 (1969)
728. Dietz, R., Larcombe, B. E.: J. Chem. Soc. *B 1970*, 1369
729. Mengoli, G., Vidotto, G.: Makromol. Chem. *150*, 277 (1971)
730. Mengoli, G., Vidotto, G.: Symp. macromolecules, Helsinki 1972, paper I-47
731. Funt, L. B., Verigin, V.: Can. J. Chem. *52*, 1643 (1974)
732. Turcot, L., Glasel, A., Funt, B. L.: J. Polym. Sci. Polym. Lett. Ed. *12*, 687 (1974)
733. Glasel, A., Murray, K., Funt, B. L.: Macromol. Chem. *177*, 3345 (1976)
734. Funt, B. L., Severs, W., Glasel, A.: J. Polym. Sci. Polym. Chem. Ed. *14*, 2763 (1976)
735. Mengoli, G., Vidotto, G.: J. Electroanal. Chem. *76*, 595 (1977)
736. Oberrauch, E., Salvatori, T., Cesca, S.: J. Polym. Sci. Polym. Lett. Ed. *16*, 345 (1978)
737. Fritz, H. P. et al.: Z. Naturforsch. *33 b*, 498 (1978)
738. Metz, D.: Adv. Chem. Ser. *91*, 202 (1969)
739. Williams, F.: Ref. [20], p. 919
740. Chapiro, A.: Ref. [22], p. 1181
741. Irie, M., Yamamoto, Y., Hayashi, K.: J. Macromol. Sci.-Chem. *A-9* 817 (1975)
742. Stannett, V.: Fourth conf. Sci. and ind. applications of small accelerators 1976, p. 127
743. Tabata, Y.: Radiat. Phys. Chem. *9*, 31 (1977)
744. Hayashi, K., Irie, M., Hayashi, K.: Europ. Symp. electr. phenomena in polymer sci., Pisa 1978, paper A 7
745. Goineau, A. M., Kohler, J., Stannett, V.: J. Macromol. Sci.-Chem. *A 11*, 99 (1977)
746. Stannett, V. et al.: Polym. Prepr. *20*, 383 (1979)
747. Tagawa, S. et al.: Macromolecules *7*, 262 (1974)
748. Tabata, Y.: Ref. [23], p. 409
749. Egusa, S. et al.: Radiat. Phys. Chem. *9*, 419 (1977)
750. Egusa, S. et al., Imamura, M.: J. Polym. Sci. Polym. Chem. Ed. *16*, 729 (1978)
751. Tagawa, S., Tabata, Y.: Polym. Prepr. *20*, 411 (1979)
752. Hayashi, K. et al.: Europ. Polym. J. *13*, 925 (1977)
753. Mehnert, R. et al.: Radiochem. Radioanal. Lett. *30*, 389 (1977)
754. Brede, O., Mehnert, R.: Zfl.-Mitt. *14*, 213 (1978)
755. Brede, O., Boes, J., Helmstreit, W., Mehnert, R.: Z. Phys. Chem. *259*, 705 (1978)
756. Brede, O., Boes, J., Mehnert, R.: Radiochem. Radioanal. Lett., in press
757. Vermeil, C. et al.: J. Chim. Phys. *61*, 596 (1964)
758. Schlag, E. W., Sparapany, J. J.: J. Am. Chem. Soc. *86*, 1875 (1964)
759. Sparapany, J. J.: ibid. *88*, 1357 (1966)
760. Viswanathan, N. S., Kevan, L.: ibid. *89*, 2482 (1967)
761. Viswanathan, N. S., Kevan, L.: ibid. *90*, 1375 (1968)
762. Lambla, M., Scheibling, G., Banderet, A.: Compt. Rend. Acad. Sci. *C 271*, 924 (1970)
763. Brendlé, M.: ibid. *C 272*, 743 (1971)
764. Koenig, R., Lambla, M., Banderet, A.: Symp. macromolecules, Helsinki, 1972, paper I-24
765. Lambla, L., Koenig, R., Banderet, A.: Europ. Polym. J. *8*, 1 (1972)

766. Ueno, K. et al.: Trans. Faraday Soc. *63*, 1478 (1967)
767. Brendlé, M., Ilvoas-Fremond, A. M.: J. Chim. Phys. *69*, 1748 (1972)
768. Brendlé, M., Ilvoas, A. M.: Ref. [21], paper C 2
769. Schmidt, W., Schnabel, W.: Ber. Bunsenges. *75*, 654 (1971)
770. Schnabel, W., Schmidt, W.: J. Polym. Sci. Symp. *42*, 273 (1973)
771. Wablat, W., Schmidt, W. F., Schnabel, W.: Makromol. Chem. *175*, 2687 (1974)
772. Wablat, W., Schmidt, W. F., Schnabel, W.: Europ. Polym. J. *11*, 113 (1975)
773. Brendlé, M. C. C., Ilvoas, A. M.: Brit. Polym. J. *1976*, 11
774. Brendlé, M.: personal communication
775. Akulov, G. et al.: Makromol. Chem. *179*, 2775 (1978)
776. Akulov, G. P. et al.: Vysokomol. Soed. *B 21*, 243 (1979)
777. Breitenbach, J. W., Olaj, O. F., Sommer, F.: Fortschr. Hochpolym. Forsch. *9*, 47 (1972)
778. Giusti, P.: Ref. [22], p. 1157
779. Giusti, P.: J. Polym. Sci. Symp. *50*, 133 (1975)
780. Europ. Symp. electric phenomena in polymer sci.: Preprints, Pisa 1978
781. Funt, B. L., Blain, T. J.: J. Polym. Sci. A-1 *8*, 3339 (1970)
782. Tidswell, B. M., Doughty, A. G.: Polymer *12*, 431 (1971)
783. Mengoli, G., Vidotto, G.: Europ. Polym. J. *8*, 661 (1972)
784. Pistoia, G.: J. Polym. Sci. Polym. Lett. Ed. *10*, 787 (1972)
785. Pistoia, G.: Europ. Polym. J. *10*, 279 and 285 (1974)
786. Pistoia, G., Scrosati, B.: ibid. *10*, 1115 (1974)
787. Ghose, A., Bhadani, S. N.: Indian J. Techn. *13*, 172 (1975)
788. Akbulut, U., Fernandez, J. E., Birke, R. L.: J. Polym. Sci. Polym. Chem. Ed. *13*, 133 (1975)
789. Funt, B. L., Blain, T. J.: J. Polym. Sci. A-1 *9*, 115 (1971)
790. Mengoli, G., Vidotto, G.: Makromol. Chem. *139*, 293 (1970)
791. Mengoli, G., Vidotto, G.: Europ. Polym. J. *8*, 671 (1972)
792. Mengoli, G., Vidotto, G.: Makromol. Chem. *153*, 57 (1972)
793. Cerrai, P. et al.: Europ. Polym. J. *10*, 1195 (1974)
794. Cerrai, P. et al.: ibid. *11*, 101 (1975)
795. Breitenbach, J. W., Sommer, F., Unger, J.: Monatsh. Chem. *107*, 359 (1976)
796. Bhadani, S. N., Baranwal, P. P.: Makromol. Chem. *178*, 1049 (1977)
797. Bhadani, S. N., Baranwal, P. P.: ibid. *178*, 2637 (1977)
798. Bhadani, S. N. et al.: ibid. *179*, 1623 (1978)
799. Cerrai, P., Guerra, G., Tricoli, M.: Europ. Polym. J. *12*, 247 (1976)
800. Cerrai, P. et al.: Ref. [780], paper B 2
801. Attalla, G. et al.: Chim. Ind. (Milan) *60*, 197 (1978)
802. Cerrai, P., Guerra, G., Tricoli, M.: Europ. Polym. J. *15*, 153 (1979)
803. Kazaryan, G. A., D'yachkovskii, F. S., Yenikolopyan, N. S.: Vysokomol. Soed. *8*, 1314 (1966)
804. D'yachkovskii, F. S., Yegyasaryan, G. M., Kazaryan, G. A.: Dokl. Akad. Nauk. SSSR *182*, 1105 (1968)
805. D'yachkowskii, F. S., Kazaryan, G. A., Yenikolopyan, N. S.: Vysokomol. Soed. *A 11*, 822 (1969)
806. Babkina, O. N., D'yachkowskii, F. S.: ibid. *A 16*, 1803 (1974)
807. D'yachkowskii, F. S., Eritsyan, M. L.: Dokl. Akad. Nauk SSSR *198*, 880 (1971)
808. Akbulut, U., Birke, R. L., Fernandez, J. E.: Makromol. Chem. *179*, 2507 (1978)
809. Crivello, J. V.: Photoinitiated cationic polymerisation. In: UV curing: science and technology (Pappas, S. P., ed.). Stamford: Technol. Mark. Corp. 1978, p. 23
810. Shirota, Y., Mikawa, H.: J. Macromol. Sci.-Rev. Macromol. Chem. *C 16*, 129 (1977-1978)
811. Irie, M., Hayashi, K.: Progr. Polym. Sci. Japan *8*, 105 (1975)
812. Hayashi, K., Irie, M., Yamamoto, Y.: Ref. [23], p. 173
813. Irie, M., Yamamoto, Y., Hayashi, K.: Pure Appl. Chem. *49*, 455 (1977)
814. Hyde, P., Ledwith, A.: Intermolecular complexes in polymerisation processes. In Ref. [64], Vol. 2, p. 173
815. Ledwith, A.: Pure Appl. Chem. *49*, 431 (1977)

816. Davidson, R. S.: Photochemical reactions involving charge-transfer complexes. In Ref. [65], p. 215
817. Hayashi, K., Irie, M., Yamamoto, Y.: Polymer prepr. *20*, 379 (1979) and refs. therein
818. Marek, M., Toman, L.: J. Polym. Sci. Symposium *42*, 339 (1973)
819. Marek, M., Toman, L.: Ref. [21], paper C 24
820. Marek, M., Toman, L., Pilar, J.: J. Polym. Sci. Polym. Chem. Ed. *13*, 1565 (1975)
821. Toman, L., Marek, M.: Makromol. Chem. *177*, 3325 (1976)
822. Pilar, J., Toman, L., Marek, M.: J. Polym. Sci. Polym. Chem. Ed. *14*, 2399 (1976)
823. Marek, M.: Ref. [23], p. 495
824. Toman, L., Pilar, J., Marek, M.: J. Polym. Sci. Polym. Chem. Ed. *16*, 371 (1978)
825. Toman, L., Pilar, J., Spevacek, J., Marek, M.: ibid. *16*, 2759 (1978)
826. Williams, F.: Ref. [23], p. 157
827. Kennedy, J. P., Diem, T.: Polym. Bull. *1*, 29 (1978)
828. Crivello, J. V., Lam, J. H. W.: Ref. [23], p. 383
829. Crivello, J. V., Lam, J. H. W.: Macromolecules, *10*, 1307 (1977)
830. Crivello, J. V., Lam, J. H. W., Volante, C. N.: J. Rad. Curing *1977* (7), 2
831. Crivello, J. V., Lam, J. H. W.: J. Polym. Sci. Polym. Lett. Ed. *16*, 563 (1978)
832. Crivello, J. V., Lam, J. H. W.: ibid. *17*, 977 (1979)
833. Crivello, J. V., Lam, J. H. W.: Polym. Prepr. *20*, 415 (1979)
834. Crivello, J. V., Lam, J. H. W.: J. Polym. Sci. Polym. Chem. Ed. *17*, 1047 (1979)
835. Crivello, J. V., Lam, J. H. W.: ibid. *17*, 1059 (1979)
836. Schlesinger, S. J.: Photog. Sci. Eng. *18*, 387 (1974)
837. Woodhouse, M. E., Lewis, F. D., Marks, T. J.: J. Am. Chem. Soc. *100*, 996 (1978)
838. Woodhouse, M. E.: personal communication
839. Ledwith, A.: Polymer *19*, 1217 (1978)
840. Ledwith, A.: Pure Appl. Chem. *51*, 159 (1979)
841. Abdul-Rasoul, F. A. M., Ledwith, A., Yağci, Y.: Polymer *19*, 1219 (1978)
842. Abdul-Rasoul, F. A. M., Ledwith, A., Yağci, Y.: Polym. Bull. *1*, 1 (1978)
843. Diem, T., Kennedy, J. P., Chou, R. T.: ibid. *1*, 281 (1979)
844. Pienta, N. J., Kropp, P. J.: J. Am. Chem. Soc. *100*, 655 (1978)
845. Gaylord, N. G.: Macromol. Revs. *4*, 183 (1970). See also Ref. [23], p. 497
846. Stille, J. K. et al.: J. Macromol. Sci.-Chem. *A 9*, 745 (1975)
847. Cesca, S. et al.: Ref. [23], p. 159
848. Jolivet, Y., Peyrot, J.: Ref. [21], paper C 18
849. Kennedy, J. P.: Ref. [23], p. 1
850. Kennedy, J. P., Feinberg, S. C., Huang, S. Y.: J. Polym. Sci. Polym. Chem. Ed. *16*, 243 (1978)
851. Seung, S. L. N., Young, R. N.: Polym. Bull. *1*, 481 (1979)
852. Bossaer, P. K. et al.: Europ. Polym. J. *13*, 489 (1977)
853. Burgess, F. J. et al.: J. Polym. Sci. Polym. Lett. Ed. *14*, 471 (1976)
854. Burgess, F. J. et al.: Polymer *18*, 719, 726 (1977)
855. Burgess, F. J. et al.: ibid. *18*, 733 (1977)
856. Plesch, P. H.: Chem. Ind. (London) *1958*, 954
857. Kennedy, J. P. (ed.): Cationic graft copolymerisation. J. Appl. Polym. Sci. Appl. Polym. Symp. 30 (1977)
858. Pary, B.: Doctorate thesis, Univers. Paris VI, 1978
859. Chapiro, A.: Ref. [23], p. 431
860. Kennedy, J. P.: Ref. [857], p. 1
861. Saegusa, T., Imai, H., Furukawa, J.: Makromol. Chem. *65*, 60 (1963)
862. Imai, H., Saegusa, T., Furukawa, J.: ibid. *81*, 92 (1965)
863. Lilley, H. S., Forster, G. L.: Nature *160*, 131 (1947)
864. Eley, D. D., Richards, A. W.: Trans. Faraday Soc. *45*, 436 (1949)
865. Salomon, G.: Rec. Trav. Chim. Pays Bas *68*, 903 (1949)
866. Walling, C. et al.: J. Am. Chem. Soc. *72*, 48 (1950)
867. Burton, H., Praill, P. F. G.: J. Chem. Soc. *1953*, 837
868. Losev, B. I., Zakhorova, I. I.: Dokl. Akad. Nauk SSSR *116*, 609 (1957)

869. Hodgdon, R. B.: J. Polym. Sci. *47*, 259 (1960)
870. Hermans, J. P., Smets, G.: ibid. *A 3*, 3175 (1965)
871. Solomon, O. F., Dimonie, M.: ibid. *C 4*, 969 (1962)
872. Solomon, O. F. et al.: Polym. Lett. *6*, 551 (1968)
873. Beletskaya, I. P., Ryabtsev, A. N.: Vysokomol. Soed. *B 17*, 75 (1975)
874. Maroy, M.: Doctorate thesis, Paris Univers. 1970
875. Cheradame, H., Gandini, A., Collomb, J.: unpublished results
876. Addecott, K. S. B., Mayor, L., Turton, C. N.: Europ. Polym. J. *3*, 601 (1967)
877. Iwasaki, K., Fukutani, H., Nakano, S.: J. Polym. Sci. *A 1*, 1937 (1963)
878. Biswas, M., John, K. J.: J. Polym. Sci. Polym. Chem. Ed. *16*, 971 (1978)
879. Biswas, M., John, K. J.: ibid. *16*, 3025 (1978)
880. Biswas, M., Mishra, P. K.: ibid. *11*, 639 (1973)
881. Biswas, M., Mishra, P. K.: Polymer *16*, 621 (1975)
882. Taninaka, T., Minoura, Y.: J. Polym. Sci. Polym. Chem. Ed. *14*, 685 (1976)
883. Taninaka, T., Minoura, Y.: Europ. Polym. J. *13*, 631 (1977)
884. Taninaka, T., Uemura, H., Minoura, Y.: ibid. *14*, 199 (1978)
885. Ino, M., Arata, K.: J. Polym. Sci. Polym. Lett. Ed. *16*, 529 (1978)
886. Biswas, M., Maiti, M. M., Ganguly, N. D.: Makromol. Chem. *124*, 263 (1969)
887. Biswas, M., Mishra, P. K.: ibid. *163*, 37 (1973)
888. Maiti, M. M., Ganguly, N. D., Biswas, M.: Proc. Indian Nat. Sci. Acad. *41*, 127 (1975)
889. Brzezinska, K. et al.: Makromol. Chem. *178, 2491* (1977)
890. Matyjaszewski, K., Kubisa, P., Penczek, S.: Macromainz Prepr. *1*, 161 (1979)
891. Sawamoto, M., Higashimura, T.: Macromolecules *12*, 581 (1979)
892. Ohtori, T., Hirokawa, Y., Higashimura, T.: Polym. J. *11*, 471 (1979)
893. Kramer, G. M.: J. Org. Chem. *44*, 2616 (1979)
894. Mirda, D., Rapp, D., Kramer, G. M.: ibid. *44*, 2619 (1979)
895. Saegusa, T.: Ref. [22], p. 1199
896. Saegusa, T.: Polymerisation of cyclic imino ethers. In: Encyclopedia of polymer science and technology. Suppl. Vol 1. New York: Interscience 1976, p. 220
897. Sangalov, Y. A. et al.: Polym. Sci. USSR *20*, 1499 (1979)
898. Villesange, M. et al.: Makromainz Prepr. *1*, 208 (1979); Polym. Bull. *2*, 131 (1980)
899. Diem, T., Kennedy, J. P.: J. Macromol. Sci.-Chem. *A12*, 1359 (1978)
900. Di Maina, M., Narducci, P., Giusti, P.: Makromainz Prepr. *1*, 397 (1979)
901. Gandini, A., Cheradame, H., Sigwalt, P.: Polym. Bull. *2*, 731 (1980)
902. Pasto, D. J., Gadberry, J. F.: J. Am. Chem. Soc. *100*, 1469 (1978)
903. Bittles, J. A., Chauduri, A. K., Benson, S. W.: J. Polym. Sci. *A2*, 1221, 1847, 3203 (1964)
904. Biswas, M., Maity, N. C.: Adv. Polym. Sci. *31*, 47 (1979)
905. Preprints of the 5th International Symposium on Cationic and Other Ionic Polymerisations, Kyoto, Japan, April 1980
906. Billingham, N. C.: in Developments in Polymerisation −1, Haward, R. N., ed., Applied Science Publ., London 1979, p. 147
907. Penczek, S.: Macromol. Chem. Suppl. *3*, 17 (1979)
908. Goethals, E. J., Schacht, E. H.: Ref. [905], p. 161
909. Griffith, T. R., Pugh, D. C.: Coord. Chem. Rev. *29*, 129 (1979)
910. Lyerla, J. R., Yannoni, C. S., Bruck, D., Fyfe, C. A.: J. Am. Chem. Soc. *101*, 4770 (1979)
911. Penczek, S.: Ref. [905], p. 90
912. Obrecht, W., Plesch, P. H.: Ref. [905], p. 32; Makromol. Chem., in press
913. Bolza, F., Treloar, F. E.: Makromol. Chem. *181*, 839 (1980)
914. Yamazaki, N., Nakahama, S., Yamaguchi, K., Kasai, H.: Ref. [905], p. 96
915. Kunitake, T., Takarabe, K.: Macromolecules *12*, 1061 (1979)
916. Kunitake, T., Takarabe, K.: ibid. *12*, 1067 (1979)
917. Sawamoto, M., Furukawa, A., Hasegawa, K., Higashimura, T.: Ref. [905], p. 34
918. Chmelir, M., Ahmad, N., Schulz, G. V.: Ref. [905], p. 26
919. Ahmad, N., Chmelir, M., Schulz, G. V.: Ref. [905], p. 28. Makromol. Chem., Rapid Comm. *1*, 369 (1980)

920. Hasegawa, H., Higashimura, T.: Polym. J. *11*, 737 (1979)
921. Sigwalt, P., Sauvet, G.: Ref. [905], p. 68
922. Sawamoto, M., Higashimura, T., Enokida, A., Okubo, T.: Polym. Bull. *2*, 309 (1980)
923. Cerrai, P., Giusti, P., Lucchesi, A., Tricoli, M.: Macromainz Preprints, Vol. I, 31 (1979)
924. Higashimura, T., Teranishi, H., Sawamoto, M.: Polym. J., in press
925. Cerrai, P., Tricoli, M.: Ref. [905], p. 36
926. Grattan, D. W., Plesch, P. H.: Makromol. Chem. *181*, 751 (1980)
927. Brüggeller, P., Mayer, E.: Z. Naturforsch. *34b*, 891 and 896 (1979)
928. Lopour, P., Pecka, J., Marek, M.: Makromol. Chem. *174*, 1 (1973)
929. Sangalov, Y. A., Nelkenbaum, Y. Y., Ponomarenko, O. A., Minsker, K. S.: Vysokomol. Soed. *21*, 2267 (1979)
930. Guenzet, J., Camps, M.: Tetrahedron *35*, 473 (1979)
931. Corno, C., Ferraris, G., Priola, A., Cesca, S.: Macromolecules *12*, 404 (1979)
932. Vilim, R., Dvořak, S., Kotas, J., Langer, V.: Chem. Prum. *21*, 21 (1971)
933. Priola, A., Cesca, S., Ferraris, G.: Makromol. Chem. *160*, 41 (1972)
934. Reibel, L., Kennedy, J. P., Chung, D. Y. L.: J. Polym. Sci.-Polym. Chem. Ed. *17*, 2757 (1979)
935. Sivaram, S.: J. Organomet. Chem. *156*, 55 (1978)
936. Balaban, A. T., Boulton, A. J.: Org. Synth. *5*, 1106–1114 (1973)
937. Collomb, J., Gandini, A., Cheradame, H.: to be published
938. Sivaram, S., Acharya, H. K., Bhardwaj, I. S.: Ref. [905], p. 22
939. Kennedy, J. P.: Makromol. Chem. Suppl. *3*, 1 (1979) and Refs. therein
940. Kennedy, J. P.: Ref. [905], p. 6 and Refs. therein
941. Kennedy, J. P., Smith, R. A.: J. Polym. Sci.-Polym. Chem. Ed. *18*, 1523 and 1539 (1980)
942. Velichkova, R. S., Toncheva, V. D., Panayotov, I. M.: Makromol. Chem. *181*, 671 (1980)
943. Heublein, G., Spange, S., Hallpap, P.: ibid. *180*, 1935 (1979)
944. Pask, S., Plesch, P. H.: Unpublished results. Pask, S.: Ph. D. Thesis, Keele University, 1980
945. Hasegawa, H., Higashimura, T.: Ref. [905], p. 100
946. Boekhoff, H. M.: Doctorate Thesis, Mainz University, 1977
947. Kingston, S. B., Pask, S. D., Plesch, P. H.: personal communication. see also Ref. [944]
948. Tagawa, S., Washio, M., Tabata, Y.: Ref. [905], p. 108
949. Tagawa, S., Washio, M., Tabata, Y., Kira, A., Arai, S.: Ref. [905], p. 110
950. Deffieux, A., Squire, D. R., Stannett, V.: Polym. Bull. *2*, 469 (1980)
951. Yamamoto, M., Gotoh, T., Tsuchida, A., Nishijima, Y.: Ref. [905], p. 114
952. Marek, M., Toman, L.: Makromol. Chem., Rapid Commun. *1*, 161 (1980)
953. Ledwith, A.: Makromol. Chem., Suppl. *3*, 348 (1979)
954. Kinstle, J. F., Tufts, T. A.: Polym. Prepr. *20* (2), 661 (1979)
955. Smets, G., Aerts, A.: Ref. [905], p. 1
956. Giusti, P.: Ref. [905], p. 14
957. Hallensleben, M. L., Möller, K.: Makromainz Preprints *1*, 401 (1979)
958. Ambrose, R. J., Newell, J. J.: J. Polym. Sci.-Polym. Chem. Ed. *17*, 2129 (1979)
959. Denes, F., Percec, V., Totolin, M., Kennedy, J. P.: Polym. Bull. *2*, 499 (1980)
960. Collomb, J., Gandini, A., Cheradame, H.: Ref. [905], p. 104
961. Collomb, J., Morin, B., Gandini, A., Cheradame, H.: Europ. Polym. J., in press
962. Collomb, J., Gandini, A., Cheradame, H.: Makromol. Chem. Rapid. Comm., in press
963. Biswas, M., Maity, N. C.: Ref. [905], p. 24
964. Kučera, M., Svábik, J., Majerova, K.: Coll. Czech. Chem. Comm. *37*, 2004 (1972)
965. Kučera, M., Kelblerovà: ibid. *44*, 542 (1979) and refs. therein
966. Zambelli, A., Sacchi, M. C., Locatelli, P.: Macromolecules *12*, 1051 (1979)
967. Zambelli, A., Allegra, G.: ibid. *13*, 42 (1980)
968. Burfield, D. R.: Ref. [905], p. 155

Received November 14, 1979
M. Gordon (editor)

XIII. Subject Index

Author Index Volumes 1-35

Crescenzi, V.: Some Recent Studies of Polyelectrolyte Solutions. Vol. 5, pp. 358–386.

Davydov, B. E. and *Krentsel, B. A.:* Progress in the Chemistry of Polyconjugated Systems. Vol. 25, pp. 1–46.

Dole, M.: Calorimetric Studies of States and Transitions in Solid High Polymers. Vol. 2, pp. 221–274.

Dreyfuss, P. and *Dreyfuss, M. P.:* Polytetrahydrofuran. Vol. 4, pp. 528–590.

Dušek, K. and *Prins, W.:* Structure and Elasticity of Non-Crystalline Polymer Networks. Vol. 6, pp. 1–102.

Eastham, A. M.: Some Aspects of the Polymerization of Cyclic Ethers. Vol. 2, pp. 18–50.

Ehrlich, P. and *Mortimer, G. A.:* Fundamentals of the Free-Radical Polymerization of Ethylene. Vol. 7, pp. 386–448.

Eisenberg, A.: Ionic Forces in Polymers. Vol. 5, pp. 59–112.

Elias, H.-G., Bareiss, R. and *Watterson, J. G.:* Mittelwerte des Molekulargewichts und anderer Eigenschaften. Vol. 11, pp. 111–204.

Fischer, H.: Freie Radikale während der Polymerisation, nachgewiesen und identifiziert durch Elektronenspinresonanz. Vol. 5, pp. 463–530.

Fujita, H.: Diffusion in Polymer-Diluent Systems. Vol. 3, pp. 1–47.

Funke, W.: Über die Strukturaufklärung vernetzter Makromoleküle, insbesondere vernetzter Polyesterharze, mit chemischen Methoden. Vol. 4, pp. 157–235.

Gal'braikh, L. S. and *Rogovin, Z. A.:* Chemical Transformations of Cellulose. Vol. 14, pp. 87–130.

Gallot, B. R. M.: Preparation and Study of Block Copolymers with Ordered Structures, Vol. 29, pp. 85–156.

Gandini, A.: The Behaviour of Furan Derivatives in Polymerization Reactions. Vol. 25, pp. 47–96.

Gandini, A. and *Cheradame, H.:* Cationic Polymerisation. Initiation with Alkenyl Monomers. Vol. 34/35, pp. 1–264.

Gerrens, H.: Kinetik der Emulsionspolymerisation. Vol. 1, pp. 234–328.

Goethals, E. J.: The Formation of Cyclic Oligomers in the Cationic Polymerization of Heterocycles. Vol. 23, pp. 103–130.

Graessley, W. W.: The Etanglement Concept in Polymer Rheology. Vol. 16, pp. 1–179.

Hay, A. S.: Aromatic Polyethers. Vol. 4, pp. 496–527.

Hayakawa, R. and *Wada, Y.:* Piezoelectricity and Related Properties of Polymer Films. Vol. 11, pp. 1–55.

Heitz, W.: Polymeric Reagents. Polymer Design, Scope, and Limitations. Vol. 23, pp. 1–23.

Helfferich, F.: Ionenaustausch. Vol. 1, pp. 329–381.

Hendra, P. J.: Laser-Raman Spectra of Polymers. Vol. 6, pp. 151–169.

Henrici-Olivé, G. and *Olivé, S.:* Kettenübertragung bei der radikalischen Polymerisation. Vol. 2, pp. 496–577.

Henrici-Olivé, G. and *Olivé, S.:* Koordinative Polymerisation an löslichen Übergangsmetall-Katalysatoren. Vol. 6, pp. 421–472.

Henrici-Olivé, G. and *Olivé, S.:* Oligomerization of Ethylene with Soluble Transition-Metal Catalysts. Vol. 15, pp. 1–30.

Henrici-Olivé, G. and *Olivé, S.:* Molecular Interactions and Macroscopic Properties of Polyacrylonitrile and Model Substances. Vol. 32, pp. 123–152.

Hermans, Jr., J., Lohr, D., and *Ferro, D.:* Treatment of the Folding and Unfolding of Protein Molecules in Solution According to a Lattic Model. Vol. 9, pp. 229–283.

Holzmüller, W.: Molecular Mobility, Deformation and Relaxation Processes in Polymers. Vol. 26, pp. 1–62.

Hutchison, J. and *Ledwith, A.:* Photoinitiation of Vinyl Polymerization by Aromatic Carbonyl Compounds. Vol. 14, pp. 49–86.

Iizuka, E.: Properties of Liquid Crystals of Polypeptides: with Stress on the Electromagnetic Orientation. Vol. 20, pp. 79–107.

Ikada, Y.: Characterization of Graft Copolymers. Vol. 29, pp. 47–84.

Imanishi, Y.: Syntheses, Conformation, and Reactions of Cyclic Peptides. Vol. 20, pp. 1–77.

Inagaki, H.: Polymer Separation and Characterization by Thin-Layer Chromatography. Vol. 24, pp. 189–237.

Inoue, S.: Asymmetric Reactions of Synthetic Polypeptides. Vol. 21, pp. 77–106.

Polymers Properties and Applications

Editorial Board: H.-J. Cantow,
H. J. Harwood, J. P. Kennedy,
A. Ledwith, J. Meißner,
S. Okamura, G. Olivé,
S. Olivé

Volume 3: A. Knop, W. Scheib

Chemistry and Application of Phenolic Resins

1979. 111 figures, 88 tables. XIII, 269 pages
ISBN 3-540-09051-7

The authors present the current theory of phenolic resin chemistry and the technical application of phenolic resins, based on day-to-day experience in research, production and marketing, and against the background of economic relevance. Where the first fully synthetic polymers (phenolic resins) stand today and what their future is are subjects of discussion. Looking back at their development, it is shown that after a wide variety of adaptions, they remain technically and economically irreplaceable products with potential for further market growth and a commensurate appreciation of their value. This book will be greatly appreciated by chemists, engineers, marketing professionals, and students.

Volume 2: H.-H. Kausch

Polymer Fracture

1978. 180 figures, 23 tables. X, 332 pages
ISBN 3-540-08786-9

"Kausch,... is well known for his work on polymer morphology and molecular mechanics as well as his research on the strength of materials. The avowed aim of this book is to connect the more conventional statistical and continuum mechanics interpretation of fracture phenomena to the newer spectroscopic studies of highly stressed polymeric chains and the kinetics of their rupture. Relating the literature on the observed modes of viscoelasticity and irreversible deformation from polymer morphology and solid-state physics, Kausch explains the behavior and rupture of polymeric materials in terms of molecular slip and breakage processes. This leads to interesting, methodical and well-thought-out interpretations of fracture toughness, crack propagation rates and fatigue of all major polymer systems... Thus, the book is an outstanding contribution to our understanding of the role of chain ruptures during mechanical failure... every student and practitioner of polymer science and engineering should find this book to be a valuable resource for his work."
Physics Today

Springer-Verlag
Berlin
Heidelberg
New York

Volume 1: B. Rånby, J. F. Rabek

ESR Spectroscopy in Polymer Research

1977. 356 figures, 29 tables. XIV, 410 pages
ISBN 3-540-08151-8

"...This book is a remarkable example for the successful combination of simplicity and clarity in its tutorial parts and of depth and width whenever and wherever is presents the state of the art...As ultimate and very gratifying reward for his investment the reader gets no less than 2519 references to the literature in excellent alphabetical order. Scientists who already work with ESR will be greatly assisted in their efforts by this book; those who do not yet use this method will have an easy time to learn and use it. All of them will be grateful to the authors for this exceptional addition to our scientific literature."
J. Polymer Science

Polymer Bulletin

Editors:
Prof. H.-J. Cantow, Makromolekulare Chemie, Universität Freiburg, Stefan-Meier-Strasse 31,
D-7800 Freiburg, West-Germany
Prof. J. P. Kennedy, Dept. of Polymer Science, The University of Akron, Akron, OH 44325, USA
Prof. T. Saegusa, Dept. Synthetic Chemistry, Kyoto University, Kyoto, 606, Japan

Editorial Board: H. Batzer, Basel; N. Calderon, Akron, OH; S. Cesca, San Donato Milanese; P. J. Flory, Stanford, CA; J. Furukawa, Tokyo; J. E. McGrath, Blacksburg, VA; H. K. Hall, Jr., Tucson, AZ; H. H. Kausch, Lausanne; T. Kelen, Budapest; M. Kryszewski, Łódź; A. Ledwith, Liverpool; E. Maréchal, Paris-Cedex; J. Meißner, Zürich; A. Nakajima, Kyoto; G. and S. Henrici Olivé, Research Triangle Park, NC; N. A. Plate, Moscow; B. Rånby, Stockholm; C. J. Simionescu, Bucureşti; S. Sivaram, Gujarat; D. H. Solomon, Melbourne; R. Steiner, Frankfurt/M.; H. Tadokoro, Osaka; M. Takayanagi, Fukuoka; I. Uematsu, Tokyo; C. Wippler, Strasbourg; H. Zahn, Aachen

Editorial Assistant: A. Heinrich, Springer-Verlag Heidelberg

Polymer Bulletin provides rapid publication of significant advances in polymer science, including biopolymers. Chemistry, physical chemistry and physics are included as well as materials science.

This journal provides

- rapid publication of papers

- no page charge

- 50 off-prints of each paper supplied free of charge

Subscription information and sample copy upon request

Springer International

Send your request to:
Springer-Verlag, Journal Promotion Dept.,
P. O. Box 105280, D-6900 Heidelberg, West-Germany